AF334100

RADIOACTIVE WASTE DISPOSAL

RADIOACTIVE WASTE DISPOSAL
Low and High Level

Edited by
William R. Gilmore

NOYES DATA CORPORATION
Park Ridge, New Jersey, U.S.A.
1977

FOREWORD

As outlined in the introduction, this book describes the technology being developed to concentrate and confine both high-level and non-high-level radioactive wastes, so that they may be stored or otherwise disposed of in a comparatively safe manner according to governmental guidelines.

Most important is a long-term solution to the problem of accumulation of wastes from the nuclear fuel cycle. Processes are under development to convert the wastes into solid and relatively insoluble forms (the so-called vitrification processes). There are also various proposals for permanent storage of solidified wastes under the ocean bed or in geological formations where they will be virtually inaccessible. Disposal of radioactive waste via rockets into deep space may seem impossible now but if a stable, non-earth-intercept trajectory or orbit can be guaranteed, extraterrestrial disposal offers a foreseeable method for the complete removal of long-lived nuclear waste constituents from the earth. Radioactive wastes in liquid form, stored in tanks, although safe enough in so-called inaccessible facilities for the present, are totally unacceptable as a long-term solution.

Perhaps there should be no commitments to large nuclear power programs or even small breeder reactors until the waste issues have been fully appreciated and weighed in the light of wide public understanding.

This book is based mainly on federally-funded studies. Advanced composition and production methods developed by Noyes Data are employed to bring these new durably bound books to you in a minimum of time. Special techniques are used to close the gap between "manuscript" and "completed book." Technological progress is so rapid that time-honored, conventional typesetting, binding and shipping methods are no longer suitable. We have bypassed the delays in the conventional book publishing cycle and provide the user with an effective and convenient means of reviewing up-to-date information in depth.

The table of contents is organized in such a way as to serve as a subject index and provides easy access to the information contained in this book. The bibliography at the end of the volume lists the highly important government reports and a source for their purchase. A valuable glossary of abbreviations of nuclear and technical terms serves to familiarize the reader with this timely topic.

CONTENTS AND SUBJECT INDEX

Contents and Subject Index

INTRODUCTION

The increasing use of nuclear energy for electric power generation, as well as the expanding application of radioisotopes in various fields, are inevitably associated with the production of growing amounts of radioactive wastes. The treatment, storage and disposal of these radioactive wastes are problems which urgently need to be solved. Considering the size of its nuclear weapons program and its current power program, it is not surprising that, of the western countries, the U.S. has stored the largest amount of waste and has expended the greatest effort on the development of the technology required to manage it.

This book describes the technology being developed to concentrate and immobilize both high-level and non-high-level radioactive wastes so that they may be stored in a comparatively safe and compact manner according to U.S. governmental guidelines. Present radioactive waste management practices at U.S. governmental nuclear facilities and commercial fuel processing plants are outlined as well as proposed schemes for waste management resulting from advances in radioactive waste treatment technology.

Although most of the book is concerned with radioactive waste treatment and storage in the United States, a brief description of procedures followed in several other countries is included to show the impact of geological features and population density on the approach to radioactive waste management.

GLOSSARY

AEC	Atomic Energy Commission; a former federal agency, disbanded by the Energy Reorganization Act of 1974.
AGNS	Allied-General Nuclear Services.
BGRR	Brookhaven Graphite Research Reactor.
BNFP	Barnwell Nuclear Fuel Plant.
BNWL	Battelle Northwest Laboratories.
BWR	Boiling-Water Reactor.
CANDU	Canadian Deuterium-Uranium Reactor.
DF	Decontamination Factor.
EPA	Environmental Protection Agency; a federal agency.
ERDA	Energy Research and Development Administration; a federal agency, successor to all nonregulatory functions of the Atomic Energy Commission.
Federal Repository	Federally operated disposal or storage facility for high-level and transuranic-contaminated wastes.
FRP	Fuel Reprocessing Plant.

HEPA	High efficiency particulate air filters. Pleated fiber glass filters with high surface area and small pore size designed to remove aerosols with a minimum efficiency of 99.97% for 0.3 micrometer particles.
HFBR	High Flux Beam Reactor.
HLLW	High-level liquid waste.
HLW	High-level wastes.
HTGR	High Temperature Gas Reactor.
ILLW	Intermediate-level liquid waste.
Interim storage	Placement of wastes in an engineered temporary surrounding that will protect the public from radioactive release while a disposal location is developed.
LLLW	Low-level liquid waste.
LLW	Low-level wastes.
LMFBR	Liquid Metal Fast Breeder Reactor.
LWR	Light-Water Reactor.
MFRP	Midwest Fuel Recovery Plant.
MOX	Mixed oxide reactor fuel (containing both plutonium and uranium).
MT	Metric tons.
MTHM	Metric tons of heavy metal.
NEC	Nuclear Energy Center.
NEPA	National Environmental Policy Act of 1969.
NFRRC	Nuclear Fuel Recovery and Recycling Center.
NFS	Nuclear Fuel Services.
NRC	Nuclear Regulatory Commission; a federal agency, successor to regulatory functions of the Atomic Energy Commission.
NW	Nuclear waste.
ORP	Office of Radiation Programs.
PWR	Pressurized Water Reactor.
rad	Radiation absorbed dose.
RCG	Radiation Concentration Guides.
rem	Radiation equivalent man.
RRF	Resin Regeneration Facility.
RSSF	Retrievable Surface Storage Facility.
RW	Radioactive waste.
SRP	Savannah River Project.
SSCC	Sealed Storage Cask Concept.
SURFF	Spent Unprocessed Reactor Fuel Facility.
SWU	Separative Work Unit.
TBP	Tributyl phosphate.
TRU	Transuranic.
UNH	Uranium hexahydrate.
WT	Waste treatment.
WTSF	Waste Treatment and Storage Facility.

SOURCES AND TYPES OF RADIOACTIVE WASTE

The material in this chapter was excerpted from following reports: ORNL-TM-4903, PB-235 804, ORNL-TM-4902, PB-235 805, PB-255 502, PB-244 928, CONF-700440-1, ORNL-TM-3965, PB-224 588 and PB-221 467. A complete bibliography appears on p 359.

URANIUM MILLING, CONVERSION AND ENRICHMENT FACILITIES

Uranium Milling Industry

In the spring of 1973 there were 15 uranium mills operating in the United States with a combined processing rate of 22,500 tons of ore per day. The 15 active mills vary in size from 350 to 5,000 tons per day, with the majority (12 mills) in the range between 900 and 2,000 tons per day.

A uranium mill extracts uranium from ore. The product is a semirefined uranium compound (U_3O_8) called yellowcake which is the feed material for the production of uranium hexafluoride (UF_6). 80% of the yellowcake is produced by a sulfuric acid leach process; the remainder by a sodium carbonate, alkaline leach process. The principal steps in the acid-leach, solvent-extraction process are:

(a) Ore is blended and crushed to pass through a 2.5 cm (1 inch) screen. The crushed ore is then wet ground in a rod or ball mill and is transferred as a slurry to leaching tanks.

(b) The ore is contacted with sulfuric acid solution and an oxidizing reagent to leach uranium from the ore. The product liquor is pumped to the solvent-extraction circuit while the washed residues (tailings) are sent to the tailings pond or pile.

(c) Solvent extraction is used to purify and concentrate the uranium.

(d) The uranium is precipitated with ammonia and transferred as a slurry.

(e) Thickening and centrifuging are used to separate the uranium concentrate from residual liquids.

(f) The concentrate is calcined and pulverized.
(g) The concentrate or yellowcake is packaged in 208 liter (55 gallon) drums for shipment.

Large amounts of solid waste tailings remain following the removal of the uranium from the ore. A mill may generate 1,800 metric tons per day of tailings solids slurried in 2,500 metric tons of waste milling solutions. Over the lifetime of the mill, about 100 hectares (250 acres) may permanently be committed to store this material.

These "tailings piles" will have a radiological impact on the environment through the air pathway by continuous discharge of radon-222 gas (a daughter of radium-226), through gamma rays given off by radon-222 and daughters which undergo radioactive decay, and finally through air and water pathways as radium-226 and thorium-230 are blown off the pile by wind and leached from the pile into surface waters.

Release of Radioactive Effluent from Uranium Mills: The radioactivity associated with uranium mill effluents comes from the natural uranium and its daughter products present in the ore. During the milling process, the bulk of the natural uranium is separated and concentrated while most of the radioactive daughter products of uranium remain in the uranium-depleted residues that are pumped to the tailings retention system. Liquid and solid wastes from the milling operation will contain low level concentrations of these radioactive materials. Airborne radioactive releases include radon gas and particles of the ore and the product uranium oxide. External radiation levels associated with uranium milling activities are low, rarely exceeding a few mrem/hr even at surfaces of process vessels.

Airborne Releases — The radiological releases from uranium milling operations include airborne particulates and vapor. Dusts containing uranium and uranium daughter products (thorium-230 and radium-226) are released from ore piles, the tailing retention system, and the ore crushing and grinding ventilation system. Natural uranium is released from the yellowcake drying and packaging operations as entrained solids.

Radon gas is released from the leach tank vents, ore piles, tailings retention system and the ore crushing and grinding ventilation system. There has been no practical method to prevent the release of radon gas from uranium mills.

Waterborne Releases — The following discussion refers to the best of current procedures for handling mill liquid wastes where these wastes plus tailings are stored in a tailings retention pond system which uses an impervious clay cored earth dam and local topographic features of the area to form the impoundment.

The liquid effluent from a mill (acid-leach process) consists of waste solutions, from the leaching, grinding, extraction, and washing circuits of the mill. The solutions, which have an initial pH of 1.5 to 2, contain the unreacted portion of the sulfuric acid used as the leaching agent in the mill process, and sulfates and some silica as the primary dissolved solids with trace quantities of soluble metals and organic solvents. This liquid is discharged with the solids into the tailings pond.

Concentrations of radioactive materials predicted in the 2,500 MT/day of waste liquor from the Highland milling plants (Converse County, Wyoming) are shown in Table 1.1. Radioactive products of radon decay may also be present in small concentrations. Since the concentrations of radium-226 and thorium-230 are about an order of magnitude above the specified Federal limits, considerable effort must be exerted to prevent any release of these materials from the site.

TABLE 1.1: CONCENTRATIONS OF RADIOACTIVE EFFLUENTS IN WASTE LIQUOR FROM A URANIUM MILL

Radionuclide	Concentration (pCi/l)
Uranium-natural	800*
Radium-226	350
Thorium-230	22,000

*About 0.001 g/ml.

Source: PB-235 804

The waste liquor is, therefore, stored in the tailings retention pond which is constructed to prevent discharge into the surface water system and to minimize percolation into the ground. This is a continuing potential problem requiring monitoring programs to insure that there is no significant movement of contaminated liquids into the environment.

If an earth-fill, clay-cored dam retention system serves as a collection and storage system for the liquid and solid process wastes generated in the mill, it will permit the evaporation of most of the contained waste liquids and serve as a permanent receptacle for the residual solid tailings. However, after the initial construction of the retention system, it is to be expected that there will be some seepage of radionuclides through and around the dam.

It has been estimated that this seepage will diminish over a period of about two years because of the sealing effect from accumulation of finer particles between the sandstone grains. On the other hand, sealing may not occur. Radium-226 is a radionuclide of concern in this case. Radium-226 levels as high as 32 pCi/liter (1) have been found in seepage from current operating mills. Assuming a seepage rate of 300 liters/min (80 gpm), the concentration of radium-226 seeping into a stream of 140 liters per second (5 ft^3/sec) is approximately 1 pCi/liter which is 33% of the current drinking water standards.

In the Humble Oil and Refining Company environmental report for the Highland Uranium Mill, a concentration of 350 pCi/l was assumed bringing the concentration of radium-226 in such a stream up to 12 pCi/l. The Highland Uranium Mill is also estimated to release to the tailings pond 22,000 pCi/l thorium-230 and trace quantities of short-lived radon daughter products. As an additional example, the analysis of plant tailings effluents for the Humeca Uranium Mill is given in Table 1.2 (2).

The radiological significance of seepage from tailings ponds will depend on the location of the pond. In arid regions, the seepage may evaporate before leaving the site, leaving the radioactivity entrained and absorbed on soil. Should the tailings pond be located near a river, minor leakage might be diluted sufficiently by the additional river water to meet relevant drinking water standards.

TABLE 1.2: ANALYSIS OF PLANT TAILINGS EFFLUENTS FROM THE HUMECA URANIUM MILL (ALKALINE LEACH PROCESS)

Radionuclide	pCi/l
Radium-226	10-2,000
Thorium-230	0.1
Uranium-238	4,000

Source: PB-235 804

Discharge of pond seepage into streams providing insufficient dilution and not under the control of the licensee would not be acceptable. In such a case, a secondary dam may be built below the primary dam to catch the seepage which may then be pumped back into the tailings ponds.

Conversion Facilities

Uranium concentrate milled from ore must be converted to the volatile compound uranium hexafluoride (UF_6) in order to be enriched by the gaseous diffusion process. Two different industrial processes are used for uranium hexafluoride production. The "hydrofluor process" consists of reduction, hydrofluorination and fluorination of the ore concentrates to produce crude uranium hexafluoride followed by fractional distillation to obtain a pure product. The wet solvent extraction process employs a wet chemical solvent extraction step at the head end of the process to prepare high purity uranium feed prior to reduction, hydrofluorination, and fluorination steps. Each method is used to produce roughly equal quantities of uranium hexafluoride feed for the enrichment plants.

The two commercial plants (3)(4) in operation in 1973 processed approximately 10,000 metric tons per year of uranium into uranium hexafluoride; 180 metric tons of uranium converted to 270 metric tons of uranium hexafluoride are required to support a GWe-yr of electricity generated by light water reactors.

The uranium concentrate feed to a conversion plant contains the equivalent of about 75 to 85% uranium oxide. The conversion process removes essentially all of the remaining impurities and produces a highly purified uranium hexafluoride product. The dry hydrofluor process separates impurities either as volatile compounds or as solid constituents of ash. The wet solvent extraction method separates impurities by extracting the uranium into organic solvent leaving the impurities behind dissolved in an aqueous solution. Therefore, the nature of the radioactive effluents from the two processes differ substantially; the hydrofluor method releases radioactivity primarily in the gaseous and solid state, while the solvent extraction method releases more of its radioactive wastes dissolved in liquid effluents.

Both plant designs stipulate virtually complete recovery of uranium, total utilization of fluorine, and high utilization of the other main reactants, (hydrogen, hydrogen fluoride, ammonia, and nitric acid). The plants are located in relatively sparsely populated areas. The range of population density in the vicinity of the two existing production facilities is 10 to 15 people per square kilometer (25 to 40 people per square mile); the region surrounding the plant using the dry hydrofluor process is the more densely populated.

Release of Radioactive Effluents from Conversion Facilities: Because no irradiated material is handled by conversion facilities, all radionuclides present also occur in nature. They are radium, thorium, uranium, and their respective decay products. Some of the decay products are delivered to the facility as impurities in the mill concentrate and others recur there by the continuing radioactive decay of the uranium. Uranium is present in the majority of the plant processes, appears in lqiuid effluents and is essentially the only source of radioactivity in the gaseous effluents. The radium, thorium, and decay products are separated from the uranium in the conversion process and thus appear in the liquid effluents or solid waste associated with specific purifying procedures.

Uranium may appear in the gaseous effluents in several chemical forms. Possible chemical species are U_3O_8, UO_2, UF_4, UF_6, $(NH_4)_2U_2O_7$ and UO_2F_2. In the conversion process employing solvent extraction, uranium is present as uranyl nitrate which may also appear in gaseous effluents. Thus, the uranium may be released as both soluble and insoluble aerosols. Measurements at one facility (4) indicate that about two thirds of the airborne uranium is in an insoluble form, and about one-third is in a soluble form. Because uranium has a low specific activity (0.7 Ci/MTU), the insoluble uranium is amenable to filtration. The discharge to the environment is through low stacks and vents.

Liquid effluents are associated with the various solvent extraction and scrubber systems so that the radionuclides in these effluents are considered to be in solution. About 1.7 metric tons of uranium, 0.03% of the material processed, appears in the liquid effluent streams.

Conversion of 10,000 MTU/yr to uranium hexafluoride produces an estimated 1,000 metric tons of solid waste (about 450 m³) that must be shipped to a commercial waste disposal burial site (5). Materials in the solid wastes are filter fines, sediments, pond muds, bed materials and miscellaneous materials. The wastes are shipped in 208 liter (55 gallon) drums. The activity disposed of per year in the solid waste is estimated as (4)(6):

Natural uranium	0.5 Ci
Natural thorium	4.0 Ci
Uranium-234	0.5 Ci
Thorium-234	0.6 Ci
Protactinium-234	0.6 Ci
Radium-226	0.5 Ci
	6.7 Ci

Uranium Enrichment Facilities

Natural uranium contains 0.7% of fissionable uranium-235. Light water nuclear power reactors, however, utilize uranium that is enriched in uranium-235 to the

range of 2 to 4%. Gaseous diffusion is the technology that has been developed
in this country for performing the enrichment operation. Uranium is enriched
by pumping the volatile uranium hexafluoride through a system of numerous
porous barriers. These barriers discriminate against the passage of the heavier
isotope of uranium by a theoretical maximum enrichment factor of 1.0043.
Existing plants would require about 1,700 barrier stages to produce a material
of 4% uranium-235. The uranium hexafluoride gas is driven through the barriers by compressors driven by electric motors.

It is the compression of the gas that generates process heat which in turn requires
cooling water that is ultimately discharged into the environment as thermal effluent. The electric motors require very large quantities of electricity, the generation of which causes additional effluents to be discharged into the environment from the electric power generating plants. The gaseous diffusion plants
also produce uranium hexafluoride depleted in uranium-235 (0.25%) which is
stored as a solid in cylinders at the plants.

As of 1973, there were three government-owned gaseous diffusion plants in the
United States (5). They were built between 1943 and 1955 and are located at
Oak Ridge, Tennessee; Paducah, Kentucky and Portsmouth, Ohio, on sites chosen
for their remote location and low surrounding population densities. The plants
average about 800 meters to their nearest site boundaries.

The complex of plants has a production capacity of about 10,000 metric tons
of separative work units per year, enough to support 90 GWe-years of electricity
generated by light water reactors. A separative work unit (SWU) is a measure of
the effort expended to separate a quantity of uranium of a given assay into two
components, one having a higher percentage of uranium-235. Separative work is
expressed in kilogram units to give it the same dimensions as material quantities.

It is planned to increase the capacity of the three-plant complex by a factor of
2.5 by 1980. This will be accomplished by improving and upgrading the present
units and will be enough to meet the projected 1980 industry demands. The
existing production capability of the plants is only partially used for commercial
production of LWR fuel. Capacity for this purpose can be increased to 10,000
MT SWU only by diverting capacity now used for other government needs.

Release of Radioactive Effluents from Enrichment Facilities: Gaseous diffusion
plants are large complexes, processing large quantities of materials, and having
many effluent streams. The effluent streams bearing radioactive materials are
limited to a few types, resulting in releases of uranium to the air and river water,
most probably as UO_2F_2. Reported release quantities are very small compared
to permissible discharge limits and consist solely of uranium. However, the large
quantities of uranium necessarily produce radioactive daughter products during
storage and processing which must be handled in some manner.

Those which decay to uranium-234, a comparatively worthless isotope, can eventually, via the uranium recovery facilities, be shipped out as uranium-234 in the
product uranium. Other radionuclides, daughters of uranium-234 and minor
contaminants of recycle fuel, must be handled through some waste disposal
system. Since they are not reported as being present in effluents, it is assumed
for the present that 100% of them go into solid waste. The effluent uranium
is reported as natural uranium which is selected as a representative isotopic mix

because in producing low-enrichment LWR fuel by gaseous diffusion, the portion of plant capacity which is processing depleted uranium is comparable to the portion which is processing enriched uranium. Effluent data for uranium enrichment plants are minimal. Solid wastes consist of sludge from onsite holding ponds. The sludge is collected and buried onsite.

REACTOR FUEL FABRICATION

The following sections describe the processing steps that produce the radwaste effluents in LWR fuel fabrication plants. Both radioactive and nonradioactive noxious components of waste must be considered in treating the radwaste effluents. The large amount of nitrates, ammonia, and fluoride in the ammonium diuranate (ADU) effluents present a significant problem in waste management.

Ammonium Diuranate (ADU) Process

ADU Process Line: Uranium hexafluoride is received at the fuel fabrication plant as a solid in a 25-ft^3 pressurized shipping vessel (Figure 1.1). The shipping vessel is placed in a sealed system where the UF_6 is vaporized and transferred to the reaction vessels. The UF_6 is hydrolyzed with water and neutralized with NH_4OH at a pH of 8 to 9 to form a slurry of ADU in an aqueous solution of ammonium fluoride and ammonium hydroxide. The ADU is recovered in a centrifuge and a clarifier and is subsequently dried and calcined to form UO_2 powder. The UO_2 powder is pressed into pellets and the pellets placed in zirconium tubes.

FIGURE 1.1: NOMINAL ADU PROCESS SYSTEM, 5 MTU/DAY

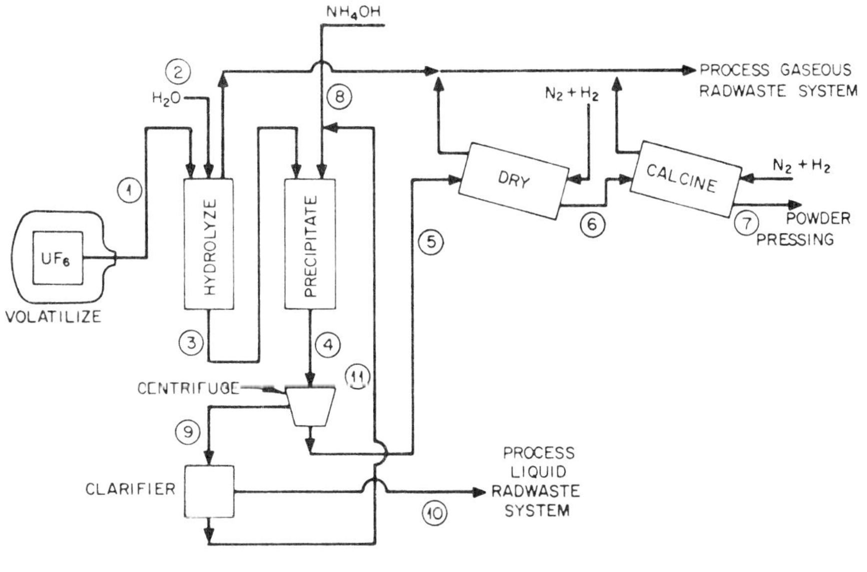

Source: ORNL-TM-4902

UO₂ Recycle System: In the UO$_2$ recycle system, the off-specification, UO$_2$ product materials, such as chipped and cracked pellets, are dissolved in nitric acid and the solution transferred to the precipitation system (Figure 1.2). The pH is raised to 3.0 by adding ammonia and uranium tetroxide is precipitated by adding hydrogen peroxide. In this system, the principal objective is to recover the uranium rather than achieve a high degree of separation from impurities, such as iron.

Consequently, the reaction is carried out at a relatively high pH where precipitation is more complete for both the uranium and the impurities, rather than at lower pH's where a better separation but lower recovery of uranium is obtained. The uranium tetroxide is separated from the mother liquor (radwaste) in centrifuge and clarifiers.

FIGURE 1.2: NOMINAL RECYCLE SYSTEM, 0.75 MTU/DAY

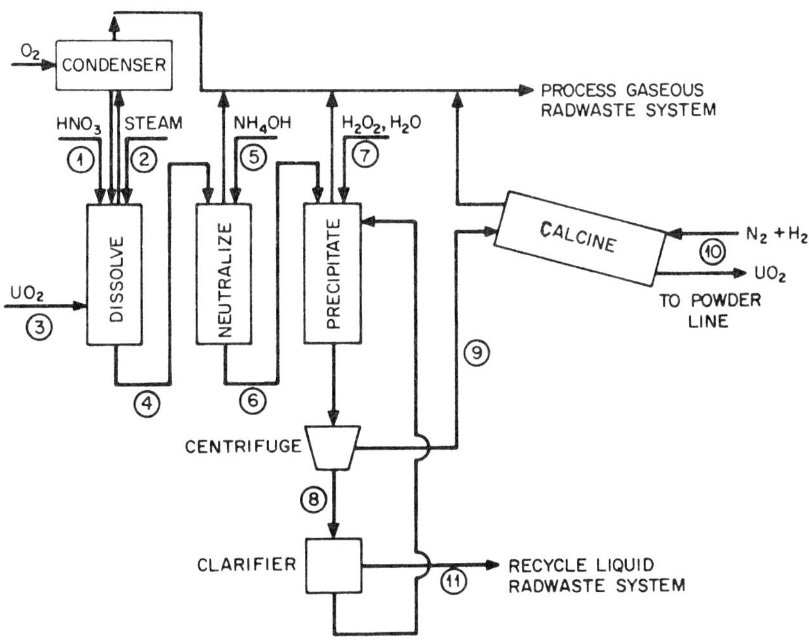

Source: ORNL-TM-4902

Scrap Recovery System: In the scrap recovery system, impure materials such as incinerator ashes and floor sweepings are treated to recover the contained uranium values (Figure 1.3). The scrap is leached with nitric acid to dissolve the uranium and the solution is filtered to remove the insoluble impurities. The solution is then processed in a solvent extraction system which is specially designed to achieve a high recovery of uranium and a high degree of separation of the uranium from impurities.

FIGURE 1.3: NOMINAL SCRAP RECOVERY SYSTEM, 0.1 MTU/DAY

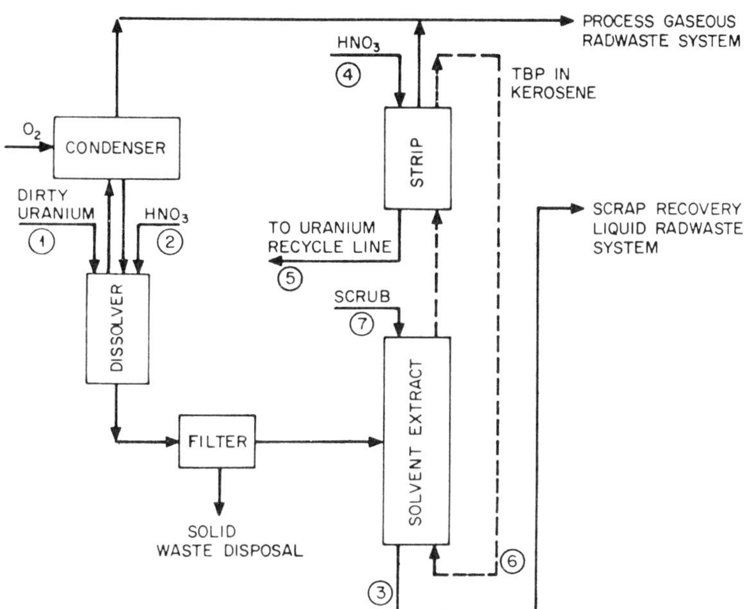

Source: ORNL-TM-4902

About 90% of the thorium is extracted with the uranium. In this system, the dissolver solution containing 3 M HNO₃ and 50 grams per liter uranium is passed countercurrent to a solvent containing 20 volume percent n-tributyl phosphate in a kerosene-type diluent in a 2-inch diameter by 20-foot high pulsed column. The solvent rises in the column and extracts the uranium from the downward flowing acidic dissolver solution. The uranium is recovered (stripped) from the solvent by contacting the solvent with 0.01 M HNO₃. The solvent is then reused.

The barren dissolver solution, or raffinate, constitutes the scrap recovery liquid waste. About once a year, the stripped solvent (a few hundred gallons) is removed from the system and is absorbed in a suitable solid material in drums. The drums of solidified waste are shipped to a licensed commercial burial ground for burial.

Liquid Scrubber: All of the process gaseous effluents in the ADU plant are passed through liquid scrubbers before release to the atmosphere. The water from these scrubbers contains uranium, NH₄OH and NH₄F. The water from the scrubbers is added to the liquid waste from the ADU line and handled in the ADU liquid radwaste treatment system.

Miscellaneous Liquid Wastes: Miscellaneous wastes are generated from the laundering of contaminated clothing, the showers for personnel working in areas where contamination is expected, the laboratory drains, and the occasional decontamination of equipment. The uranium in this stream is primarily UO₂ suspended in detergent solution.

Process Gaseous Effluent: The processing units in the ADU production and waste treatment systems are connected to the process gaseous effluent treatment system. The gaseous effluents contain small amounts of UF_6 and HF gases and an aerosol of droplets of solution which contains a complete spectrum of the radioactive and nonradioactive materials in the process liquids. On drying, these droplets form radioactive particles containing uranium, thorium and protactinium. Additional particulates are derived from the drying and calcining operations and from the radwaste treatment units. The process gaseous flow rate is about 69,000 scfm.

Ventilation Gaseous Effluent: The air from the operating areas contains small amounts of radioactive particulates of uranium, thorium, and protactinium compounds. The suspension of these particles occurs during operations such as loading transfer containers with UO_2 powder, loading the pelletizing machines with UO_2 powder and pellet grinding, or from the leakage and drying of process solutions from pipes or pump-packing glands. The ventilation gaseous flow rate is about 90,000 scfm.

Direct Conversion (DC) Process

The flowsheet for the direct conversion process for the production of UO_2 from UF_6 is shown in Figure 1.4. Cylinders of UF_6 are placed in steam-heated cabinets to vaporize the contained UF_6. The UF_6 gas enters into a bed of UO_2F_2 particles which is fluidized by steam. The gas reacts with the steam on the hot, wet surface of the particles to form a coating of UO_2F_2.

FIGURE 1.4: NOMINAL DC PROCESS SYSTEM, 5 MTU/DAY

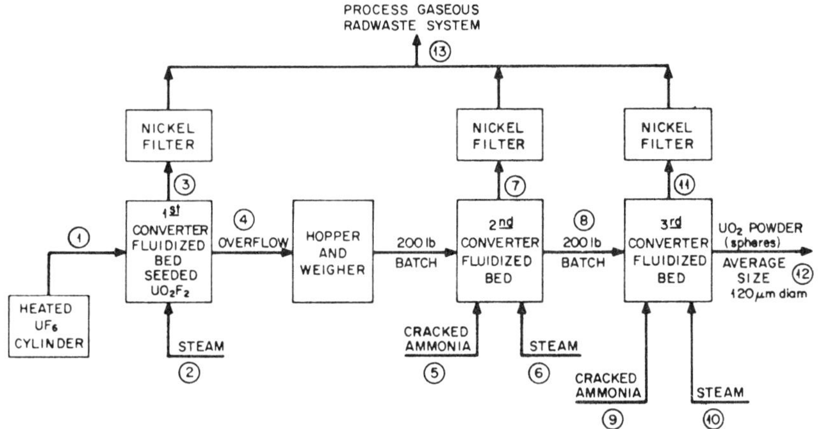

Source: ORNL-TM-4902

The reaction is $UF_6 + 2H_2O \rightarrow UO_2F_2 + 4HF$. The particles of UO_2F_2 overflow to a product hopper. The particles at this point are approximately 120 μm in diameter. After a given amount is accumulated, the batch is transferred to the next vessel where the bed is fluidized by steam and cracked ammonia. A second reaction yields $UO_2F_2 + H_2 \rightarrow UO_2 + 2HF$. A high percentage of the UO_2F_2 is converted to UO_2 in the second reactor, but the product goes into a third reactor where, by the same process, the reaction is carried to completion. The gaseous effluent from each of the three converter vessels passes through a sintered nickel filter in the top of each vessel before going to the gaseous effluent treatment system where HF and particulates are removed from the off-gas stream. The process gaseous flow rate is ~19,000 cfm. The ventilation gaseous flow rate is ~90,000 cfm. There is no liquid effluent from this process.

NUCLEAR POWER REACTORS

Present estimates of electrical power growth indicate a substantial increase in the growth of nuclear powered generating stations. By the year 2000 approximately 65% of the U.S. electrical generation is expected to come from nuclear energy.

In order to meet this projected demand, approximately 1,200 nuclear reactors with a capacity of 1-gigawatt each (1 GWe = 1,000,000 kilowatts) will be required. Projections of future technology indicate that the Liquid-Metal Fast Breeder Reactor (LMFBR) is expected to account for a substantial portion of this forecasted capacity. Based upon these projections only about 500 GWe will be from light-water-cooled reactors. However, at present, most nuclear power stations employ light water reactors.

The capacity of individual reactors has increased from 50 to 200 MWe in the early 1960s to 1,110 to 1,200 MWe (1.1 to 1.2 GWe) for advanced reactors presently being ordered by utilities.

Light Water Reactors (LWR)

A light-water-cooled nuclear power station operates on the same principle as a conventional fossil-fueled (oil or coal) power station except that the heat generation is by nuclear fission rather than combustion. The heat liberated in either process is used to convert water into steam. The steam enters a three-stage turbine consisting of one high-pressure stage and two low-pressure stages.

The turbine consists of a common central shaft attached to a circular array of curved blades. The steam impacts on these blades turning the rotor at high speeds. The turbine shaft is connected to a wire wound armature in the generator. This armature rotates in an applied magnetic field producing alternating electric current.

After passing through the turbine the low pressure steam passes through a condenser where the steam transfers its remaining heat to the condenser cooling water and is condensed back into water and is recycled to the boiler. The heated condenser cooling water may be released directly to the environment in a single-pass open-cycle cooling system. This heated water may have an adverse impact on aquatic organisms and the use of open-cycle systems is decreasing in favor of augmented cooling systems. This is important for nuclear power plants as they have a lower thermal efficiency (32%) than fossil-fueled plants (40%) and, consequently, discharge about two-thirds of their heat output to the environment.

There are two basic types of light-water-cooled nuclear reactors: the pressurized-water reactor (PWR) or indirect cycle and the boiling water reactor (BWR) which operates on a direct cycle. At the present time, pressurized-water reactors comprise approximately two-thirds of the light-water generating capacity committed through 1982. This 2:1 PWR:BWR ratio has been assumed to continue through the year 2000.

The fundamental difference between these two designs is evident from a comparison of the two reactor systems which are shown in Figures 1.5 and 1.6. In the pressurized-water reactor (Figure 1.5), the coolant is maintained at a high pressure (~2,250 pounds per square inch) which inhibits boiling. Steam is produced by allowing the heated primary coolant to transfer heat to a secondary coolant which is at a lower pressure (~1,000 pounds per square inch) where boiling can occur. Because the steam production is separated from the heat generation source, this mode of operation is termed an indirect cycle.

The boiling water reactor operates on a direct cycle where the process steam is generated directly in the reactor vessel. This is possible because of a lower reactor coolant pressure (~1,020 psi) than in the pressurized-water reactor. The steam generated in the reactor vessel is separated from excess moisture and passes directly to the first stage (high-pressure) turbine. The principal components of the boiling-water reactor system are shown in Figure 1.6.

Sources of Radioactive Discharges to the Environment: Nuclear power reactors generate radioactive materials as a consequence of the fissioning of uranium and by neutron absorption in the coolant and in structural materials which leads to induced radioactivity in these components.

The products of uranium fission comprise a large number of elements and include both stable and radioactive isotopes of these elements. Among the more important radionuclides produced by uranium fissioning are isotopes of the noble gases krypton and xenon, the alkali metals cesium and rubidium, the alkaline earths barium and strontium, and the halogens iodine and bromine.

The capture of the neutrons liberated in fission by the nuclei of stable elements often results in the production of radioactive activation products. The coolant activation products are generally gases such as argon-41, fluorine-18, nitrogen-13, nitrogen-16, and oxygen-19 which have short half lives in the range of several seconds to a few hours.

The induced activities in the structural materials may have considerably longer half lives and comprise a much wider range of elements including zirconium, manganese, nickel, iron, carbon, chromium, cobalt and copper. These radionuclides usually remain fixed in the structural materials but can enter the coolant as a consequence of corrosion and erosion in the pumps and other moving components.

Nuclear power reactors are constructed with multiple barriers for isolating these radionuclides from the environment. The principal barriers are: (1) the fuel cladding, (2) the reactor systems, and (3) the reactor and auxiliary buildings. Release of radioactive material to the environment occurs principally as a consequence of the penetration of one or more of these barriers. This penetration can occur due to the presence of structural defects, leakage from pumps or other components, or intentionally as a consequence of the particular plant design.

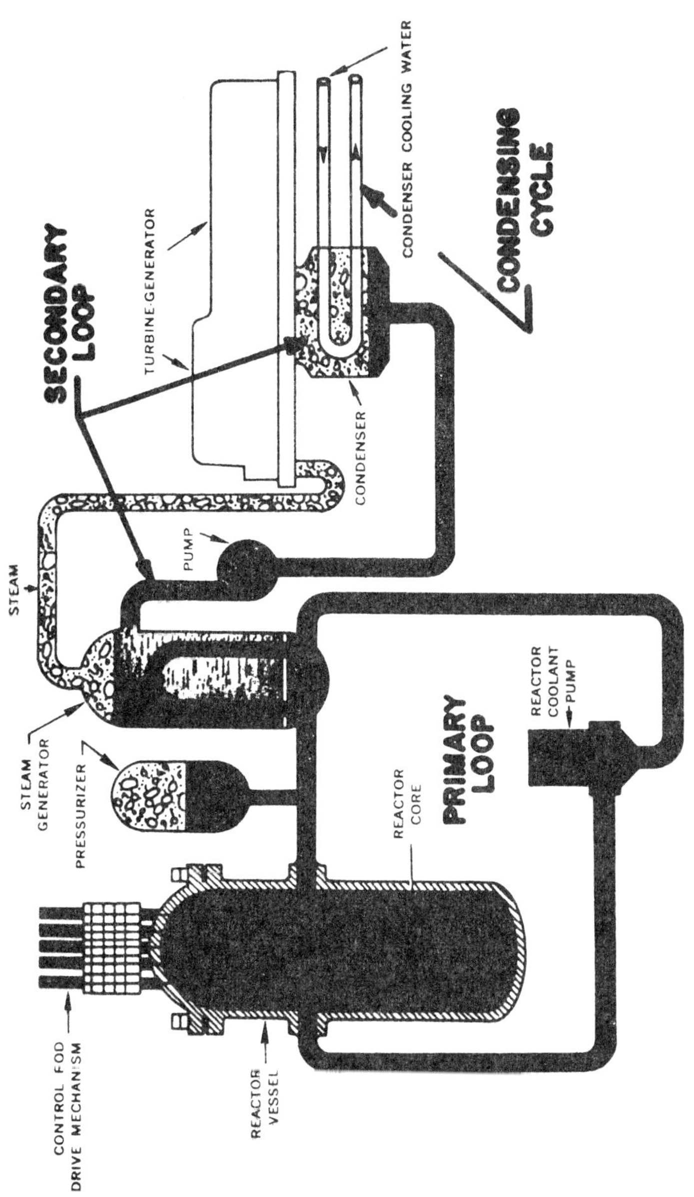

Source: PB-235 805

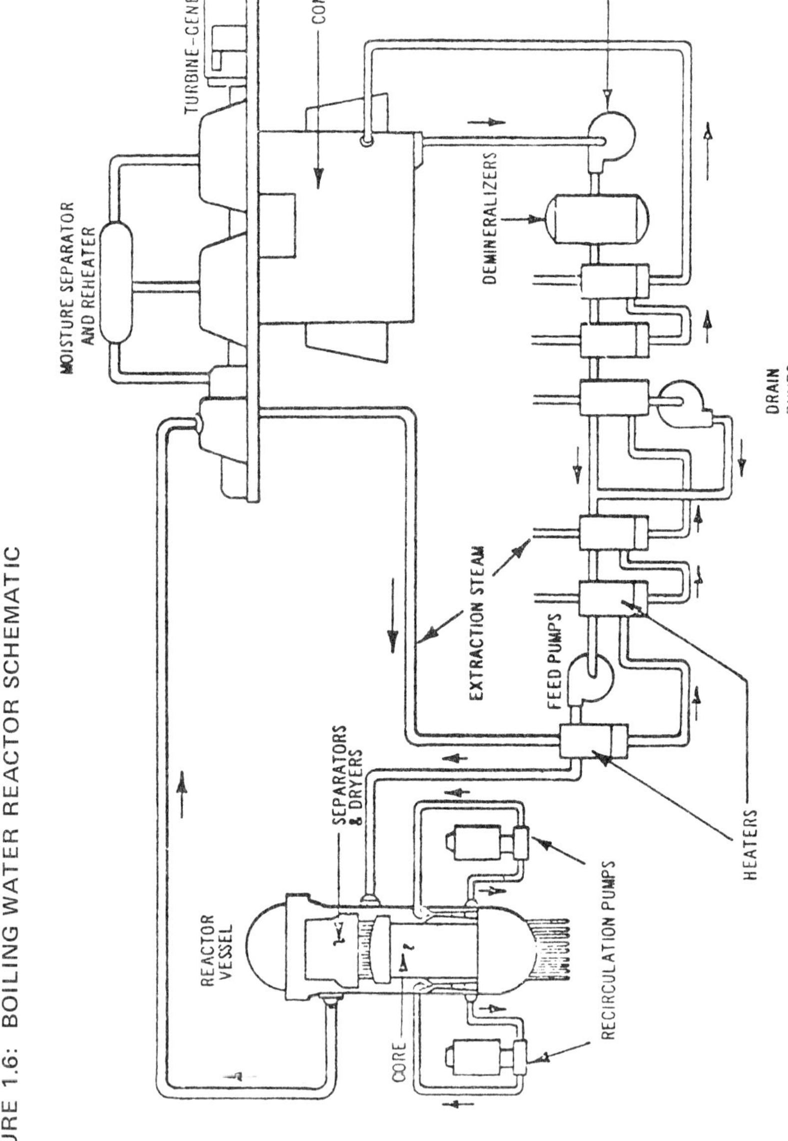

FIGURE 1.6: BOILING WATER REACTOR SCHEMATIC

Source: PB-235 805

Fuel Cladding Defects — The primary barrier for isolating radioactive fission products from the environment is the fuel rod cladding. Within a 1,000-MWe nuclear reactor there are millions of curies of radioactive isotopes; the iodine-131 alone can amount to over 70 megacuries. The major fraction of these fission products is retained within the ceramic matrix of the uranium dioxide fuel pellet. However, the more volatile elements such as the halogens and noble gases can diffuse through this matrix into the space between the fuel pellet and the cladding.

This diffusion process is accelerated by the high fuel temperatures and the presence of cracks and fissures produced by thermal stresses so that appreciable amounts of these elements accumulate in the fuel-cladding gap. Nonvolatile elements such as strontium, barium, and cerium also accumulate there as they are daughter products produced by the radioactive decay of short-lived noble gas precursor radionuclides. These radionuclides will be contained within the fuel rod as long as the thin metallic cladding remains intact.

In this situation, the quantity of radioactive material reaching the coolant will be limited to fission products arising from small traces of uranium which remain on the outer surface of the fuel as a consequence of the manufacturing process, activation products which arise from neutron induced reactions with water and air, and traces of metallic elements which enter the coolant as a result of corrosion of the reactor vessel, piping, and other structural components.

The fuel cladding is approximately 0.025 inch in thickness for pressurized water reactors and approximately 0.033 inch for boiling water reactors and is subjected to thermal stresses as the reactor power level is changed and mechanical stresses from the high pressure and velocity of the coolant or from physical contact with the fuel as it expands. These stresses, combined with variations in the cladding thickness or other irregularities in manufacture, can result in small pinholes or defects in the fuel cladding which allow the volatile radionuclides in the cladding gap to escape into the coolant.

Under severe conditions large failures could occur in the cladding which would permit the coolant to contact the fuel and leach out the less volatile fission products. However, these occurrences are not common and cladding failures of the small pin-hole type are more usual.

The extreme conditions imposed in the reactor core on the fuel cladding together with the difficulties of producing large quantities of thin, near-perfect tubing for the large number of fuel rods (approximately 40,000) make it extremely difficult to eliminate such fuel cladding failures. As a result, nuclear reactor systems are designed to accommodate the equivalent of 1% of the gap activity (contained in all of the fuel rods) escaping to the coolant through cladding defects.

Table 1.3 shows the relationships between the total core inventory, the fuel plenum (gap) inventory, and the primary coolant inventory for a representative 1,000 MWe light-water reactor. Even with defective fuel, the primary coolant activity remains a small fraction of the total inventory within the reactor. The coolant purification systems are responsible for removing most of the primary coolant activity so that low levels are maintained in the circulating coolant. Typical primary coolant radionuclide concentrations for a reactor having 1% failed fuel are shown in Table 1.4.

TABLE 1.3: VOLATILE RADIONUCLIDE INVENTORY IN A 1,000 MWe NUCLEAR POWER PLANT

Parameters: 3,040 MWt PWR, operating at full power for 500 days with 1% of the fuel rods having cladding defects*.

| | | | Total Activity In: | |
| | | Reactor Core (mega-curies)* | Fuel-Cladding Gap (mega-curies)* | Primary Coolant (curies) |
Radionuclide	Half-life			
Iodines				
I-131	8.05d	74.9	0.76	465
I-132	2.3 h	114.0	0.14	186
I-133	21. h	171.0	0.64	766
I-134	52. m	206.0	0.12	117
I-135	6.7 h	158.0	0.34	420
Kryptons				
Kr-85	10.8 y	0.66	0.067	334
Kr-85m	4.4 h	33.5	0.95	439
Kr-87	76. m	64.4	0.076	261
Kr-88	2.8 h	93.0	0.149	775
Xenons				
Xe-133	5.3 d	164.0	4.17	52,290
Xe-133m	2.3 d	4.0	0.019	692
Xe-135	9.2 h	43.6	0.084	1,488
Xe-135m	15.6 m	46.4	0.016	42

*Converted from 1,721.4 MWt to 3,040 MWt and core volume adjusted to scale.

Source: PB-235 805

BWR Condenser Air Ejector Off-Gas — The boiling-water reactor operates on a direct cycle and the contaminated coolant passes directly through the turbine. Entrained radioactive gases, air which has leaked into the condenser, and hydrogen and oxygen which result from the radiolytic dissociation of water are removed from the main turbine condenser by the steam jet air ejector which is used to maintain a vacuum in the condenser.

These gases are removed at a rate of about 300 cubic feet per minute. Approximately 230 cubic feet per minute represents the dissociated hydrogen and oxygen, 5 to 20 cubic feet per minute represents air in-leakage and the remainder is water vapor; the radioactive gases contributing negligible volume.

TABLE 1.4: ESTIMATED REACTOR COOLANT SPECIFIC FISSION PRODUCT AND CORROSION PRODUCT ACTIVITIES (AT 578°F) FISSION PRODUCT REACTOR COOLANT CONCENTRATIONS CORRESPONDING TO 1% FAILED FUEL

Isotope	µCi/cc	Isotope	µCi/cc
Noble Gas Fission Products		**Fission Products**	
Kr-85	1.11	Br-84	3.0×10^{-2}
Kr-85m	1.46	Rb-88	2.56
Kr-87	0.87	Rb-89	6.7×10^{-2}
Kr-88	2.58	Sr-89	2.52×10^{-3}
Xe-133	1.74×10^2	Sr-90	4.42×10^{-5}
Xe-133m	1.97	Y-90	5.37×10^{-5}
Xe-135m	0.14	Y-91	4.77×10^{-4}
Xe-138	0.36	Sr-92	5.63×10^{-4}
Total Noble Gases	187.3	Y-92	5.54×10^{-4}
		Zr-95	5.04×10^{-4}
Corrosion Products		Nb-95	4.70×10^{-4}
Mn-54	4.2×10^3	Mo-99	2.11
Mn-56	2.2×10^{-2}	I-131	1.55
Co-58	8.1×10^{-3}	Te-132	0.17
Fe-59	1.8×10^{-3}	I-132	0.62
Co-60	1.4×10^{-3}	I-133	2.55
Total Corrosion Products	3.7×10^{-2}	Te-134	2.2×10^{-2}
		I-134	0.39
		Cs-134	7.0×10^{-2}
		I-135	1.4
		Cs-136	0.33
		Cs-137	0.43
		Cs-138	0.48
		Ce-144	2.3×10^{-4}
		Pr-144	2.3×10^{-4}
		Total Fission Products	12.8

Source: PB-235 805

In the absence of appreciable failed fuel, the principal contributor to the radioactive emission is nitrogen-13. When there is significant failed fuel, the noble gas fission product releases dwarf the activation gas releases as shown in Table 1.5.

TABLE 1.5: ESTIMATED AIR EJECTOR OFF-GAS RELEASE RATES FOL-LOWING 30-MINUTE HOLDUP; 1,064-MWe BWR WITH 0.25% FAILED FUEL, 18.5 SCFM IN-LEAKAGE

Radionuclide	Half-Life	Emission Rate pCi/sec	Annual Discharge Ci/yr
Nitrogen-13	10 min	340	8,580
Krypton-83m	1.9 hr	2,537	64,000
Krypton-85m	4.4 hr	5,700	143,800
Krypton-85	10.8 yr	7.5	189
Krypton-87	76 min	15,700	396,000
Krypton-88	2.8 hr	17,367	438,000
Krypton-89	3.2 min	262	6,610
Xenon-131m	11.8 days	15	378
Xenon-133m	2.3 days	188	4,743
Xenon-133	5.3 days	5,100	128,700
Xenon-135m	9.1 hr	8,000	202,000
Xenon-135	15.6 min	17,367	438,150
Xenon-137	3.9 min	860	21,700
Xenon-138	17.5 min	26,500	668,600
Total		100,000	2,523,000

Source: PB-235 805

PWR Gaseous Radwaste System — In the operation of a PWR, boron is added to the primary coolant to act as a neutron absorber. In the beginning of the fuel cycle its concentration is approximately 1,000 ppm. As the reactor produces power, less and less boron is required. In order to remove this boron a small portion of the coolant purification flow is typically "bled" to the boron recovery system. Radioactive gases evolved at the gas stripper are routed to the waste gas system for treatment. Table 1.6 provides an estimate of the radio-activity releases from a waste gas system providing 45 days of holdup for these gases.

TABLE 1.6: REPRESENTATIVE ESTIMATED GASEOUS RELEASES ASSOCIATED WITH PRIMARY-TO-SECONDARY LEAKAGE (20 GALLONS PER DAY)

Radionuclide	Containment Purge	Waste Gas Processing System	Steam Generator Leakage
	- - - Annual Activity Release to the Environment - - - (curies per year) from		
Krypton-85	13.0	791	2.0
Krypton-87	0.04	–	3.0
Krypton-88	–	–	10.0
Xenon-131m	10.0	63	3.0
Xenon-133	1,005.0	1,500	682.0
Xenon-135	0.018	–	3.0
Xenon-138	0.007	–	2.0
Iodine-131	0.018	–	0.62

Source: PB-235 805

Liquid Radioactive Waste Treatment Systems — In BWRs and PWRs, various
sources of liquid waste are handled by liquid waste treatment systems. Each
reactor type provides for the purification of the reactor coolant. In BWRs this
system is simply referred to as the reactor water cleanup system (RWCS). In
PWRs, coolant purification (and chemical adjustment) is provided by the chemical and volume control system (CVCS) which itself may be classified into two
subsystems, the reactor coolant cleanup subsystem and the boron recovery subsystem.

The boron recovery subsystem of the CVCS in a PWR, and RWCS in a BWR,
may contribute radioactive liquids to the respective liquid radioactive waste
treatment systems in each type of reactor. These liquid radioactive waste treatment systems handle the miscellaneous radioactive liquids generated by plant
operation as well as those liquids from the coolant purification systems. Table
1.7 compares the magnitude of PWR liquid radioactivity releases from the CVCS,
the liquid radwaste system, and steam generator blowdown during a postulated
20 gallon per day primary-to-secondary leak.

**TABLE 1.7: COMPARISON OF ESTIMATED RADIONUCLIDE LIQUID DIS-
CHARGES TO THE ENVIRONMENT WITH APPRECIABLE
PRIMARY-TO-SECONDARY LEAKAGE (20 GALLONS PER DAY
LEAKAGE AND 10 GALLONS PER MINUTE BLOWDOWN)**

Major Radionuclides	- - - Activity (curies per year) Discharged from - - -		
	Chemical and Volume Control System	Waste Treatment System	Steam Generator Blowdown
Molybdenum-99	0.005	0.018	5.51
Technetium-99m	0.004	0.016	0.61
Iodine-130	0.002	0.006	0.009
Iodine-131	0.59	2.06	8.1
Iodine-132	0.056	0.19	0.12
Iodine-133	0.56	1.92	3.46
Cesium-134	0.004	–	7.1
Iodine-135	0.14	0.45	0.62
Cesium-136	0.001	0.005	2.05
Cesium-137	0.003	0.012	6.06
Total	~1.4	~4.7	~35

Source: PB-235 805

Primary-to-Secondary Leakage in PWRs — In pressurized water reactors, the secondary coolant system is isolated from the primary reactor coolant by virtue
of the tubing in the steam generators. If these tubes remain intact, the secondary
system would be free of radioactive material. The number of these tubes can
be about 4,000 per steam generator, depending upon the reactor make and plant
design power level, and the number of steam generators per plant may range
between 2 and 4, depending on the power rating and reactor vendor.

Due to the large number of tubes present in a plant, the possibility of defects
in the tubing either due to manufacturing errors, or operating conditions

(corrosion, burn-out, or stress) is enhanced. If holes develop in the steam generator tubing, the primary reactor coolant water will leak into the steam generator as a consequence of the higher primary system pressure. This water will contain radioactive materials at the concentration present in the primary coolant and consequently contaminate the secondary coolant system.

Volatile radionuclides which enter the secondary coolant system as a result of primary-to-secondary leakage may be discharged to the environment via two pathways: (1) the condenser air ejector and (2) the steam generator blowdown flash tank. The condenser air ejector removes entrained gases from the secondary coolant and will extract radioactive noble gases, gaseous activation products, and some of the halogens (iodines). In many designs, the steam is generated from evaporation of water in the steam generator.

Solids build up in the steam generator and may impair heat transfer. To counteract this buildup, typically 5 to 15 gallons of the steam generator bottom liquid are withdrawn per minute. This hot liquid is bled into a flash tank at a lower pressure than the secondary coolant system. The low pressure and high temperature cause the rapid evaporation (flashing) of 30 to 40% of the liquid into steam which, in older plant designs, was released to the atmosphere. The volatile radionuclides would also be carried over in the steam and, consequently, released.

The relative magnitude of these gaseous releases from the steam generator leakage are shown in Table 1.6. The major contribution from these releases is the additional iodine-131 which can be considerably greater than that from other sources.

In older plants, the blowdown liquid remaining in the flash tank would be discharged to the condenser coolant without any cleanup. Any radionuclides which were in the steam generator blowdown as a result of primary-to-secondary leakage would be released to the environment. If appreciable (20 gallons per day) primary-to-secondary leakage is present, concurrent with discernible fuel cladding perforations, the unprocessed steam generator blowdown could be the major source of liquid radionuclide discharges as illustrated in Table 1.7.

Operating experience has shown that pressurized water reactors eventually develop some primary-to-secondary leakage. Generally, with only a few tubes having defects, this leakage may amount to only a few gallons per day. However, several plants have experienced long periods of operation with leakage rates of 50 gallons per day or more. Because of the additional solids contributed from the boric acid in the primary coolant, high leakage rates cannot be tolerated for more than a few days. These high leakage rates require corrective action which usually means plugging up the defective tubes by sealing them off with plugs and small explosive charges or by welding.

System Leakage to Building Atmosphere — Release of radioactivity from reactor and waste treatment systems can occur via leakage directly from system components. Much of this leakage is from coolant pump or valve seals and is generally returned directly to the reactor coolant system. Other leakage paths include smaller valve seals and releases associated with chemical and radiological analysis. Most of the liquid released will be collected by plant drain systems and be processed by the waste treatment system.

The volatile elements, including the noble gases and halogens, will be released to plant buildings such as the containment building where they are available for leakage or discharge to the environment. In pressurized water reactors, this leakage may average between 0.2 to 0.3 gallons per minute and account for 0.4 to 1.0% of the coolant volume per day. Similar leakage rates may be expected at boiling water reactor plants. Thus appreciable quantities of the volatile elements may accumulate in the reactor containment building and auxiliary building atmospheres.

Containment Purging — Radioactive halogens and noble gases which escape to the reactor containment from the reactor system may be discharged to the environment during containment venting or purging. The containment atmosphere is vented or purged in order to test the containment isolation system on a periodic basis, reduce containment temperature and activity levels prior to and during maintenance involving entry into the containment, and also to reduce containment pressure if excessive system leakage exists. The purges for testing are typically of one to several minutes in duration and may occur on a monthly schedule. The other purging intervals may last several hours or more and may occur 1 to 10 times per year. Estimated releases of gaseous radionuclides via containment purging are shown in Tables 1.6 and 1.8 for the two types of light-water reactors.

TABLE 1.8: ESTIMATED GASEOUS RELEASES OF THE PRINCIPAL RADIONUCLIDES FROM MISCELLANEOUS EFFLUENTS AT A BWR STATION (CURIES PER YEAR PER UNIT)

Nuclide	Turbine Gland Seal	Turbine Building	Reactor Building and Containment	Mechanical Vacuum Pump
Krypton-83m	41	10	–	–
Krypton-85m	69	16	–	–
Krypton-85	–	–	–	–
Krypton-87	200	49	–	–
Krypton-88	220	53	–	–
Krypton-89	490	17	–	–
Xenon-131m	–	–	–	–
Xenon-133m	4	1	–	–
Xenon-133	120	29	–	1,445
Xenon-135m	320	82	–	–
Xenon-135	350	84	–	215
Xenon-137	900	290	–	–
Xenon-138	1,020	260	–	–
Iodine-131	0.041	0.547	0.012	–
Iodine-133	0.214	2.54	0.041	–

Source: PB-235 805

Gland Seal Leakage — Equipment with external moving parts such as valves and the coolant pumps contain a soft packing to retard the loss of fluid and steam from the reactor system. This packing does not provide total isolation and is a major source of the coolant system leakage described previously.

In a boiling water reactor, a similar condition exists with regard to the turbine generator shaft. As the steam passing through the turbine was generated in the reactor vessel, it contains volatile radioactive fission and activation products such as the noble gases and iodines. In order to reduce the loss of these volatile nuclides from the turbine, process steam is bled into the outer portions of the turbine seals and removed via a gland steam condenser. The nonvolatile radionuclides are condensed and the volatile radionuclides pass through a 2-minute delay line to the gaseous release discharge point.

Small quantities of the steam from the seal together with the entrained volatile radionuclides also escape to the turbine building atmosphere and are released to the environment unprocessed via the turbine building exhaust. Estimates of the total discharge of volatile radionuclides from the gland seal system are shown in Table 1.8.

Other Sources of Leakage — Leakage from the coolant purification and waste treatment systems will be collected by the auxiliary building floor drains. Volatile radionuclides which remain entrained in the liquids will be released to the auxiliary building ventilation system and roof vents on the reactor building to the environment. In addition, gases will be released from radiochemical fume hoods during sample analysis and from tank venting and purging operations. These releases are highly variable and depend on system design parameters, construction techniques, maintenance, sampling and venting frequencies, etc. Estimates of the releases from these sources are shown in Table 1.6 for the PWR and Table 1.8 for the BWR.

During reactor start-up, it is necessary to initially depressurize the cooling-water condenser. As a vacuum is drawn, the coolant present in the condenser will be partially degassed and the noble gases and a fraction of the halogens will be released. The number of start-ups and the intervals between shutdown and start-up (which represents a decay period) are highly variable. Estimates of this frequency are about 2 to 10 cold start-ups per year. The estimated gaseous releases for 2 to 3 startups per year are shown in Table 1.8 under the column "mechanical vacuum pump."

Atmospheric Steam Dumps from PWRs — In order to relieve high pressures in the secondary system from various abnormal operations (e.g., load rejection), PWR designs include a provision for relieving steam directly to the atmosphere through atmospheric steam dump valves. Of particular interest is the steam release which would accompany runback operations (i.e., rapid reduction of reactor power from 100% to a level at least high enough to supply the unit auxiliary load) via the main steam relief valves. The magnitude of this release can be on the order of many tens of thousands of pounds of steam in one minute. With primary-to-secondary leakage and failed fuel, radioactivity as well as steam would be released directly to the atmosphere. Although noble gases and some particulates would be released, the main concern would focus upon the release of radioiodine. Table 1.9 shows the estimated release resulting from the actuation of the main steam valves for a one-minute period.

Another direct atmospheric pathway for secondary system steam exists via the feedwater heater relief valve discharge. Radioactivity release is again predicted on the concurrent presence of failed fuel and primary-to-secondary leakage.

TABLE 1.9: PWR RADIOACTIVITY RELEASES VIA ATMOSPHERIC STEAM
DUMPS (Ci/yr FOR 20 gpd PRIMARY-TO-SECONDARY LEAKAGE
AND 0.2% FAILED FUEL IN A 3,358 MWt PWR)

| | - - - - - - - - - - Curies Discharged per Year - - - - - - - - - | |
Radioisotope	Main Steam Relief Valves (one minute release)	Feedwater Heater Relief Valves (one minute release)
Noble Gases		
Kr-85	9.06×10^{-5}	–
Kr-88	1.30×10^{-4}	–
Xe-133m	1.09×10^{-4}	–
Xe-133	9.78×10^{-3}	–
Xe-135m	3.18×10^{-4}	–
Xe-135	2.44×10^{-4}	–
Radioiodine		
I-131	6.60×10^{-4}	1.28×10^{-4}
I-132	8.66×10^{-5}	1.68×10^{-5}
I-133	6.44×10^{-4}	1.25×10^{-4}
I-134	9.22×10^{-6}	1.78×10^{-6}
I-135	1.83×10^{-4}	3.54×10^{-5}
Particulates		
Mo-99	1.18×10^{-4}	2.28×10^{-5}
Tc-99m	7.32×10^{-5}	1.41×10^{-5}
Te-132	6.02×10^{-6}	1.16×10^{-6}
Cs-134	6.78×10^{-6}	1.31×10^{-7}
Cs-137	3.54×10^{-5}	6.64×10^{-6}
All others	3.0×10^{-4}	4.0×10^{-7}
Total	1.27×10^{-2}	3.56×10^{-4}

Source: PB-235 805

Table 1.9 also shows the estimated releases resulting from the actuation of the
feedwater heater relief valves for a one minute period. On the order of ten
thousand pounds of steam may be released during this procedure.

Fusion Power Systems

Within the past few years, fusion research has moved from the realm of experimental physics into the stage where power reactor designs are being seriously
discussed. A national fusion power development program has come into being,
funded in excess of $200 million per year. Goals are being established for the
orderly development of fusion devices which are aimed at the operation of a
power-producing demonstration reactor in the 1990s.

Among the conceptual designs for large fusion power systems published in the
United States, the Wisconsin and Princeton designs represent the most complete
and best documented of the tokamak (closed toroidal) power system designs.
Both machines are deuterium-tritium fueled and breed tritium from lithium. The
main features of a conceptual tokamak power system are presented in Figure 1.7.
System features and operating characteristics for the two reference designs are
given in Table 1.10.

FIGURE 1.7: TOKAMAK FUSION POWER PLANT SCHEMATIC

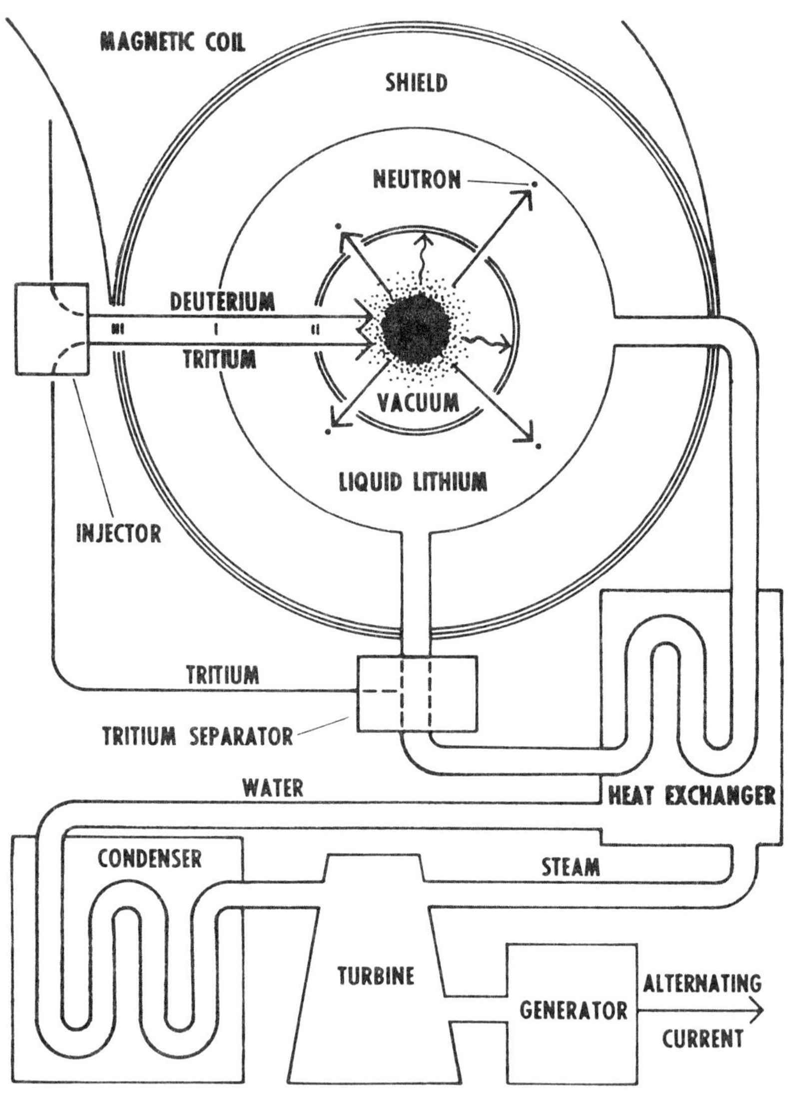

Source: PB-255 502

TABLE 1.10: TOKAMAK REACTOR REFERENCE DESIGN FEATURES

	UWMAK-I	PPPL
Power: MWt	4,660	5,305
MWe (net)	1,473	2,030
Net Electrical Efficiency (%)	32	38
Fuel Cycle	(D-T) Li	(D-T) Li
Toroidal Vacuum Chamber:		
Major Radius (meters)	13	10.5
Minor Radius (meters)	5	3.25
Vacuum - First Wall Structural Material	316 - stainless steel	PE-16 - austenitic nickel alloy
Shielding Materials	Pb, SS, B_4C	high-density concrete, Pb, B, polyethylene
Main Coolant	Lithium	Helium
Moderator--Tritium-Breeding Material	Lithium	$(LiF)_2BeF_2$ eutectic mixture (flibe)
Magnets: Super-Conducting Material	Nb Ti + Cu	Nb_3 Sn
Power Cycle	Li-Na-Steam	He-Steam
Plasma Burn Cycle:		
Burn Time (min)	90	100
Recovery Time (min)	6.5	3
Thermal Flywheel	Sodium storage loop	flibe diverted from blanket

Source PB-255 502

Like fission plants, fusion power plants can be expected to produce gaseous, liquid, and solid radioactive wastes. While fusion plants will not produce fission products and actinides, tritium and a variety of activation products will constitute the sources of radioactive waste. The majority of wastes associated with routine operation will be produced in the various tritium-handling and corrosion product removal systems in the blanket and plasma exhaust clean-up systems.

An idea of the magnitude of the tritium-handling problem can be gained from an examination of Figures 1.8 and 1.9. Figure 1.8 shows the principal fuel flows in the Princeton Plasma Physics Laboratory (PPPL) design and Figure 1.9 shows the tritium flows and environmental release sources for the Wisconsin design (UWMAK-1). For example, in the Wisconsin design, there are a minimum of 36 blanket and shield coolant loops from which tritium is recovered. In addition, there are tritium recovery systems in the vacuum systems and in the divertor system cooling loop.

Another major source of radioactive waste will be activated corrosion products
in the various coolant loops, since these regions lie within the zone of significant
neutron fluxes in the blanket. Details of treatment and handling systems and
procedures for this category of waste are not available in the designs under consideration.

FIGURE 1.8: PRINCIPAL FUEL FLOWS IN PRINCETON TOKAMAK

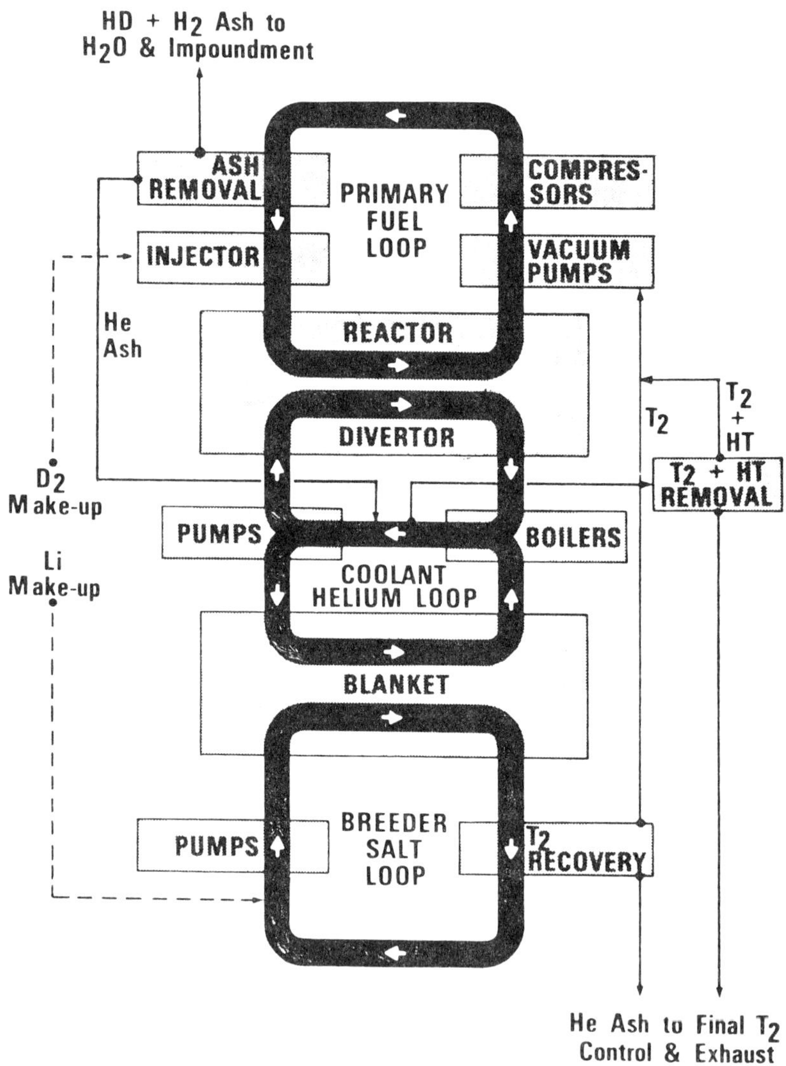

Source: PB-255 502

FIGURE 1.9: TRITIUM FLOW IN WISCONSIN TOKAMAK

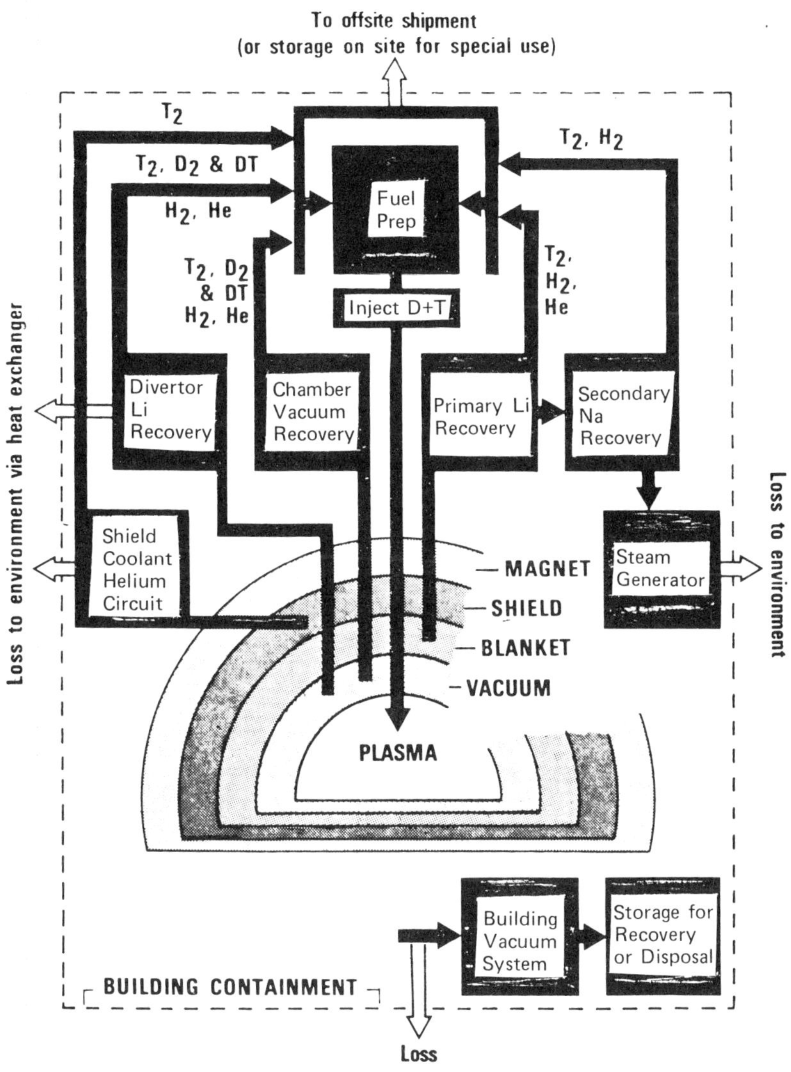

Source: PB-255 502

The Wisconsin design analysis includes a discussion of corrosion product formation in the coolant and the neutronic analysis of activation product formation in stainless steel structural materials. What is currently lacking is the coupled analysis to predict coolant system radioactivity inventories which are necessary to determine coolant-radioactivity clean-up system design requirements. It is expected that liquid metal cooled fast breeder reactor (LMFBR) design experience will be of value in this area.

Discussions of solid waste treatment systems are limited to general discussions of sources of activated components, such as vacuum and divertor wall replacement, failed pumps and equipment, and solids or sludges from the various tritium and corrosion product clean-up systems. Estimates can be made of amounts of solid waste expected and several modes of handling can be considered.

Gaseous wastes are expected from various tritium leakage paths, and neutron activation of air and coolant gas contaminants. Both conceptual designs include the use of containment-building vapor barriers around the main torus to control gaseous wastes. In fact, the Wisconsin design includes an evacuated primary containment building to limit tritium out-leakage.

Building ventilation systems and secondary sources of gaseous wastes, such as cooling system off-gassing, and turbine condenser air ejectors must be included in plans for general plant gaseous waste clean-up and treatment systems. Several hundred mols of gases are produced each day in the Princeton flibe blanket by neutron transmutations. The bulk of this is He, but significant amounts of ^{16}N, ^{19}O, ^{18}F and ^{20}F are produced. With the exception of ^{18}F, all have half lives of less than 30 seconds. The equilibrium inventory of 110 minute ^{18}F, however, is 6.6×10^8 Ci.

A similar situation exists with respect to liquid waste treatment systems. The balance of plant designs do not include integrated liquid waste treatment systems to handle the various expected sources of liquid radioactive wastes.

Radiological Effluents: Tritium, primarily, and additional amounts of activation products are expected to constitute the main radiological effluents. From routine operations, it is expected that both gaseous and liquid releases will occur. While expected release rates are unavailable, some discussion has occurred as to what practicable design tritium release rates might be. Inventories of various isotopes estimated for Wisconsin and Princeton Tokamak designs are summarized in Table 1.11. In comparison with light water reactors, the estimated tritium release rates for these fusion designs are about one to two orders of magnitude higher. To achieve this level of release requires that leak rates be one to two orders of magnitude lower than currently achieved in light water reactors. For example, a boiling water reactor (BWR) leaking about 100 Ci of tritium per year corresponds to a leakage fraction of about 10^{-5} of the reactor coolant inventory per day (7). Tritium diffusion rates through some fusion reactor candidate materials are not well known.

Gaseous effluents are expected from routine operation as indicated previously. Candidates for potential release include the large inventory of ^{18}F (7×10^8 Ci) evolved at a rate of 70 mols per day of fluorine from the flibe blanket in the Princeton Tokamak. Fluorine is very reactive chemically, and it is expected that most of the fluorine evolved will appear in the flibe as various compounds, such as fluorides, some of which may be gases.

It is apparent that there will be large potential sources for tritium release as HTO vapor and T and D-T gas. Another source of gaseous effluent is ^{41}A from neutron activation of air and cover or off-gases in various liquid metal systems. In the Princeton design, a major potential source of this isotope is the several hundred grams per hour of argon that is circulated through the plasma as an impurity for burn control and quenching.

TABLE 1.11: SUMMARY OF RADIOLOGICAL QUANTITIES ASSOCIATED WITH TOKAMAK DESIGNS

	UWMAK	PPPL
Tritium inventory, kg	13.5*	2.6
Estimated tritium environmental release rate (Ci/day)	10	6
Fractional tritium leakage of total inventory per day	0.7×10^{-7}	2.3×10^{-7}
Major isotope inventories in reactor blankets (Ci)	Mn-56 (2.0×10^8)**	F-18 (6.6×10^8)***
	Fe-55 (6.5×10^8)	Fe-55 (1.5×10^9)
	Co-58 (1.4×10^8)	Co-57 (2.5×10^9)
	Mn-54 (1.1×10^8)	Mn-54 (1.5×10^8)
	Co-60 (2.2×10^7)	Co-60 (1.5×10^8)
	V-49 (3.1×10^6)	V-49 (1.7×10^9)
	Ni-57 (5.2×10^6)	
Estimated annual quantities of solid waste (m^3/yr)		
First wall replacement	31.4	20.7 (from first wall and blanket)
Annual equivalent from replacement of blanket	62.6	
Corrosion products removed from coolant	0.3	45 (from flibe and structural material)

*Estimates of tritium inventory for UWMAK vary, other reports indicate higher values than the design report.

**First wall inventory only, based on 10-yr operation.

***Entire blanket inventory calculated for 30-yr operating history.

Source: PB-255 502

Sources of solid radioactive wastes from routine reactor operation include: sludges from various coolant clean-up systems, activated materials and components from wall and blanket replacement and failed equipment, and miscellaneous contaminated small tools and clean-up wastes.

In the Wisconsin design, an estimated 1,500 to 2,000 kg of stainless steel is expected to be dissolved in the lithium coolant per year (8). This material will have to be removed continuously to avoid fouling of heat transfer surfaces. An additional source of radioactive materials in coolants is from neutron sputtering of first wall materials into the coolants. It has been estimated that this mechanism could contribute up to 50% as much radioactivity as chemical corrosion alone (8).

Equilibrium Li coolant corrosion product inventory total radioactivity is estimated to be about 5×10^7 Ci. This indicates that the specific activity of coolant-borne corrosion products will be on the order of curies per gram. These materials from coolant clean-up systems will have to be handled as high-level radioactive waste requiring biological shielding, and perhaps cooling as well.

A major source of solid waste generated during routine operation is the activated structural material from periodic replacement of first walls and blankets. Some volume estimates obtained from the Wisconsin and Princeton design reports are included in Table 1.11. This material will also have a specific activity on the order of curies per gram, particularly from first wall material (9).

The solid waste handling problems posed by these materials are unique to fusion designs in that large volumes of high specific activity solid wastes are not associated with fission reactor routine operation. The conceptual fusion designs include provisions for remote handling of this material via manipulators and hot cells. It is likely that the volumes of this material that will actually have to be handled have been underestimated in the design reports, as it does not appear to be amenable to compaction. The importance of minimizing blanket and structural radioactivity has been recognized by fusion reactor designers and it is expected that future designs will improve on the Princeton and Wisconsin designs in this regard.

While fusion reactors do not require off-site shipment of highly radioactive spent fuel as with fission reactors, it appears that shipments of high-level radioactive waste will be required unless it is decided to locate fusion plants such that on-site, long-term storage of these wastes is feasible.

It can also be expected that an additional several thousand cubic feet per year of solid radioactive waste will be generated from miscellaneous sources as in any large nuclear power plant. These could include: filters from the heating and ventilation system, deactivated coolant traps as in an LMFBR, analytical laboratory and liquid waste treatment residues, contaminated small tools and parts, clean-up wastes such as towel wipes and plastic bags, and protective clothing.

Additional materials unique to fusion plants would include: vacuum system traps, tritium clean-up system beds, tritium getters, tritided metals used to contain tritium fuel, and failed fuel system components. Decommissioning of fusion power plants at the end of design lifetime would be expected to present waste disposal problems similar to those associated with LWR power plant decommissioning.

Tritium Production

The control of tritium is of increasing concern to the power reactor industry. Leakage of fission tritium, production from chemical shims, from control rods, from n-capture by water, from impurities in construction materials—can bring the concentration up to the $\mu Ci/ml$ level in reactor coolant and the release to the environment is a potentially serious adverse consequence of the extensive use of nuclear fission and fusion power sources. The present trend appears to be toward zero release to the environment. However, this approach substitutes an in-plant problem and only postpones the environmental one and either an acceptable means must be found for releasing tritiated water or a method for separation (concentration) and permanent storage must be developed. The problem, of course, is also an important one for the fuel reprocessing plants which handle 50 to 90% of the tritium produced in the reactors.

The projected growth of thermal power reactors indicates that in about two decades a production equal to the world's natural tritium inventory will be reached. In addition, the development of fusion power will add greatly to the tritium inventory.

J.W. Ray (10) has summarized the reactions occurring in a thermal power reactor which lead to tritium production. Chief among these is ternary fission (11) of U or Pu. In the case of ^{235}U the T yield is 0.87×10^{-4} per fission. Other significant sources are: n,T reactions with core construction materials, the ^6Li (n, α)T reaction on trace lithium purities, the n,T reaction with deuterium in the coolant and the production from burnable poisons such as boron. The contribution from photonuclear reactions in power reactors is small, about 0.01% of that produced by fission (12).

Thus, there exists considerable variability in tritium production in power reactors. For example, the tritium yield from ^{239}Pu is about twice that from ^{235}U (11). The production from the normal deuterium content (0.015% of the hydrogen atoms) of water is less than 0.1% of the fission yield. However, for heavy water moderated reactors the yield would be several times that from fission.

In the pressurized water reactor (PWR) a very significant source is the boron used as a "chemical shim", i.e., a chemical added to the coolant to level out the reactivity as a function of time. The rate of tritium production varies with the length of time the fuel has been in service and the coolant turnover rate. An average value would place the contribution from this source about equal to that from fission (12). This tritium becomes a part of the primary coolant. Boron used in control rods produces significant quantities, but the tritium will be largely fixed within the rods.

It is difficult to estimate the effect of controlled thermonuclear reactor (CTR) development on tritium production. As it is a fuel as well as a product, containment rather than disposal becomes the major problem. The overall tritium hazard does become magnified, however, due to the large quantities involved. On an energy basis it has been estimated that the production rate may be 10^5 times greater than from fission (11), and a single 3,000 MWe reactor could burn nearly 10^7 Ci/day, or a fair fraction of the present world inventory. Table 1.12 presents rough approximations of tritium yields associated with different nuclear reactors.

TABLE 1.12: TRITIUM YIELDS—3,000 MWe REACTOR

	Curies per Day
Thermal fission, U-235	50
Reactivity control (chemical shims), PWR	10-60
Reactivity control (chemical shims), BWR	2-15
1 ppm Li impurity in primary system	25
Deuterium in light water*	0.03
Heavy water moderated	80-150
Fast breeder	50-100
Thermal fusion reactor (CTR)	10^6-10^7

*Assumes doubling of natural D content by n reactions with light water.

Source: PB-244 928

Prediction of the total environmental tritium level is thus subject to considerable uncertainty. Jacobs (11) lists annual production rates of 2×10^6 Ci in 1980, 6×10^6 in 1990 and 15×10^6 in the year 2000 leading to accumulated values of 6×10^6, 32×10^6 and 96×10^6 Ci, respectively. These data are apparently Cowser's estimates (13). Petersen (10) includes contributions from all the sources discussed above and leads to production rates about a factor of 3 greater.

It would seem that Petersen's estimates of 2×10^7 Ci in 1980, 10^8 in 1990 and 3×10^8 in the year 2000 might be within a factor of two of the real values assuming (1) power reactors as the major source, and (2) no further H-weapon testing.

There are several aspects that must be considered in assessing the impact to the tritium production: the total production and world inventory, the concentration levels reached in process liquids, e.g., the primary coolant in a thermal reactor, the amounts actually released to the environment and the localization with respect to the environment. With thermal reactors which are the immediate sources of the increased tritium, most of the fission tritium will remain in the fuel until the chemical reprocessing step.

With Zircaloy-clad fuels probably only 0.1% escapes from the fuel elements (12) (14). Diffusion through the cladding or effusion through pinholes appear to be the primary mechanisms. The losses are higher by a factor of 10 if stainless cladding is used (14).

In any case, the reprocessing plant will be handling most of the tritium, probably 50 to 90%, associated with both thermal reactors and fast breeder reactors. If direct dissolution of fuel is used then the bulk of the tritium equilibrates with water and process chemicals, e.g., nitric acid and organic solvents. Thus, prudent tritium control in reprocessing dictates removal before the dissolution step. Otherwise, the disposal of waste H_2O could become a serious problem.

NUCLEAR REACTOR FUEL REPROCESSING

In atomic power reactors of the type expected to predominate at least through the 1980s, the fission process is carried out in sealed small diameter metal tubes designed to function as miniature pressure vessels. Inside these pressure vessels, commonly called cladding, uranium or plutonium fuel in the very stable oxide form is exposed to a flux of neutrons causing fission to occur with resultant release of energy within the fuel tube.

The fission energy is transferred in the form of heat to the water surrounding the tube and, directly or indirectly, produces steam to drive turbine-generators. By maintenance of an appropriate balance of design and operating parameters, the fission rate is controlled in a manner such that the fuel and cladding temperatures are kept in a range that assures the integrity of the fuel cladding throughout the operating life of the fuel. Data covering operation of nuclear power reactors demonstrates quite clearly that containment of fission product radioactivity within the fuel tubes has been very effective.

Over a period of several years the reactivity of the fuel decreases gradually thereby necessitating that the fuel be replaced. Upon removal from the reactor, the depleted fuel is stored under water for several months to permit the short-lived fission products to decay. At reactor shutdown, the fission product decay energy is approximately 5% of the rate of energy release during reactor operation. After 3 months of decay the fission product decay energy has decreased by a factor of approximately 75, and thus corresponds to less than 0.08% of the reactor operating energy. At this point the fuel, containing essentially all the fission product activity remaining from the power reactor operation, is shipped in U.S. Atomic Energy Commission-licensed casks to a reprocessing plant for recovery of unused uranium and plutonium and for encapsulation of residual radioactivity.

Essentially all the radioactivity leaving a nuclear power reactor is transferred to the fuel reprocessing plant. Commercial reactor fuel reprocessing has been developed from the U.S. government's atomic energy program spanning the last 25 years. In general, reprocessing consists of dismantling reactor fuel in a manner that permits dissolution of the core material without dissolving the corrosion-resistant cladding. The resulting solution is subsequently treated by several cycles of solvent extraction or ion exchange to recover, separate and purify the residual uranium and plutonium.

The fission product activity is consolidated, concentrated and stored for a prolonged period to permit the radioactivity to decay. In addition to this highly radioactive waste, varying waste streams containing smaller amounts of radioactivity may also arise from the cladding, wastewater and from chemical treatment of the solvent, water and off-gas.

Purex and Aquafluor Processes

The reprocessing of spent nuclear fuel will be performed for some years to come by either the Purex process or the Aquafluor process. Inasmuch as these operations require the processing of large quantities of more radiotoxic material than all other operations of the nuclear power cycle or nuclear industry in general, a brief description of these processes is included to identify the sources and the means of handling the radioactive wastes.

In both the Purex and the Aquafluor processes, the spent fuel, transported in heavily shielded casks from the reactor facilities to the reprocessing plant, is unloaded underwater and stored temporarily underwater until there is an elapsed time of about 150 days from the time of withdrawal from the reactor. This storage time allows the decay of many short, half-life fission products in the fuel. After suitable storage, the fuel assemblies are processed remotely in heavily shielded concrete cells.

To start, the fuel assembly (up to 15 feet long) is dismantled to separate the stainless steel or Zircaloy fuel-bearing tubes from the remainder of the assembly. The stainless and Zircaloy tubes are then chopped into small segments to expose the fuel and these segments are placed in nitric acid which dissolves the fuel and its fission products and leaves the tubing segment (hull) intact. This "chop-leach" operation produces contaminated fuel assembly hardware and hulls as solid waste, and it releases fission product gases which undergo treatment to remove radioiodines and particulates before discharge.

Next, the dissolved fuel is processed by means of solvent extraction to separate the uranium and plutonium from each other and from the fission products (high-level wastes). In the Purex operations, the fission product wastes from solvent extraction are concentrated to recover nitric acid and are stored in acid form in stainless steel tanks for up to 5 years before they have to be converted to solids in an acceptable form for ultimate disposal at a Federal Repository. In the Aquafluor process, the fission product wastes will not be stored temporarily as liquids, but will be calcined to a solid form. The calcined solids will be sealed in stainless steel cylinders which will be stored underwater to remove fission product heat.

Up through the first solvent extraction operation, the Purex and Aquafluor processes are very similar, and they vary only in subsequent purification of uranium and plutonium. The Purex process uses additional solvent extraction operations to purify both the uranium and plutonium, and the Aquafluor process purifies the uranium by fluorination.

Low-level liquid wastes in the reprocessing operation are primarily the effluent from nitric acid recovery. This and other low-level aqueous waste streams are treated by scavenging, precipitation, and ion exchange before discharge to a watercourse. In the following description of fuel reprocessing wastes, the wastes of three types of nuclear reactors are considered: LWR, HTGRs and LMFBRs.

The light water reactor is based upon the Diablo Canyon PWR. It was concluded from separate studies that the high-level wastes from BWR fuel will closely approximate those from PWRs on the basis of equal exposure. Plutonium recycle in the PWR assumes self-sustaining recycle that uses a mixture of assemblies that contain either enriched uranium (PWR-U) or plutonium mixed with natural uranium (PWR-Pu). These assemblies are assumed to be shipped and reprocessed together, such that the plutonium product and wastes will be blended.

At steady state: (1) 1.21 atoms of fissile plutonium replace 1.00 atom of ^{235}U in the plutonium-fueled assemblies; (2) 32.4% of the assemblies in the reactor are fueled with plutonium; and (3) the plutonium used in the plutonium-fueled assemblies has an initial fissile content of 61.3% and a fissile content at discharge of 55.4%. The plutonium product produced in the enriched uranium elements (70.64% fissile) is blended with the product of the plutonium elements (55.4% fissile) in order to prepare the plutonium for the next refueling (61.3%).

The model HTGR is based upon design studies of an 1,160-MWe HTGR that uses annual refueling (at 80% capacity factor) of approximately one-fourth of the core. Recycle of ^{233}U is assumed to begin in the third reload. The ^{235}U from the fissile makeup particles is assumed to be reprocessed separately from the ^{233}U-thorium particles and recycled one time before discard. This recycle of partially burned ^{235}U particles is also assumed to begin in the third reload.

The assumed properties of the Atomics International and General Electric Follow-On LMFBR designs are taken from normalized data that were developed by the Argonne National Laboratory.

High-Level Wastes and Cladding

High-level wastes are defined in federal regulations (15) as "those aqueous wastes resulting from the operation of the first cycle solvent extraction system, or equivalent, and the concentrated wastes from subsequent extraction cycles, or equivalent, in a facility for reprocessing irradiated reactor fuels." These wastes contain virtually all of the nonvolatile fission products, several tenths of 1% of the uranium and plutonium originally in the spent fuels and all the other actinides of the uranium formed by transmutation of the uranium and plutonium in the reactors. They can be generally characterized by their very intense, penetrating radiation and their high heat-generation rates. Regulations call for these wastes to be solidified within 5 years after they are generated and for the resultant stable solids to be shipped to a federal repository within 10 years after the liquids are generated.

Cladding wastes are the residual Zircaloy and stainless steel cladding and structural components of the fuel assemblies that remain after the fuel cores have been dissolved. Although their radioactivity arises mainly from neutron-induced isotopes, the hulls are similar in some respects to high-level waste in that they may contain up to 0.1% of the plutonium originally in the spent fuel, need biological shielding equivalent to several inches of lead, and have heat-generation rates of 50 to 100 W/ft^3.

LWR Fuels: In the reprocessing of LWR fuels, a chop-leach head-end method is used, followd by a standard Purex solvent extraction procedure for separating the uranium and plutonium and subsequently purifying the uranium, followed by an amine extraction for the final purification of plutonium.

Compositions, after concentrations, of the four liquid wastes from LWR fuel reprocessing are given in Table 1.13. The iron, nickel, and chromium in the concentrated fission product and alpha-active wastes are present as corrosion products. The caustic scrubber waste volumes and concentrations are based on the loss of 1% of the HNO_3 that was charged to the dissolver.

TABLE 1.13: COMPOSITIONS OF LWR AQUEOUS REPROCESSING WASTES

	Concentrations* (g/l)			
Component	Fission Product Waste	Alpha-Active Waste	Solvent Clean-up Waste	Caustic Scrubber Waste**
U^{6+}	0.80	0.018	2.6	–
Pu^{6+}	0.008	0.16	0.008	–
H^+	0.95	1.01	–	–
Na^+	–	–	159	190
Fe^{3+}	0.6	0.2	–	–
Ni^{2+}	0.08	0.02	–	–
Cr^{3+}	0.08	0.02	–	–
Fission products***	21.1	Trace	Trace	Trace
Actinides	4.43	Trace	Trace	Trace
CO_3^{2-}	–	–	18.2	–
OH^-	–	–	–	70
NO_3^-	98.0	62.9	388	255
PO_4^{3-}	0.1	–	Trace	–

*Concentrations are based on volumes, in gal/metric ton, of 330 for first column, 15 for second, 30 for third and 10 for fourth.
**Based on 1% HNO_3 loss from process.
***Based on a fuel exposure of 33,000 MWd/metric ton.

Source: ORNL-TM-3965

HTGR Fuel: The data for reprocessing HTGR fuel is based on methods being developed at Gulf General Atomic, Allied Chemical Corporation at Idaho Falls, Idaho, and ORNL (16). The fuel block is crushed prior to burning the graphite in a fluidized bed. The burning step is carried out in two parts: (1) an exothermic burning mode with heat evolution, and (2) an endothermic mode with heat addition. The residual fuel particles are separated from each other by screening and the ^{235}U (recycled) particles constitute a solid which is not processed further.

The fertile ThO_2-$^{233}UO_2$ (recycle) particles are dissolved, and after clarification, are purified by a modified Thorex solvent-extraction process (17). The ^{235}U (makeup) particles are crushed in a roll grinder to fracture the silicon carbide coating, and the residual carbon is burned off in an exothermic burner. After dissolution and clarification, the ^{235}U is recovered using a tributyl phosphate (TBP)-25 solvent extraction process flowsheet (18).

Compositions of the principal liquid and solid wastes from this process are given in Table 1.14. The combined fission-product wastes contain about 74% by weight of the fission products in the original fuel, the two solid wastes contain about 10%, and the remainder of the fission products appear as off-gases.

TABLE 1.14: COMPOSITIONS OF HTGR REPROCESSING WASTES

| | Weight (kg/metric ton) | | Concentration* (g/l) | | |
| | (1) | (2) | (3) | (4) | (5) |
Component	^{235}U Recycle Particle Waste	Combined Clarifier A and B Wastes	Combined Thorex and TBP Fission-Product Wastes	Combined Thorex and TBP Alpha Wastes	Solvent Cleanup Wastes
C	15	2.3	–	–	–
SiC	18	27.3	–	–	–
Si	–	–	0.05	–	–
Fe	–	–	0.40	–	–
Ni	–	–	0.11	–	–
Cr	–	–	0.07	–	–
Li	–	–	0.04	–	–
B	–	–	0.26	–	–
Mo	–	–	0.01	–	–
Al	–	–	1.68	–	–
Cu	–	–	0.01	–	–
Na	–	–	–	59.1	124
H^+	–	–	1.01	2.5	–
NO_3	–	–	115	31	305
SO_4	–	–	0.28	245	–
F	–	–	0.49	–	–
CO_3	–	–	–	–	15.1
PO_4	–	–	–	23.9	–
U	7.5	0.04	0.01	2.31	0.86
Th	–	0.5	0.23	8.78	19.1
Pu	0.5	0.002	0.28	1.04	0.018
Fission products	4.0	6	19.2	Trace	Trace
Actinides	0.7	Trace	0.38	Trace	Trace

*Concentrations based on volumes, in gal/metric ton, of 1,000 for (3); 25 for (4); and 25 for (5).

Source: ORNL-TM-3965

LMFBR Fuels: A number of alternatives are considered for reprocessing LMFBR mixed core and blanket fuel. Following a mechanical disassembly of the fuel element and separation of hardware that has only induced activity associated with it, the fuel tubes are sheared and heated to about 400°C in an oxygen atmosphere for removal of volatile fission products. The oxide fuel is then leached from the stainless steel cladding with HNO_3.

Laboratory tests show that from 1% to more than 10% of the cladding may dissolve during leaching. Accordingly, concentrations of stainless steel in the solvent extraction feed corresponding to these two percentages are considered. Primary and alternative dissolution and solvent extraction procedures may be used. Both utilize a TBP partitioning cycle, and TBP and amine purification cycles for plutonium, but in neither case is the uranium purified to an extent necessary for fuel recycle.

The primary method relies on concentration and geometry for criticality control and on ferrous ion for reduction of the plutonium in the TBP cycles. The alternative procedure assuming about 300 grams of uranium and plutonium in the feed, uses boron as a soluble poison for criticality control and U(IV) plus hydrazine for reduction of the plutonium. The compositions of the concentrated wastes, taking into consideration the alternative procedures are given in Table 1.15.

Characteristics of High-Level Solidified Wastes and Cladding: Estimated characteristics of the residues obtained from evaporation and heating aqueous wastes to 900°C are given in Tables 1.16, 1.17 and 1.18. In addition to the solids obtained from solidifying the individual waste streams, solidification of several combinations of these streams is considered. For convenience, the solid waste components are generally expressed as oxides; however, in a few instances some elements are expressed as tetraborates, sulfates, and fluorides.

Those solidified wastes in Tables 1.16 and 1.18 that contain 70% voids are calcines which may be incorporated in borosilicate or other glasses to yield void-free products that contain 50 weight percent oxides from waste. Formation of such glassy products doubles the weight of waste per metric ton of fuel; however, the volume of waste per metric ton remains the same and the thermal conductivity of the glassy product is about four times that of the calcine. Similarly, the solid wastes of Table 1.17 that contain 50% voids are powders which may be incorporated in glasses. This would double their weights, increase their volumes by 10 to 20% and increase their thermal conductivity by factors of 4.4 to 4.9.

The solidified wastes in Table 1.16, 1.17 and 1.18 which are described as dispersions of waste solids in sodium tetraborate (or in sodium sulfate and phosphate) may also be made into glasses. This would increase their weights by factors of 1.3 to 1.4, increase their volumes by factors of 1.4 to 1.5, and decrease their thermal conductivities by factors of 1.1 to 1.2. The weights and volumes of cladding wastes from LWR and LMFBR fuels, given in Tables 1.16 and 1.18, were calculated from fuel element specifications, assuming that the cladding and its associated hardware is compacted, on the average, to 70% of theoretical density (19).

Recommended Volumes of Solidified High-Level Wastes: Based on the above analysis, combined volumes of solidified reprocessing wastes (excluding cladding) should lie in the neighborhood of 1 to 2 cubic feet per ton of LWR fuel, 4 to

TABLE 1.15: COMPOSITIONS OF LMFBR AQUEOUS REPROCESSING WASTES (MIXED CORE AND BLANKETS)

	Concentration* (g/l)							
	Fission Product Wastes				Alpha-Active Wastes			
	Without Boron		With Boron					
	(1)	(2)	(3)	(4)	(5)	(6)	(7)	(8)
Component	1% Cladding Dissolved	10% Cladding Dissolved	1% Cladding Dissolved	10% Cladding Dissolved	With $Fe(NO_3)_2$ Reductant	With U^{4+} Reductant	Solvent Cleanup Wastes	Caustic Scrubber Wastes**
Fe^{3+}	2.10	21.0	2.10	21.0	105	—	—	—
Cr^{3+}	0.55	5.54	0.55	5.54	—	—	—	—
Ni^{++}	0.26	2.6	0.26	2.62	—	—	—	—
U^{6+}	0.69	0.69	0.69	0.69	—	0.23	0.76	—
Pu^{6+}	0.08	0.08	0.08	0.08	0.35	0.35	0.18	—
H^+	1.01	1.01	1.01	1.01	1.01	1.01	—	—
Na^+	—	—	—	—	—	—	136	192
H_3BO_3	—	—	82.5	82.5	—	—	—	—
$CO_3^=$	—	—	—	—	—	—	16.1	—
NO_3^-	109	195	109	195	413	62.3	332	258
OH^-	—	—	—	—	—	—	—	71
Fission products***	23.1	23.1	23.1	23.1	Trace	Trace	Trace	Trace
Actinides***	0.64	0.64	0.64	0.64	Trace	Trace	Trace	Trace

*Concentrations are based on volumes, in gal/metric ton, of 330 for fission product wastes [(1), (2), (3), (4)], 15 for alpha-wastes [(5), (6)], 30 for (7) and 30 for (8).

**Based on 1% HNO_3 loss from process.

***Based on an average exposure of 30,000 MWd/metric ton.

Source: ORNL-TM-3965

TABLE 1.16: CHARACTERISTICS OF LWR SOLIDIFIED AQUEOUS REPROCESSING WASTES AND CLADDING

Characteristic	(1) Fission Product Waste	(2) Alpha- Active Waste	(3) Solvent Cleanup Waste	(4) Caustic Scrubber Waste	(5) Cladding	Combinations (6) (1) + (2) + (3)	(7) (1) + (2) + (4)	(8) (1) + (2) + (3) + (4)
Composition, kg/metric ton								
U_3O_8	1.17	0.001	0.348	—	1.1[*]	1.52	1.17	1.52
PuO_2	0.011	0.011	0.001	—	0.01[*]	0.023	0.022	0.023
$Na_2B_4O_7$	—	—	79.3	31.6	—	79.3	31.6	111
Fe_2O_3	1.07	0.016	—	—	—	1.09	1.09	1.09
Cr_2O_3	0.15	0.002	—	—	—	0.15	0.15	0.15
NiO	0.13	0.001	—	—	—	0.13	0.13	0.13
P_2O_5	0.09	—	Trace	—	—	0.09	0.09	0.09
FP oxides	31.5	Trace	Trace	—	0.02[*]	31.5	31.5	31.5
Actinide oxides	6.28	Trace	Trace	—	—	6.28	6.28	6.28
Zircaloy-4, Inconel	—	—	—	—	270	—	—	—
Weight, kg/metric ton	40.4	0.031	79.6	31.6	270	120	72.0	152
Volume, ft^3/metric ton	0.74[**]	0.001[**]	1.18[***]	0.47[***]	2.12†	1.41[***]	0.69[***]	1.88[***]
k_e, Btu/hr/ft/°F at 500°C	0.26	0.26	0.55	0.55	3.6††	1.02	1.48	0.96
Density, g/ml	1.92[**]	2.07[**]	2.37[***]	2.37[***]	4.5†	3.01[***]	3.67[***]	2.86[***]

[*]Assumes 0.1% Pu and U, and 0.05% of the fission products, are lost to cladding.
[**]Product assumed to be 70% voids.
[***]Product assumed to be void-free dispersion of solids in sodium tetraborate.
†Cladding compressed to 70% of theoretical density.
††Conductivity at 100°C.

Source: ORNL-TM-3965

TABLE 1.17: CHARACTERISTICS OF HTGR SOLIDIFIED REPROCESSING WASTES

Composition, kg/metric ton	(1) ^{235}U Recycle Particle Waste	(2) Combined Clarifier A + B Wastes	(3) Combined Thorex and TBP Fission Product Wastes	(4) Combined Thorex and TBP Alpha-Wastes	(5) Solvent Cleanup Wastes	. . . . Combinations. . . . . (6) (3) + (4) + (5)	(7) (1) + (2) + (3) + (4) + (5)
C	15	2.3	—	—	—	—	17.3
SiC	18	27.3	—	—	—	—	45.3
$Na_2B_4O_7$	—	—	113*	—	51.4**	155***	240†
Na_2SO_4	—	—	—	34.3††	—	34.3	34.3
Na_3PO_4	—	—	—	3.90	—	3.90	3.90
U_2SO_4	—	—	1.21	—	—	1.21	1.21
AlF_3	—	·	2.73	—	—	2.73	2.73
Al_2O_3	—	—	10.4	—	—	10.4	10.4
Fe_2O_3	—	—	2.16	—	—	2.16	2.16
NiO	—	—	0.53	—	—	0.53	0.53
Cr_2O_3	—	—	0.39	—	—	0.39	0.39
SiO_2	—	—	0.40	—	—	0.40	0.40
MoO_3	—	—	0.06	—	—	0.06	0.06
CuO	—	—	0.05	—	—	0.05	0.05
UO_2	8.52	—	—	—	—	—	8.52
U_3O_8	—	0.047	0.045	0.258	0.096	0.398	0.446
ThO_2	—	0.569	0.991	0.964	2.06	3.99	4.56
PuO_2	0.567	0.002	1.20	0.112	0.002	1.31	1.88
FP oxides	4.80	7.2	91.30	Trace	Trace	91.30	103.3
Actinide oxides	0.794	Trace	1.633	Trace	Trace	1.633	2.427
Weight, kg/metric ton	47.7	37.4	226	39.5	53.5	309	480
Volume, f^{-3}/metric ton	1.14†††	0.79†††	2.41§	0.51§§	0.77§	3.56§	5.79§
k_e, Btu/hr/ft/°F at 500°C	0.41	0.41	1.42	1.5	0.60	1.13	1.39
Density, g/cc	1.48	1.68	3.30	2.73	2.44	3.07	2.93

*1.05 kg of sodium per metric ton added to balance boron; an additional 103 kg of $Na_2B_4O_7$ per metric ton added to give 50 wt % waste solids in tetraborate.

**35.5 kg of B_2O_3/metric ton added to balance sodium.

***10.7 kg of B_2O_3/metric ton added to balance sodium; an additional 135 kg of $Na_2B_4O_7$/metric ton added to give 50 wt % waste solids in tetraborate.

†10.7 kg of B_2O_3/metric ton added to balance sodium; an additional 220 kg of $Na_2B_4O_7$/metric ton added to give 50 wt % waste solids in tetraborate.

††5.50 kg of sodium per metric ton added to balance sulfate and 1.64 kg sodium per metric ton added to balance phosphate.

†††Assumed to be 50% voids.

§Assumed to be void-free dispersion of solids in tetraborate.

§§Assumed to be void-free dispersion of solids in sodium sulfate and phosphate.

Source: ORNL-TM-3965

TABLE 1.18: CHARACTERISTICS OF LMFBR SOLIDIFIED AQUEOUS REPROCESSING WASTES AND CLADDING

| Characteristic | Fission Product Wastes | | | | Alpha-Active Wastes | | | | Cladding | | Combinations | | | |
| | Without Boron | | With Boron | | | | | | | | | | | |
	(1) 1% Cladding Dissolved	(2) 10% Cladding Dissolved	(3) 1% Cladding Dissolved	(4) 10% Cladding Dissolved	(5) With $Fe(NO_3)_2$ Reductant	(6) With U^{4+} Reductant	(7) Solvent Cleanup Wastes	(8) Caustic Scrubber Wastes	(9) Alpha-Contaminated	(10) Induced Activity Only	(11) (1)+(5)+(7)+(8)	(12) (2)+(5)+(7)+(8)	(13) (3)+(5)+(7)+(8)	(14) (4)+(5)+(7)+(8)
Composition, kg/metric ton														
Fe_2O_3	3.75	37.5	3.75	37.5	8.52	—	—	—	—	—	12.27	46.02	12.27	46.02
Cr_2O_3	1.01	10.1	1.01	10.1	—	—	—	—	—	—	1.01	10.1	1.01	10.1
NiO	0.42	4.17	0.42	4.17	—	—	—	—	—	—	0.42	4.17	0.42	4.17
PuO_2	0.11	0.11	0.11	0.11	0.02	0.02	0.02	—	0.03*	—	0.15	0.05	0.05	0.15
U_3O_8	1.02	1.02	1.02	1.02	—	0.02	0.10	—	0.3*	—	1.12	1.12	1.12	1.12
$Na_2B_4O_7$	—	—	86.5	86.5	—	—	67.9	95.8	—	—	164	164	164	164
FP oxides	34.2	34.2	34.2	34.2	Trace	Trace	Trace	Trace	0.02*	—	34.2	34.2	34.2	34.2
Actinide oxides	0.91	0.91	0.91	0.91	Trace	Trace	Trace	Trace	—	—	0.91	0.91	0.91	0.91
Stainless steel	—	—	—	—	—	—	—	—	650	700	—	—	—	—
Weight, kg/metric ton	41.4	88.0	128	175	8.54	0.04	68	95.8	650	700	214	261	214	261
Volume, ft^3/metric ton	0.81**	1.84**	1.53***	1.84***	0.19**	0.0003**	1.01***	1.43***	4.17†	4.50†	2.74***	3.05***	2.74***	3.05
k_e, Btu/hr/ft/°F at 500°C	0.26	0.26	1.02	1.41	0.26	0.26	0.55	0.55	9.0††	9.0††	0.90	1.13	0.9	1.13
Density, g/ml	1.81**	1.69**	2.95***	3.35***	1.56**	2.90**	2.37***	2.37***	5.50†	5.50†	2.75***	3.02***	2.75***	3.02***

*Assumes 0.03% Pu and U, and 0.05% of the fission products are lost to the cladding.

**Product assumed to be 70% voids.

***Product assumed to be void-free dispersion of solids in sodium tetraborate.

†Cladding compressed to 70% of theoretical density.

††Conductivity at 100°C.

Source: ORNL-TM-3965

6 cubic feet per ton of HTGR fuel, and 1 to 3 cubic feet per ton of LMFBR
fuel. The actual volumes will, of course, depend on many factors that cannot
be resolved without benefit of actual operating experience.

Other Reprocessing Wastes

Noble Gases: The noble gas fission products consist principally of stable and
short-lived isotopes of krypton and xenon. Typically, from 5 to 6 kg is present
in each ton of spent fuel from LWRs and LMFBRs, and while xenon comprises
about 95% of the weight of the mixture, the only radioisotope remaining after
150 days decay is 10.8-y ^{85}Kr. These gases are discharged to the atmosphere
following interim holdup for decay.

Over 99% of them are released at reprocessing plants, and the remainder are discharged at power stations as a consequence of their leakage through defective
fuel cladding. Although off-site radiation exposures have been small as compared with current guidelines for population exposure, the noble gases will probably be recovered in the near future as a result of stated policies of maintaining
radioactive releases to the environment at the lowest practicable levels.

Several processes are available or are under development for recovery of these
gases from plant off-gas streams (20). Initially, it is likely that the mixed gases
will be collected under high pressure in gas cylinders and then shipped to a large
remote site and stored under conditions that will promote long-term integrity
of the cylinders.

Iodine: Iodine is a semivolatile fission product which, because of its complex
physical and chemical properties and its high biological significance, has always
required special attention to ensure adequate safety in its management. About
0.01 gram of iodine isotopes is formed per MWd (thermal), and the isotope of
greatest concern during reactor operation and fuel reprocessing is 8.05-d ^{131}I.
However, the species of consequence to longer-term waste management is 1.6 x
10^7-y ^{129}I, which comprises about 75% of the weight of the fission product iodine
isotopes.

Research and development work aimed at reducing iodine releases from fuel
reprocessing plants to "near zero" levels shows promise of removing at least
99.9% of it from the other fuel constituents by volatilization at the head-end
of the process (21). Once separated, it would subsequently be trapped and retained separately from the other waste streams. The final form into which the
iodine may be processed for packaging, shipment, and disposal has not been
defined. Such concentrates of ^{129}I must be stored permanently because of its
very long half-life. As a consequence, disposal by isotopic dilution and dispersal,
or by transmutation, may ultimately be adopted.

Tritium: Tritium wastes are generated at nuclear power stations and at fuel reprocessing plants. Tritium in light water reactor wastes arises principally from
neutron reactions with light elements such as lithium and boron that may be
present in the primary coolants. New power stations may have the capability
for recycling the coolant until the tritium concentration reaches about 5 μCi/ml,
and then to maintain that concentration by additions of fresh water to the coolant as required. Operating experience with this technique is lacking; however,

it is estimated that, on the average, about 20,000 gallons of tritiated water per year will be bled from the primary circuit of each 1,000-MW LWR. This waste can be stored in tanks at the power stations until it is shipped to disposal sites.

About 0.02 to 0.03 Ci of ^{3}H per MWd (thermal) or 700 to 1,200 Ci/ton of fuel, is produced in fission and appears in wastes from fuel reprocessing. A few percent may be associated with the fuel cladding, but most is present with the core materials and is eventually released as water vapor to the atmosphere. Future plants having head-end operations like "voloxidation" (22) should be able to separate and recover the tritium in a relatively small volume. This concentrate could then be converted into an appropriately stable, solid form [such as $Ca(OH)_2$, perhaps], packaged, and shipped to a designated storage or disposal site.

There is evidence that tritium may act differently in LMFBRs than in other reactor types. As much as 95% may diffuse through the stainless steel cladding during reactor operation and appear as a sodium tritide sludge in the primary-coolant cold traps. If experience bears out this behavior, processes for recovery and packaging of the tritium from LMFBR fuels will be installed at the power stations rather than at the reprocessing plants.

Alpha Waste: The so-called "alpha" wastes are defined as those solid materials that contain plutonium or other long-lived alpha emitters in concentrations greater than 10 μCi/kg, and yet have sufficiently low external radiation levels that they can be handled directly without supplementary shielding (surface dose rates of $\leqslant$10 mrem/hr). The criterion of 10 μCi/kg is adopted as a lower activity density for alpha wastes because this corresponds to the upper range of alpha-emitting isotopes in naturally occurring deposits. It is reasonable that wastes containing less alpha activity than 10 μCi/kg be regarded as disposable in carefully selected burial grounds, whereas alpha wastes must be stored in special repositories that offer maximal assurance of permanent containment.

Alpha wastes arise principally at fuel preparation and fabrication plants, and to a lesser extent at fuel reprocessing plants (23)(24). They consist of a wide assortment of solid materials including items made of paper, cloth, wood, plastic, rubber, glass, ceramic and metal, as well as salts and sludges that arise in the treatment of liquid waste streams and filters from clean-up of off-gas.

The transuranium-element contents of these wastes range from trace amounts to several grams per cubic foot, and average about 0.25 gram per cubic foot. The densities of the uncompacted wastes vary from 2 to 200 lb/ft^3. About one-half to two-thirds of these wastes (by volume) are combustible and can be reduced via incineration by factors of about 50 and about 20 in volume and weight, respectively. About one-half to three-fourths of the wastes (by volume) can be reduced in volume by factors of 2 to 10 through compaction.

A survey of operations at AEC laboratories and production facilities indicates that future large plants will generate about 10,000, 20,000, and 4,000 cubic feet of uncompacted alpha waste per ton of plutonium or ^{233}U processed in fuel preparation, fabrication, and reprocessing, respectively (23). Thus, an average plutonium or ^{233}U concentration in the waste of 0.25 gram per cubic foot represents fuel losses of 0.25% in preparation, 0.5% in fabrication and 0.1% in reprocessing.

Alpha-Beta-Gamma Wastes: Alpha-beta-gamma wastes are defined as those solid materials, other than high-level and cladding wastes, which contain long-lived alpha activities greater than 10 μCi/kg and have gamma radiation levels sufficient to require biological shielding and remote handling techniques (typical surface dose rates lie between 10 and 1,000 mrem/hr). They arise at fuel reprocessing plants and consist of an assortment of materials similar to alpha wastes. About 10,000 ft^3 of this waste is generated per ton of plutonium or ^{233}U processed, and it contains an average of 0.025 gram of plutonium or uranium per cubic foot (23).

Beta-Gamma Wastes — These diverse solid wastes are common to all facilities handling radioactive materials. They range from concentrates of the radionuclides generated in the decontamination of plant effluent streams to almost every conceivable type of contaminated solid refuse from plant operations. They normally require minimal shielding, and since they contain less than 10 μCi of long-lived alpha activity per kilogram, they are disposable in surface burial grounds. Based on operating experiences at AEC sites, it is estimated that 40,000, 80,000 and 16,000 ft^3 per ton of plutonium or ^{233}U will be generated at fuel preparation, fabrication and reprocessing plants, respectively (23). The radioactivity level will average 0.001 Ci/ft^3.

RADIONUCLIDES USED FOR MEDICAL AND INDUSTRIAL APPLICATIONS

Carbon-14, Cobalt-60, Iridium-192 and Radium-226

These four radionuclides are representative of the radioisotopes of commercial interest which are generally produced, distributed, and used by the private sector of the economy. With the exception of radium-226, these radioactive materials are either directly, or indirectly, subject to federal regulations. In many states individuals can purchase radium without proof of competence to handle it safely. It is interesting to note that two-thirds of the companies selling radium devices are located in nonagreement states and account for 91% of the total devices sold (25).

Radium-226 is a naturally occurring radionuclide first discovered in 1898. It has the longest history of use of any radioactive material and is also one of the most hazardous radionuclides. The ingestion of the luminous dial paint prepared from radium-226 was the cause of death of many of the early dial painters before the hazard was fully understood. Much of what is known of the biological effects of ionizing radiation on man is based on the effects of radium ingested by these early watch dial painters. It is estimated that over 3 million timepieces containing radium are sold annually and the exposure to the total population is probably greater than that from all other consumer products containing radioactive material.

Radium-226 has a half-life of 1,602 years and decays by the emission of high energy, 5.68 Mev alpha particles. Carbon-14, cobalt-60, and iridium-192 are produced in nuclear reactors by the bombardment of a target material with neutrons. Their principal use is in the fields of nuclear medicine, radiation detection and control equipment, and in nuclear power sources. Carbon-14 has a half-life of 5,730 years and emits only a 0.156 Mev beta particle.

Radium-226 has been and still is used in a wide variety of products and applications. In the past 10 years there has been a decrease in radium usage made possible by the increased availability and acceptance of use of other less hazardous radioactive materials. Radium-226 is used in producing timepieces, electron tubes, record player brushes, gauges, fire detectors, and in various self-luminous products. It is used in medicine for the treatment of tumors, superficial skin lesions, lymphoid tissue and other diseases. It is estimated that 330 curies of radium-226 contained in 50,000 sources are used in medical applications at 2,300 facilities which provide approximately 85,000 medical treatments per year (25).

Carbon-14 is used in medicine to study metabolic diseases and in radioisotope gauges. Iridium-192 is widely used in the field of medicine in radiographic units. Cobalt-60 is used in isotopic power devices, in teletherapy units to treat cancer, and in radiation processing applications. Presently, there are approximately 1,830 cobalt-60 teletherapy units in use with each unit containing about 3,000 curies of cobalt-60 (26).

In almost all applications, these materials are used in the solid form. They are generally used as a sealed source and the material is retained within the encapsulating material. When radium-226 is used in self-luminous compounds, it is retained within the crystalline radium salt; however, there is some release of radium from luminous compounds. These materials are generally distributed throughout the country and can be found in hospitals, commercial facilities, and in households.

Other Radionuclides

The medical profession is rapidly developing diagnostic and therapeutic uses for both reactor-produced and cyclotron-produced radioisotopes. These uses produce relatively small quantities of radioactive waste. In diagnostic applications, whereby a patient ingests radioactive material to permit a direct measurement of an area of concern, the radioactive material is of short half-life to minimize radiation exposure of the patient. On the other hand, in the case of therapeutic treatments, radiation energy is utilized so the radioisotopes used are of high radiation energy and are long half-life well encapsulated materials, used to the very maximum before they are replaced. Replaced encapsulated materials are sometimes reused for other applications or are of such low activity they are sent to one of the six burial sites for disposal which also serves nuclear power facilities. Table 1.19 presents some of the radionuclides used in medical and industrial applications.

TABLE 1.19: RADIONUCLIDES USED FOR MEDICAL AND INDUSTRIAL APPLICATIONS

Radioisotope	Emission	Half Life	
Used in medical diagnostic work			
Carbon-11	β^+	20.5	m
Nitrogen-13	β^+	10.1	m
Oxygen-15	β^+	2.1	m
Fluorine-18	β^+	1.87	m
Sulfur- 37	β^-,γ	5.0	m
Calcium-47	β^-,γ	5.8	d

(continued)

TABLE 1.19: (continued)

Radioisotope	Emission	Half-Life
Chromium-51	γ	26.0 d
Gallium-68	β^+,γ	68.0 m
Selinium-73	β^+,γ	7.0 h
Technetium-99m	β^-,γ	5.9 h
Used in doubly encapsulated form of therapy work		
Cobalt-60	γ	5.25 y
Strontium-90	β^-	28.0 y
Cesium-137	γ	30.0 y
Plutonium-238	α	92.0 y
Used in doubly encapsulated form for industrial applications		
Cobalt-60	γ	5.25 y
Strontium-90	β^-	28.0 y
Cesium-137	γ	30.0 y
Thulium-170	γ	127.0 y
Iridium-192	γ	75.0 d
Americium-241	α	475.0 y

Source: PB-221 467

REFERENCES

(1) U.S. Environmental Protection Agency, *Evaluation of the Impact of the Mines Development, Inc. Mill on Water Quality Conditions in the Cheyenne River,* EPA Region VIII, Denver, Colorado 80203, September 1971.

(2) U.S. Atomic Energy Commission, *Draft Detailed Statement on the Environmental Considerations Related to the Proposed Issuance of a License to the Rio Algom Corp. for the Humeca Uranium Mill,* Docket No. 40-8084, Fuels and Materials, Directorate of Licensing, U.S. Atomic Energy Commission, Washington, D.C., 20545, December 1972.

(3) Kerr-McGee Corporation, "Uranium Hexafluoride Plant," Supplemental Applicant's Environmental Report, USAEC Docket No. 40-8027, June 1972.

(4) Allied Chemical Corporation, Special Chemicals Division, Docket File, Attachment to Form AEC-2, Application for Amendment to Source Material License No. SUB-526, USAEC Docket No. 40-3392, August 21, 1970.

(5) U.S. Atomic Energy Commission, *Environmental Survey of the Nuclear Fuel Cycle,* Directorate of Licensing, Fuels, and Materials, U.S. Atomic Energy Commission, Washington, D.C. 20545, November 1972.

(6) U.S. Public Health Service, *Process and Waste Characteristics at Selected Uranium Mills,* Technical Report W62-17, Robert A. Taft Sanitary Engineering Center, HEW, Cincinnati, Ohio 45268, 1962.

(7) T.H. Pigford, M.J. Keaton and B.J. Mann, *Fuel Cycles for Electric Power Generation,* Teknekron Report No. EEED 101 EPA Contract No. 68-01-0561, January 1973.

(8) R.W. Conn and G.L. Kulcinski, "Technological Implications for Tokamak Fusion Reactors," in *Proceedings of the First Topical Meeting on the Technology of Controlled Nuclear Fusion,* April 16-18, 1974, San Diego, California, CONF-740402-PI Vol. 1, pp. 56-69.

(9) The University of Wisconsin Fusion Feasibility Study Group, *UWMAK-I, A Wisconsin Toroidal Fusion Reactor Design,* UWFDM-68, Vol. II, May 1975.

(10) H.T. Petersen et al., *Environmental Tritium Contamination from Increasing Utilization of Nuclear Energy Sources,* Hearing before the Joint Committee on Atomic Energy, Congress of the United States, Oct. 28-Nov. 7, 1969.

(11) D.G. Jacobs, *Sources of Tritium and Its Behavior Upon Release to the Environment,* Oak Ridge National Laboratory, TID-24635, 1968.

(12) J.W. Ray, *Reactor and Fuel Reprocessing Technology,* 12, No. 1, 19, 1968-69.

(13) K.E. Cowser et al., ^{85}Kr *and Tritium in an Expanding World Nuclear Power Industry,* Oak Ridge National Laboratory, ORNL-4007, 1966.

(14) C.L. Weaver et al., *Tritium in the Environment from Nuclear Power Plants,* Public Health Report, 1969.

(15) "Siting of Fuel Reprocessing Plants and Related Waste Management Facilities," *Federal Register,* 35, No. 222, 17530, November 14, 1970.

(16) Oak Ridge National Laboratory and Gulf General Atomic, *National HTGR Fuel Recycle Development Program Plan,* ORNL-4702, August 1971.

(17) L. Küchler, L. Schäffer and B. Wottech, "The Thorex Two-Stage Process for Reprocessing Thorium Reactor Fuel with High Burnup," *Kerntechnik,* 13, *Johrgeng,* No. 7/8, 1971.

(18) J.W. Codding, W.O. Haas, and F.H. Heumann, *Equilibrium Data for Purex Systems,* KAPL-602, November 1951.

(19) J.O. Blomeke and J.J. Perona, *Storage, Shipment and Disposal of Spent Fuel Cladding,* ORNL-TM-3650, January 1972.

(20) J.P. Nichols and F.T. Binford, *Status of Noble Gas Removal and Disposal,* ORNL-TM-3515, August 1971.

(21) W.E. Unger et al., *Aqueous Fuel Reprocessing Quarterly Report for Period Ending December 31, 1972,* ORNL-TM-4141, April 1973.

(22) J.H. Goode and S.D. Clinton, *Aqueous Processing of LMFBR Fuels—Technical Assessment and Experimental Program Definition, Sections 4.4 and 5.4,* ORNL-4436, Supplement 1, June 1972.

(23) H.W. Godbee and J.P. Nichols, *Sources of Transuranium Solid Wastes and Their Influence on the Proposed National Radioactive Waste Repository,* ORNL-TM-3277, January 1971.

(24) H.H. Van Tuyl et al., *A Survey of Alpha Waste Generation and Disposal as Solids in the U.S. Nuclear Fuel Industry,* BNWL-B-34, December 1970.

(25) G.L. Pettigrew, E.W. Rodinson and G.D. Schmidt, *State and Federal Control of Health Hazards from Radioactive Materials Other than Materials Regulated Under the Atomic Energy Act of 1954,* U.S. Department of Health, Education and Welfare, Rockville, Maryland, June 1971.

(26) Atomic Energy Commission, *The Nuclear Industry, 1971,* Washington, U.S. Government Printing Office, WASH-1174-71, 1971.

HIGH-LEVEL RADIOACTIVE
WASTE TREATMENTS

The sources of the material in this chapter were the following reports: BNWL-2092, BNWL-2138, ICP-1088, PB-221 467, BNWL-SA-5757, BNWL-SA-5895, BNL-21571, DP-MS-76-66, BNWL-1876, BNWL-1913 and PB-254 737. A complete bibliography appears on p 359.

Since high-level wastes (HLW) are generally considered to be those liquid wastes resulting from the first cycle solvent extraction system, or concentrated wastes from subsequent extraction cycles, in reactor fuel reprocessing, this chapter is primarily concerned with the problem of converting these liquid wastes to a solid form. The treatment of cladding wastes is also included since they may contain up to 0.1% of the plutonium originally in the spent fuel and have heat generation rates of 50 to 100 W/ft^3.

There are a number of characteristics which are desirable in the final solidified waste form. Good thermal conductivity in the waste form is desired to prevent excessive internal temperatures during the initial few years of storage, remembering that the waste may be packaged as monoliths with volumes up to 1,200 liters. A coherent ceramic waste form has a thermal conductivity which is 3–4 times higher than loosely consolidated powdery oxide waste form.

The relative leachability of a given waste form is a function of both surface area and specific reaction rate with water, thus a monolith having a low reaction rate with water may be preferable to a high surface area material with an even lower leach rate. It is of course most important that neither the surface area or leach rate of the waste form be increased deleteriously by thermal or radiation effects at any time during storage or transportation.

Mechanical ruggedness is important, particularly during transportation and, since the canister is the first line of protection, it is essential that no interactions occur between the canister and the waste form which will weaken the canister.

Finally, from a practical plant operating standpoint, certain processing characteristics are desirable. The solidified waste should occupy the minimum possible volume. Generally a volume reduction factor of 7-10 is achieved during solidification and solidification should be done at minimum cost. Since the solidifying of HLW must be done remotely behind 4-5 ft thick concrete walls, process reliability and simplicity will have a great deal to do with minimizing cost.

SPRAY CALCINATION/IN-CAN MELTING

Federal regulations, Appendix F of the Code of Federal Regulations, Title 10, Part 50, require that High-Level Liquid Waste (HLLW) be converted to a solid and placed in a sealed container prior to transfer to a Federal repository. The HLLW must be solidified within 5 years after generation at a commercial nuclear fuel reprocessing plant and shipped to a Federal repository within 10 years of origination.

The regulations indicate that the dry solids should be chemically, thermally, and radiolytically stable to the extent that the waste container will maintain its integrity until it is placed in a disposal site. It is the intent of the regulations to maximize the public safety by converting the HLLW to a form that will minimize radionuclide mobility and toxicity under normal and abnormal conditions. The solids are also to be isolated from the biosphere. The United States Government has conducted research and development for about three decades to provide the appropriate technology.

Solidifying radioactive wastes poses unique problems requiring specialized considerations due to the radiological, chemical, and physical properties of the waste. Under sponsorship of the United States Atomic Energy Commission (AEC) and more recently the Energy Research and Development Administration (ERDA) extensive research and development has taken place.

The associated program goals were to prepare a suitable waste product form and develop processes well adapted to remote operation and maintenance. Much of the work was carried out at Battelle Pacific Northwest Laboratories (PNL) in the Waste Solidification Engineering Prototypes (WSEP) program. In this program, promising processes developed in the United States up to about 1967 were pilot plant tested and compared, using radioactive waste, except for fluidized bed calcination, which was demonstrated at the Idaho National Engineering Laboratory.

The WSEP program ended in 1971 after radioactive demonstrations were completed with the pot, spray, and phosphate glass solidification processes. A total of 33 demonstrations were completed during which more than 50 million curies of radionuclides were processed and collected in 33 canisters ranging from 6 to 12 inches in diameter by 8 ft tall. In 1972 the Waste Fixation Program (WFP) started at PNL and has continued to the present. A major objective of the WFP is to assure that suitable processes and solidified waste forms are available to industry for the fixation of HLLW.

In most countries developing nuclear power, HLLW solidification and storage programs are in progress. The more promising processes ultimately incorporate the toxic waste constituents into a highly nondispersible borosilicate glass.

These processes include:

In-Canister Drying and Melting	England
Rotary Kiln Calcination with a Continuous Metallic or Ceramic Melter	France
Spray Calcination with a Continuous Metallic or Ceramic Melter	Germany

In the United States during recent years, spray and fluidized bed calcination with in-can or continuous ceramic melting to a borosilicate glass have been developed further than other methods. After extensive evaluation of the available solidification processes by PNL, the Spray Calciner (SC) and In-Can Melter (ICM) process was selected as the system most ready for demonstration as a production plant process. This process is now being designed for vitrification of commercial fuel reprocessing waste generated in the near future.

In 1975 PNL undertook a joint program with Nuclear Fuel Services, Inc. (NFS) to provide equipment design and a process for use by a commercial nuclear fuel reprocessor to produce solidified waste in a canister acceptable at a Federal repository. PNL was to provide the process and basic equipment design. NFS and their architect-engineering firm designed the facility together with the associated services for location at the West Valley Fuel Reprocessing Plant. This cooperative program resulted in the development of a conceptual design (1) and flowsheet (2) for a commercial HLLW vitrification facility for NFS.

Although the PNL information is presented using the West Valley Plant in New York for the bases, it is applicable to other fuel reprocessing plants. These include the higher capacity Allied General Nuclear Services, Barnwell plant, and the proposed Exxon plant. The NFS flowsheet consists of basically four steps: calcination, in-can melting, canister sealing, handling, and inspection, and off-gas treatment. Figure 2.1 is a flow diagram showing these steps.

FIGURE 2.1: HIGH-LEVEL LIQUID WASTE VITRIFICATION FLOW DIAGRAM

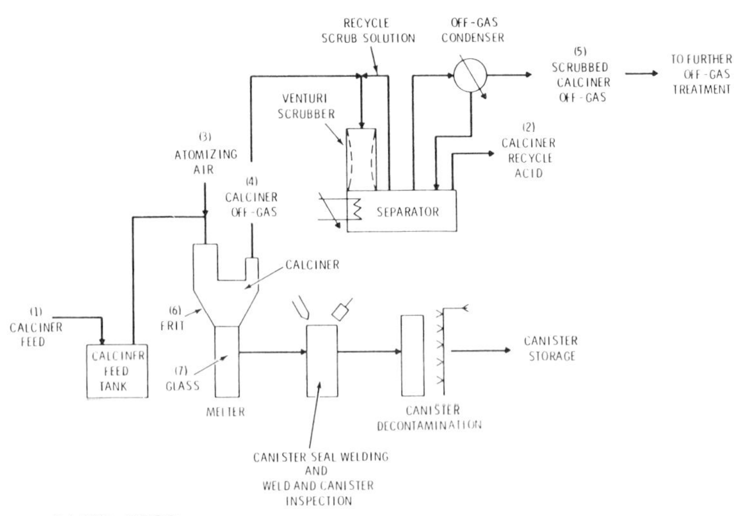

Source: BNWL-2092

Spray Calcination

In spray calcination, liquid waste is pumped and metered to the spray calciner at a pressure of 1 to 2 atm. The waste is pneumatically atomized in an internal mix nozzle and sprayed into the top of a cylindrical calcination chamber. The atomized liquid waste as ~70-μm-diameter droplets is sequentially evaporated, dried, and calcined as it falls through the calcination chamber.

The chamber walls are heated to about 700°C by an external multizone resistance furnace. Inside the chamber heat is transferred to the droplets by combined radiation and convection. This produces calcine containing typically less than 0.5% moisture. Periodic operation of side-mounted vibrators prevents significant depositions on heat transfer surfaces within the calciner. On passing through the chamber, water vapor and atomizing air entrain 15 to 50% of the calcine particles.

During normal operation only 0.1% of the total particulates pass through the porous stainless steel filters. The entrained powder collects on the filters and is periodically removed by a pulse of blowback air. After this it falls into the lower calciner cone and is discharged along with that calcine falling directly from the calcination chamber.

The simplicity of the spray calciner makes it easily automated to minimize operator attention and facilitate remote operation. Successful remote operation and maintenance of a heated-wall spray calciner has been demonstrated while processing actual high-level waste (3)(4). The calciner can be started up and shut down in less than an hour and produces no additional waste. Also during operation radionuclide volatilization is acceptably low (4).

Feed System: The feed system to the calciner consists of the following subsystems: feed tank, pump, flow control, and spray nozzle. The feed system has been developed to deliver properly atomized HLLW of uniform composition to the calciner at a controlled rate. The development of these subsystems is described here. Figure 2.2 shows the system schematically.

FIGURE 2.2: SIMPLIFIED SPRAY CALCINER FEED SYSTEM

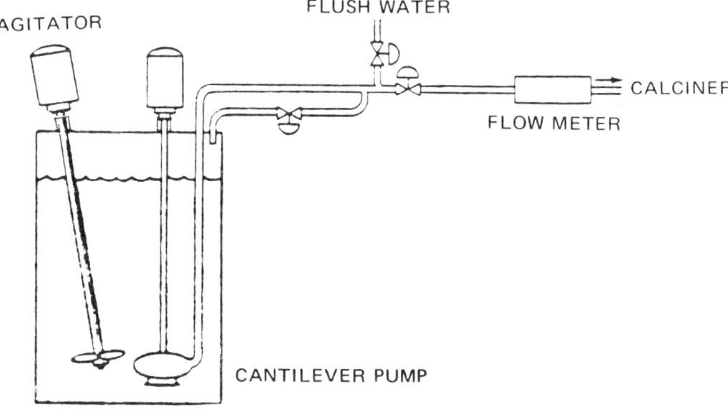

Source: BNWL-2092

Feed Tank — During normal plant operation, waste will likely be batch transferred to the calciner feed tank. If a canister of waste solid such as glass is to be produced in a batch or low-holdup melter, the liquid waste volume may be sized to produce one canister of solidified product. By this technique added assurance of canister contents can be gained because liquid sampling is normally more accurate and easier than solids sampling. Since chemical adjustment of the feedstock probably will not be necessary, a single feed tank can be used by simply adding the next waste batch as the liquid level reaches a lower limit. This technique was routinely used in the WSEP and WFP. A tank with a working volume of 1,000 gal would fulfill this need for the design bases flowsheet.

Lag storage of waste slurries is complicated by the need to prevent settling of nondissolved species. Accurate melt formulation may be dependent on a known amount of solids entering the calciner and subsequently, the melter. Also deposition of solids in waste tanks may result in plugging, eventual hot spots and accelerated corrosion. No difference in calciner performance has been observed when operating at waste concentration from 200 to over 650 liters of waste per tonne of fuel. Experience (5) has shown that waste concentrations and/or composition may change drastically during operation without requiring corrective action.

An electric motor-driven propeller agitator for feed tank agitation has proven successful for maintaining homogeneity in waste slurry tanks. The agitator can be designed to provide the degree of mixing desired. In the WSEP, use of electric motor-driven agitators has proved not only effective for solids suspension but also capable of sustained operation in high-level waste service. If desired, the agitator motor's radiation resistance can be increased by shielding the motor or other potential failure areas. Studies with simulated material have shown that agitation by air sparging is not effective in maintaining homogeneity throughout the tank. Accurate liquid sampling is thus virtually impossible.

In the absence of proper agitation, suspended solids which predominate in the lower part of the tank are drawn into the calciner first, thus complicating accurate melt formation. Furthermore, pluggage of air holes in the sparge ring has been a continuing problem. A pulse type of agitation technique developed at Windscale, England, has been effective in preventing accumulation of solids in the bottom of large storage tanks. Applying this technique to feed tanks, however, can result in somewhat less than optimum performance because of the predominance of solids near the tank bottom, the frequent pulsation required, and the continually changing liquid level.

Liquid level monitoring dip tubes are accurate and effective for determining tank volumes. In the WSEP program, dip tube plugging in solutions with high solids content was negated by periodically discontinuing the air purge to the dip tube. This allowed liquid to enter the dip tube and dissolve deposits formed at the air/liquid interface. Plugging of waste tank dip tubes has not been encountered in WFP simulated nonradioactive wastes. Liquid level monitoring serves to confirm the feed rate measured by electromagnetic flowmeters on the calciner feed piping.

Maintaining the waste temperature at 40°C or lower reduces feed drying and blockage at the spray calciner atomizing nozzle. Thus, to counteract the self-heating of the calciner feeds, the feed tank(s) should have a cooling system.

However, the cooling required to maintain acceptable limits will fluctuate with the quantity of feed and the specific feed heat-generating rate. Manual cooling water flow control systems can be used to compensate for fluctuating feed tank temperatures. However, an automatic control system capable of maintaining the feed within ± 10°C of a set value should be considered for increased reliability and for economical use of manpower and cooling water. Neither the absolute temperature of the feed nor feed temperature fluctuations significantly affect the operation of the calciner. However, the feed should be kept cool to minimize evaporation in the feed tank and minimize corrosion of equipment.

Pump – Many types of transfer devices have been evaluated for pumping radioactive slurries. Air lift techniques have been developed and are routinely used for metering and transferring dilute suspensions. But handling concentrated slurries at low flow rates by this technique has normally not been successful. A pressure head is useful in preventing pipe and nozzle blockage and is required for some atomizing nozzles. Continuous control valve cycling normally required with air lift systems in this service is unnecessary. In the WSEP, radioactive slurries were pumped with disposable in-line centrifugal pumps. Unfortunately, their short service life limits their usefulness.

On the other hand, Savannah River Laboratory and other sites have found that concentrated radioactive slurries can be routinely pumped using vertical cantilevered shaft pumps. With this type pump all seals and bearings are above and out of the liquid. This permits long-term operation without seal replacement. Experience at Savannah River with 35 cantilever pumps in a radioactive environment has shown that in high-level waste service this type pump can be expected to operate 5 to 10 years without maintenance. These pumps typically permit high flow rates and are potentially capable of supplying feed tank mixing from a jet of recycled liquid waste.

In storage of concentrated liquid waste, at the air-liquid interface, caking on vessel walls often occurs. Dislodged caked solids in tanks and lines pose potential line blockage problems. This may be intensified in denitration vessels if the sugar denitration process is used. Restrictions in the feed lines ($^3\!/_{16}$ inch diameter), the electromagnetic flowmeter, and the atomizing nozzle ($^1\!/_4$ inch orifice) are potential plugging points that could require shutdown if appropriate provisions are not made. To prevent blockage at these points, a solids filter may be installed in the feed system. A strainer is planned to be tested on the inlet to the cantilever pump.

Since the amounts of solids material larger than $^1\!/_4$ inch are likely to be very small, strainer cleaning is not expected to be necessary. Collected solids could be dissolved during system downtime. If desired, a basket type filter containing the accumulated oversize solids could also be used. In a vitrification facility the collected solids could be chemically or physically reduced in size, or it may be possible to periodically place the baskets in a canister to be filled with glass.

Flow Control – The spray calciner/in-can melter is relatively insensitive to waste feed rate fluctuations with short-term variations of as much as ± 50% being acceptable. Since the frit flow rate is controlled so that it is proportional to the waste feed rate, proper glass formulation on melting is possible with significant feed rate fluctuations. But such variations are unnecessary since the control system described has an accuracy of about ± 1% of the set point. The design bases feed rate is about 80 liters/hr, but for process flexibility, the flow control system should have a usable range of from about 30 to 150 liters/hr.

A recirculation loop from the feed pump back to the feed tank, shown in Figure 2.2 is recommended. This accomplishes several things: it minimizes settling of undissolved solids in the feed lines when the feed rate to the calciner is low or shut off completely, and it enables the feed pressure to be adjusted to optimize the performance of the feed flow control valve. To minimize the settling of solids in the feed line, the spray nozzle, the flow sensor, the feed flow control valve, and the feed tank selector valves should be coupled as closely together as is practical, and each feed tank selector valve should be coupled as closely as practical to the respective feed recirculation loops.

A pneumatically operated flow control valve, remotely operated from a manual control station, is indicated in the feed recirculation loop for each feed tank. This valve and the recirculation loop piping should be so sized that adjusting the position of the recirculation valve permits varying the feed pressure ahead of the calciner feed flow control valve so as to keep the valve in a 60 to 80% open condition while maintaining the desired feed flow rate. This avoids the flow rate fluctuations caused by undissolved solids accumulating in the control valve when operated in its more nearly closed position.

The recirculation loop control valve could conceivably be replaced with a fixed orifice if it could be assured that solids would not block the orifice and that the required calciner flow rate range is narrow enough that the feed flow control valve does not have to routinely operate outside of its 60 to 80% open condition to satisfy this flow rate range. The calciner feed flow control valve should have a linear spline plug trim, with a single tapered spline on the plug, similar to the Hammel-Dahl design. This type of trim realizes the desired flow control characteristics, and it presents the largest passage possible for undissolved solids for any given feed flow rate.

An electromagnetic flowmeter may be considered for the calciner feed flow rate sensing device. With a 1/10 inch diameter vitreous enamel-lined tube and platinum electrodes, indicated flow rates will be accurate to within ± 1% of the actual flow rate, from 20 to 230 liters/hr. The ¼ inch diameter tube decreases the change of plugging with a solid particle, but the lower flow detection limit is 30 liters/hr. The accuracy from 30 to 110 liters/hr is ± 2%, and above 110 liter/hr the accuracy improves to ± 1%. A feed metering system using the atomizing nozzle orifice as the flowmeter is being developed to potentially eliminate the need for an electromagnetic flowmeter.

It is recommended that a flush line be connected to the calciner feed line immediately ahead of the calciner feed flow control valve. This will enable the flow control valve, the flow sensor, and the spray nozzle to be flushed to the calciner, and the feed pump and recirculation line to be flushed to the feed tank when the feed to the calciner is shut down.

In addition, if the flush supply is not in an adequately shielded area, it is recommended that the flush line be pressurized at all times to a value slightly in excess of the feed supply pressure. This will reduce the possibilities of feed being forced into the flush water piping system, past check valves and flow control valves that can potentially leak, or be inadvertently left open.

A conventional Bourdon tube-type pressure indicator is suggested to measure the feed supply pressure ahead of the feed flow control valve. However, since the standard glass face cover will darken quickly from cell radiation, making the gage unreadable, the glass will have to be removed or replaced with radiation-resistant glass. If the gage cover glass is removed, the gage movement will be exposed to the probably corrosive environment of the cell, and gage replacement will probably be frequent. To avoid an identical problem, in indicating the feed pressure at the spray nozzle, the feed pressure can be implied by the feed flow rate and an empirical relationship developed from simulated feed compositions and prototypical nozzle assemblies.

Spray Nozzle — Proper atomization is the key to successful spray calcination. Experience (6) at PNL has shown that a droplet mass median diameter of $\leqslant 70$ microns is sufficient for successful operation with waste compositions, chamber configuration and temperatures studied.

The required particle size may be obtained by using a commercial nozzle supplied by the Spraying System Company. The air cap has been modified by addition of an insert made of 96% alumina to avoid abrasive wear. A nozzle tested has logged 1,000 hr of use without detectable wear. Alumina air caps are more susceptible to damage by thermal shock than are caps of stainless steel. These shocks are prevented if atomizing air flows through the nozzle during heatup to the operating temperature before the water flush is turned on, and the feed and air should not be turned off simultaneously while the calciner is hot. The atomizing air flow should also be maintained while the calciner is at the operating temperature whether or not liquid is flowing. To obtain the required atomization, a Spraying System Setup No. 72 or the equivalent can be used in a reprocess plant.

Steam, hot air, and ambient temperature air were evaluated as atomizing gases during the design verification testing (DVT) and WSEP. The steam and pre-heated air caused nozzle plugging because they baked waste in the orifices. Therefore a cleanout needle was designed into the nozzle assembly. At Karlsruhe, Germany, where superheated steam is used as the atomizing gas in their nuclear waste spray calciner, severe nozzle plugging problems were recently solved by cooling the liquid line and nozzle (7). Nozzle plugging has been virtually eliminated at PNL by using ambient temperature atomizing air, by placing a sieve with 0.04 inch openings in the liquid waste line before the nozzle, and by flushing feed line and nozzle with water before each startup and after each shutdown.

The only treatment given the atomizing air is filtration to remove entrained oil introduced by the compressor. Plugging is not anticipated with the Spraying System Co. Setup No. 72 nozzle because the minimum diameter orifice is ¼ in. instead of the ⅛ in. nozzle used in the early tests. A remotely operated nozzle cleanout needle may not be required on a production calciner.

Atomizing air flow control is not critical in the operation of the spray calciner. To produce drops in the desirable range of 40 to 70 microns, a volumetric air-to-liquid flow rate ratio of about 300 should be used. To achieve these droplet sizes the atomizing air pressure at the nozzle should be about double the fluid feed pressure at the nozzle for feed flow rates above 40 liters/hr. Below 40 liters/hr the nozzle air pressure should be about triple the nozzle fluid pressure. To eliminate an in-cell measurement of the atomizing air pressure at the nozzle, the atomizing air pressure could be measured at some appropriate point in the out-of-cell atomizing air piping, and the piping from this point to the in-cell nozzle could be sized so as to minimize the pressure drop between this pressure measurement point and the nozzle.

A pressure switch should be interlocked with the feed pump switchgear in such a way that at low atomizing air pressure values the feed pumps will be turned off. This switch would be activated by the out-of-cell nozzle atomizing air pressure and set at a pressure slightly below the minimum anticipated atomizing air pressure. This will minimize the probability of forcing feed material past potentially leakable check valves and into the out-of-cell atomizing air piping. In addition, antibackflow devices such as check valves or filters which will not pass liquid should be installed in the atomizing air line along with a radiation alarm.

Calciner: The calciner comprises three main components: the chamber, the furnace, and the off-gas filter system. The technology for these systems is described here.

Calciner Chamber — The calcination chamber is a large, externally heated vertical tube. Atomizing liquid waste is sprayed into the top center. Combined radiant and convective heat transfer within the calcination chamber allows evaporation and calcination at a high rate. Under normal operating conditions spray particles dry before striking the wall and thus do not adhere to the wall. Calcine produced in the spray chamber falls into a conical section together with calcine blown back from the filters. It then flows by gravity to the melter below. Experience at PNL with other geometries for combining the two calcine streams has shown that the filters should be as close to the calcination chamber as possible. All surfaces are sloped $\leqslant 60°$ from horizontal to avoid calcine accumulation in the cone.

Both Hastelloy and stainless steel have been used as spray chamber material. Presently, the rolled and welded stainless steel chamber has been in nonradioactive operation for over 800 hr. Spray calciner wall thickness measurements on the stainless steel chamber failed to show a detectable decrease in wall thickness. Since the instrument used was accurate to within 5 μm (0.0002 in.), a corrosion/erosion rate of less than 75 μm (0.003 in.) per year is expected (8). The ¼ inch wall thickness in use is designed for strength with minimum concern for corrosion.

A recent nonradioactive capacity run, using the two-zone (50 kW/zone) furnace, 21 in. i.d. spray chamber heated to 700°C, and simulated clean waste, calcined waste at flow rates up to 75 liters/hr. Higher feed rates were not possible because of the limited capacity of the available calciner furnace, but scale-up calculations show that over 150 liters/hr can be processed in the full-scale unit. This unit can operate at the design bases flowsheet rate.

The calciner chamber size required, both diameter and length, has not been established. Thus a conservative size, 3 ft dia by 10 ft tall, is specified to perform the design bases flowsheet. It is anticipated that optimization will reduce the size required.

Scale (calcine and corrosion) deposits on the calciner wall can form an insulating layer which reduces calciner capacity. Experience (9) has shown that scaling increases when wastes containing high-sodium concentrations are calcined under conditions where moist particles impinge on the wall. However, HLLW wastes containing over 2 M sodium have been satisfactorily calcined without feed additives.

Under normal operating conditions wall scale is minimal and extended operation is possible without significant scale buildup. But to ensure the flexibility and operability of the unit, vibrators have been mounted on the calciner to minimize the possibility of scale buildup. Experience (10) has shown that a vibrator located on the vertical center of the chamber and mounted on the furnace shell has prevented scaling even when using high-sodium-containing waste or when atomization of the feed was intentionally changed to introduce scaling conditions. Mounting the vibrator on the vertical center of the furnace outer shell has greatly reduced the required vibrator maintenance. Frequent replacement was required when the vibrator was located on the calciner top flange where it was exposed to high temperatures.

To help determine the effect of vibration on the spray calciner a 100 hr test was conducted in which the vibrators were activated frequently with the calciner at operating temperature to simulate up to 4,000 hr of operation (11). The diameter of the spray chamber across the impact pads and at 90° to the pads was measured before and after 4,000 vibrator cycles acting on 700°C calciner walls. No measurable difference in calciner shape was found when measuring to 25 μm (0.001 in.). However, the calciner must be designed to minimize the undesirable effects of vibration.

A thermal shock descaling technique for use during calciner shutdown has proved rapid and effective for removing even the most stubborn scale. A small amount of water sprayed onto the hot (500°C) calciner wall produces sufficient thermal shock to dislodge the scale. The descaling operation can be carried out with less than 40 liters (10 gal) of water, and nearly all water sprayed from a properly designed nozzle can be vaporized by the hot wall. Should it ever be needed, this descaling process requires 5 to 10 min and can be done with remotely operated equipment.

The vibrator cycle timer and solenoid operated air control valve should be adjustable for a pulse duration range of 3 to 30 sec with a 5 to 60 min cycle, depending on potential feed scaling characteristics.

Furnace — Multizone resistance furnaces have been used successfully at PNL to calcine HLLW. A ceramic fiber insulated furnace used for radioactive demonstration during the WSEP program contained three independently controlled heating zones. The heating capacity of each zone was 15 kW. The electrical power to each heating zone was controlled by an automatic on-off controller which operated over a temperature range. A refractory brick insulated furnace used for nonradioactive demonstrations contained two independently controlled heating elements. The maximum heating capacity of each zone was 50 kW. The electrical power to each heating zone was controlled by a silicon controlled rectifier.

The recommended spray calcination chamber is heated by a resistance furnace divided into several horizontal zones. A maximum of four zones is required in the production unit, with a heating capacity of about 70 kW per zone. The electrical power to each furnace zone is controlled with a zero-crossover, silicon controlled rectifier power supply, using a thermocouple attached to the furnace side of the calcination chamber as the reference temperature. The temperature control system for each zone should be capable of maintaining the zone furnace temperature at any value within the range of 500° to 1000°C, with an accuracy of ± 10°C. The reference temperature should be recorded, and the alarm should be activated outside the control range to detect malfunctions of the control system. At least one redundant thermocouple for each zone should be mounted on the furnace side of the calciner opposite from the reference thermocouple. These redundant thermocouples should also be recorded and alarmed above the furnace set points to detect possible scaling within the calciner over the reference thermocouple locations.

Thermocouples should be positioned inside the calciner above the feed nozzle spray pattern, in the powder collection cone, above the powder diverter valve, and in the off-gas filter zone to monitor the calciner operation. These temperatures should be recorded to detect long-term changes.

The furnace should be encased in a material resistant to vapors and to potential spills of process solutions. Brackets should be mounted on the furnace to hold the vibrators. To reduce head room requirements the furnace may be designed capable of disassembly by zones for removal from the calciner.

Radial and longitudinal temperature uniformity does not appear to be critical to calciner operation. Longitudinal temperature differences of over 100°C have been observed at times during operation. Since the furnace will likely be held at a constant temperature of about 700°C during operation, thermal cycling will occur only at startup and shutdown. The heatup rate will likely be limited by the thermal shock resistance of furnace materials, not by the calciner.

Filter System — Early experiments (12) with various filter types for dust removal led to the choice of sintered stainless steel filters. Various filter media tested included porous alumina, fine woven wire, fine wound wire, sintered fiber metal, and sintered metal powder. Sintered powder filters appear to exhibit the best compromise between filter efficiency and ease of cleaning by pulsed blowback air. Alumina filters have 10 mm (³⁄₈ in.) thick walls and pore sizes ranging from 10 to 100 μm (0.0004 to 0.004 in.). They are fragile and not effectively cleaned by pulsed blowback. Wire wound and woven wire filters are not as effective as sintered filters, and sintered fiber metal filters are expensive and require high differential pressure.

The filters used in previous radioactive demonstrations as well as the present sintered filters being used in nonradioactive development testing have a mean pore size of 65 microns. They have been very effective in filtration of the calcine. However, due to the large filter pore size relative to the calcine particle size, some calcine can penetrate the filters before a layer accumulates on the filters. Thus, to avoid passage of calcine through the filtering media and particles becoming permanently lodged on the filter, filter candles having a mean pore size of 10 microns will be tested. Ten-micron seamless filters will be tested on the full-scale development spray calciner.

Pressure drop across the filters increases as the calcine accumulates on the filter candles. Pressure drop as high as 20 in. of water across the filters has shown no effects on filtration. Thus the blowback cycle is a function of the available vacuum to maintain the required 10 in. vacuum in the spray calciner unit. Accumulated filter cake is periodically dislodged from the sintered stainless steel off-gas filters by directing a short pulse of high-pressure filtered air into each filter element through a venturi nozzle in opposition to the normal off-gas flow through the unit. This blowback air is routed to each filter element venturi nozzle through a separate line controlled by a solenoid operated valve. The blowback system is analogous to a jet ejector. The venturi draws calciner off-gas from the filter off-gas plenum back into the filters to provide additional blowback air volume.

For efficient calcine cake removal and minimum calciner pressure surges, a blowback pulse of 60 to 100 psig air for ¼ sec is used. This removes most of the cake but leaves a thin calcine layer. The layer acts as filtering media and reduces the particles that penetrate and permanently lodge in the filter.

The air to the solenoid control valves is supplied from a single manually operated pressure control valve. A cycle timer controls the operation of the solenoid valves. The cycle timer is designed so that (1) the solenoid valves are operated sequentially, (2) the period each valve remains open is adjustable from 0.25 to 5 sec, (3) the cycle time for blowing back the total complement of filters is adjustable from 1 to 100 min, and (4) any filter element may be blown back manually at any time. The time between filter blowbacks may be controlled by the filter differential pressure. The blowback piping, the solenoid valves, and the manual pressure control valve should be sized so that the pressure at the venturi nozzle will be maintained at 60 psig or greater for the duration of the blowback pulse.

The differential pressure across the sintered stainless steel off-gas filter elements should be measured and recorded to observe trends in filter performance, effects of process variations, and the effectiveness of remedial measures.

After passing through the sintered stainless steel filters, the off-gases pass directly to the off-gas cleanup system. The off-gas temperature should be measured immediately on leaving the filter plenum, and recorded, to observe trends in the calciner performance and the effects of process changes.

Calciner vacuum is normally controlled at 10 in. of water. However, during filter blowback the vacuum may only be 1 to 2 in. Internal vacuum is influenced by the interrelationships of: pressure drop across the off-gas filters, calciner feed flow rate, pressure drop through the off-gas cleanup system, and fluctuations in the volume of other gases contributing to the total flow through the vacuum source. It is recommended that the differential between the calciner operating pressure and the cell pressure be measured and recorded. The recorder should be provided with alarm contacts to warn of impending pressurizing conditions.

In-Can Melting

An empty can with a gasket in the canister mouth is transported into the hot cell where vitrification is done. As assembly consisting of the lid for the retort/susceptor (R/S), the insulating plug for the top of the furnace, and the can suspension hooks is lowered onto the top of the can with the overhead crane, and the hooks are engaged in the ears of the can. This assembly and the can are then raised by the crane and lowered into the R/S of an empty furnace.

The lid assembly rests on the top flange of the R/S, and the can is suspended by the ears from the lid assembly in the R/S. Suspending the can from the top while it is in the furnace permits expansion and contraction of the can to occur during processing without affecting the connection to the calciner or the can content weight monitoring system. Suspending the can in the furnace will probably also minimize can distortion during the in-can melting process.

This process designates the furnace to be at 750°C when the empty can is placed into it, but it could be at ambient temperature. Cycling the cell furnace between 750°C and the processing temperature, rather than between ambient temperature and the processing temperature, should increase the life of the furnace and significantly reduce the time required to bring an empty can to the processing temperature. However, it will also contribute to the total heat load which the cell ventilation system must remove. An additional insulating cap for the furnace will also be required for those periods when there is no can in the furnace and the furnace is idling at 750°C. Placing a can made from 304L stainless steel in the furnace at 750°C without the cover gas should not create a spall problem because scaling does not occur below 900°C.

When the can and lid assembly are in place on the furnace, the crane hook is disconnected and the furnace is rolled into position below the coupling section. Pneumatic cylinders then lower the coupling section into the top of the can to make a dust-tight seal at the gasket located there. As the cone valve is opened and lowered into the neck of the can, the pressure in the can lowers to that of the calciner. Normal operation is at a vacuum of 10 in. of water. If the can was not already suspended on load cells when it was lowered into the R/S, this is also accomplished at this time by the action of pneumatic cylinders.

The retort is purged with the selected cover gas to prevent canister spall formation, and the can is heated to the processing temperature. About 0.1 lb of scale per square foot of can exterior surface forms when a can made of 304L stainless steel is held at temperatures above 1000°C for more than 8 hr in an air atmosphere. Thus a 20 in. diameter by 10 ft tall can could be coated by an oxide layer of up to 8 lb if adequate air were available to form the spall during processing. The spall can be reduced or eliminated by minimizing the air volume in the R/S or employing an inert purge gas.

If the can is removed from the furnace at temperatures above 750°C for quenching, the scale is still loosely attached to the can, and the majority is deposited in the quench tank rather than in the furnace R/S. This spall is a waste to be disposed of if it is not prevented. It could be collected and placed in each canister, or could be handled as spent equipment. Alternatives to the use of a cover gas to prevent spall include (a) designing the R/S to collect spall for later disposal, (b) placing a protective coating on the can exterior such as plasma sprayed zirconia (13), and (c) making the can from alloys which do not form scale at the processing temperature.

The processing temperature range is between 1050° and 1100°C. The actual can surface temperature which the furnace must achieve depends on the batch (calcine and frit mixture) composition, the heat generation rate of the waste, and the thermal conductivity of the melt and fin design. Present batch compositions must be heated above 1025°C to melt well. WSEP experience with in-can melting showed that melt centerline temperatures of 1040°C could be obtained with self-heating waste in 8 inch diameter cans without fins by heating the can walls to between 900° and 950°C (14). The use of larger diameter cans containing fins will require that the temperature of the can wall be much closer to the processing temperature. The maximum processing temperature is determined by the strength of the can material and its resistance to attack by the melt. The minimum temperature is determined by the batch composition and desired melting rate. The viscosity of the melt as a function of temperature is an important consideration.

When using batch compositions and 304L can material, the best melting rates and products are obtained when temperatures are maintained as high as practical, between 1025° and 1100°C. If the cans are made from nickel base alloys, the processing temperatures may be raised to 1150°C or more. The optimum time to hold each zone at the melting temperature has not been determined. But, based on WSEP experience, it will probably not exceed by more than 2 hr the time required to fill the zone. A short time at temperature is desirable to minimize the loss of volatile species and settling of dense phases.

The rate of temperature rise of the empty can will most likely be determined by the limits of the furnace construction materials. If the furnace is an all metal induction unit, it may be heated very fast (500°C/hr). If it is a resistance-heated furnace insulated with the fiber ceramics, the resistance elements may be the rate limiting material. It is expected that the cans used will be large enough to allow adequate cycle time for a conservative heatup rate not to exceed 200°C/hr. At a melting rate of 30 kg/hr a 20 inch diameter by 10 ft long can will take over 50 hr to fill. By assuming 4 hr to top up and fine (remove gas bubbles) the previous can, 2 hr to cool it to 750°C, and 8 hr to change cans in the furnace, over 40 hr remain to heat the empty can to the processing temperature if desired. Thus ample time is available for the auxiliary parts of the melting process.

Carbide precipitation at grain boundaries in stainless steel alloys results in a phenomenon known as sensitization. This increases the metals' susceptibility to intergranular attack by halides. The degree of sensitization is a function of the carbon content of the metal, the temperature, and the time held at temperature. The 750°C temperature is considered by some sources to be the upper bound of the sensitization range of 304L stainless steel. The lower bound of the range is near 480°C.

However, if the metal is cooled rapidly enough that transition through this range takes less than 1 hr, minimal sensitization should occur. Quenching the can of melt appears to minimize both sensitization and stress by causing the can to contract while the melt is still fluid. In developmental work the canister is immersed in flowing, cold (5° to 15°C) water. But more sophisticated spray or agitated quench systems can more easily assure uniform cooling to avoid localized stresses in the can.

When the empty can and furnace are at the processing temperature and the can in the other melter located below the calciner is nearly full, calcined waste and glass-forming frit are directed by the diverter valve down the coupling section into the empty in-can melter. The diverter is used because the heated-wall spray calciner is best operated continuously, but in-can melting is a batch process requiring periodic changing of cans in the furnace. The calciner, however, is capable of intermittent operation if desired.

Calcine is produced at a rate of about 10 kg/hr, and frit is metered into the cone below the calciner and above the diverter at a rate proportional to the calcine production rate to provide the desired batch composition. Compositions require two parts frit to one part calcine by weight, so that the melter charging rate is about 30 kg/hr. This charging rate does not exceed the melting capacity of an 8 inch or 12 inch diameter can containing eight fins.

The in-can melter charge rate should not exceed the melting rate for most satisfactory operation. It was initially believed that the melting rate could be increased for a given diameter can by having the batch in contact with more of the hot can surface. Therefore, cans were precharged with batch or very rapidly charged with batch to the full level and then charged at a rate which maintained that level as melting proceeded. This charging method was called the batch melting method. It was found that in the relatively short, single-zone development furnace the rising level melting method resulted in higher melting rates.

The batch melting rate is believed to be lower than the rising level melting rate because the NO_x and other gases released from the calcine as the batch is heated may create an insulating foam in the static batch; however, these gases are readily released at the surface of a rising level melt. The batch melting process also has been known to leave voids in the vitrified product as large as 2 inches in diameter.

As melt accumulates in the can, the temperature of the melt begins to rise above the processing temperature because of the heat generated by the waste. Power to the zones of the furnace below the melt level is switched off, and cooling of the zones is required. Forced air cooled the furnaces in the WSEP and will be used to cool the full-scale demonstration furnace. An air vacuum system gives better zone control, but a forced air system is better for hot cell application because the blowers can be located outside the cell and the hot air can either be quenched or dumped directly from the furnace to the cell.

The off-gas from the in-can melting process rises from the melter through the coupling section to the sintered metal filters of the calciner off-gas extraction system. Although characterization of the in-can melter off-gas is not yet complete, preliminary results show that cesium, boron, sodium, zinc, and molybdenum are not present, and only traces of ruthenium, zirconium, tellurium, lanthanum, praseodymium, and neodymium have been detected in the vapors to the filters.

It is believed that the counter flow of batch and off-gas in the coupling section, the scrubbing effect of the calcine-coated filters, and the relatively cool, small surface of the melt in the in-can melter are responsible for the low radionuclide level of effluents from the process.

When the melt has filled the can to the desired level, the diverter directs the batch into the other melter. As the melt is fined (gas bubbles are removed), the melt level drops. If necessary to maintain the desired level, the diverter can add small quantities of batch periodically, or the melter can be overfilled initially if sufficient space is available in the heated portion of the can. Any solids accumulation in the canister section above the furnace heating sections is not completely melted due to the inability to achieve an adequate temperature.

During the fining phase of the process, off-gas from the melter must still be vented into the calciner even though the batch is being directed into the other melter. This can be accomplished within the diverter, or a bypass line can be provided around the diverter.

Following the fining phase of the process, the inside surfaces of the unfilled portion of the can, the coupling section, and cone valve are coated with a partially melted or sintered layer of batch rich in calcine and species which have risen from the melt as volatiles and have condensed on these cooler surfaces. This coating spalls as the can is cooled following the melting process and has a leach rate about four times that of the glass. Depending on how stringent the solidified waste specifications are, this coating may have to be removed or totally vitrified. Attempts made to dislodge this material by thermal shock have not been totally successful. Other suggestions are to fill the can to the top with a glass containing no waste or to wash out the top of the can before making the closure.

After the melting is completed, the furnace and can of vitrified waste are cooled to about 750°C. The cone valve is closed, and the coupling section is disconnected from the canister. The furnace is rolled away from the coupling section to permit access to the top of the can, and the cap is placed on the can. The can is then removed from the furnace and can be water-quenched to avoid sensitization and to reduce residual stress caused by the difference in thermal expansion of the glass and steel (15). The quench must remove enough residual heat from the can and product that self-heating will not raise the can surface temperature above 480°C (the sensitization range).

Following the quench, the R/S lid assembly is removed from the can. The can is transferred to the closure weld station and sealed. Leak testing and decontamination of the can are then performed.

At this stage in the in-can melting process the canister may be put into water basin storage. Or it may be heat treated further at 480°C to improve the canister corrosion resistance by relieving residual stresses. At this temperature the borosilicate glass will increase in density with time, and it is anticipated that as the glass volume decreases, residual stress in the can walls will also decrease. This feature is being evaluated at this time.

The rising level method is compatible with direct coupling to a continuous calciner. To operate with the rising level melting method, the rate at which the calcine and frit enter the melter should not exceed the rate at which they are being melted. Therefore, the melter fill rate is determined by the melting rate. Some of the variables that have been determined to increase the melting rate are:

> increased furnace temperature
> increased can diameter
> internal fins
> optimum batch feed rate to melter
> addition of reducing agent to the batch
> decreased moisture content of batch
> decreased nitrate content of calcine
> increased temperature of calcine entering melter.

Table 2.1 presents the melting rates demonstrated using a single-zone furnace and simulated HLLW calcine. To date both 8- and 12 inch canisters have been tested.

TABLE 2.1: MELTING CAPACITIES OF THE IN-CAN MELTER

CAN DIAMETER, INCH	MELTER CHARGING METHOD	NUMBER OF FINS	MELTING RATE		TIME TO MELT 100 KG, HR
			KG/HR	KG/HR-M	
8	BATCH	0		13.5	8.2
	DUMP	8		29	3.3
	RISING	0	14		7.1
	LEVEL	8	30		3.3
	RISING LEVEL CONCENTRATE	8	12		8.3
12	BATCH	0		18	6.0
	DUMP	8		21	4.2
	RISING	0	14.5		6.9
	LEVEL	8	50		2.0
	RISING LEVEL CONCENTRATE	8	13 - 19		5.3 - 7.7

Source: BNWL-2092

The results are not directly comparable because some of the variables listed above were not constant for all demonstrations. But some general conclusions can be made. Inclusion of fins in the can and use of the rising level method each had the effect of doubling the melting rate.

The effects on the melting rate of the addition of 1.5 weight percent powdered silicon metal to the batch has not been quantitatively evaluated, but it has been observed to reduce the power required by the furnace to melt an equivalent amount of batch. Between 225 and 360 cal/g were required to melt batch without the additive while only 120 to 200 cal/g were required to melt the same batch with the silicon added.

The melting rates reported in Table 2.1 are considered conservative because they were established using nonradioactive, simulated waste which has no internal heat generation. High-level nuclear waste will generate from 15 to 150 W/liter of vitrified product, depending on the burnup of the fuel processed, the age of the waste, and the waste loading in the host glass.

During the radioactive in-can melting demonstrations at PNL in 1970, centerline temperatures reached 1040°C during melting in 8 inch diameter cans without fins while the walls were heated to only 900° to 950°C by the furnace. Even without the aid of this self-heating the melting rate of 30 kg/hr required for the waste from the NFS fuel reprocessing plant can be obtained with an 8 inch diameter can containing a drop-in fin assembly when coupled to a heated-wall spray calciner. This 30 kg/hr rate was determined assuming a frit-to-calcine ratio of 2:1.

By this procedure, plant HA Column Wastes, Low-Level Wastes, and various HLLW process recycle streams could be combined to produce a borosilicate glass with these properties: calcine content, 33 weight percent; density, 3.0 g/cm^3; volume, 3.3 ft^3/MTU; heat generation rate, $\leqslant$ 60 W/liter; thermal conductivity, 1.0 W/m-°K at 200°C and leach rate, $\sim 10^{-6}$ g/cm^2-day.

Canister Handling and Inspection

Canister handling and inspection consists of movement of the canister within the cell, canister closure welding, canister closure inspection, canister decontamination, and canister in-cell storage. These operations are described below.

Canister Movement: Canister movements between operations in the cell will be performed by the crane. At this time it is expected that ears or a holding knob will be included on the canister as part of the cap or head. A yoke or a grappling device will be placed on the cell crane hook for attachment to the canister. At this time final canister design has not been completed and evaluated to the extent that a detailed design can be designated.

Canister Closure Welding: The filled canisters are closed with a threaded closure. This closure is then seal welded with a ⅛ inch fusion weld fillet using a standard water-cooled TIG welder. The welder power supply and controls are located out of cell. The welding torch, which is in-cell, is mounted on a rotating holder. During welding this holder is mounted on and indexed off a reference center point on the canister closure. Rotation is accomplished with a fractional horsepower, variable speed dc motor.

This method of remote seal welding of canister closures was developed and used in the WSEP program. As with any welding, it was found that weld joint preparation (cleanliness) is very important. This is assured by proper design of the canister neck and closure. During processing the canister is sealed below the weld joint, thus preventing contaminants from contacting or being present at the joint. With these precautions a good quality weld can be made.

Leak Test of Canister Closure Weld: It is currently planned to verify the integrity of the canister closure seal weld by a helium mass spectrometer leak test or by ultrasonic techniques. Three basic methods exist for performing the helium leak test: (1) Place a slow-release helium source in the canister and use a sniffer type of detector after closure weld. (2) Evacuate the canister through the detector, and flood the exterior of the canister with helium. (3) Place a slow-release helium source in the canister, and place the canister top in a leak chamber after closure weld.

Method 1 may not suffice due to inadequate detection sensitivity for this application. Method 2 was used for leak testing the WSEP canister closure welds. Though adequate, this method involves using an additional canister closure which must also be leak checked by another means. Outgassing the canister can be very time consuming if scale, calcine, etc., exist.

A modification of Method 3 is preferred. With this method a slow-release helium source is placed in the canister; the lid is placed on the canister and seal welded; a metal "bell jar" is placed over the seal weld to form the "leak chamber." A high-temperature silicone gasket is placed between the chamber and a flat on the canister neck to effect a seal. The vacuum created by the leak detector is used as the clamping force for the gasketed seal. With this method outgassing prob-

lems associated with metal scales, etc., are minimized because the only portions
of the canister in the leak chamber are the closure, which has not been subjected
to process heat, and the weld joint. Leak detection capability of this method
is limited only by the accuracy of the mass spectrometer. The mass spectrometer is placed outside the hot cell for ease of operation and maintenance. The
only equipment contained in the cell is the leak detection head. Other methods
considered to assure good closure weld integrity will be:

> (1) visual inspection ($\geqslant$7X)
> (2) dye penetrant or magnetic particle test if the
>   cap can be cooled
> (3) stringent welding procedures and sample
>   testing.

But at this time the helium leak check or ultrasonic techniques appear to be the
most practical and reliable. Radiograph methods will not be practicable because
of the high amount of radiation generated by the waste glass.

Canister Decontamination: Canister decontamination and smear sampling techniques and equipment have been developed at PNL as part of a program to
assist in establishing waste canister acceptance criteria at the Federal repository
(16)(17).

Canister decontamination is achieved by repetitively passing the canister vertically
past a medium-pressure spray ring where the canister is alternately sprayed with
100 psig steam or 70 psig water. In recent tests this alternate steam/water procedure took about 140 min and used 310 gal of water and 40 lb of steam. A
steam cleaning procedure required 40 min and consumed 20 lb of steam. A
total of four WSEP canisters were decontaminated using the spray ring with both
steam and water as cleaning agents (17). Two additional WSEP canisters were
decontaminated using the spray ring and steam only. Nearly the same level of
cleaning can be done with steam in 40 min as with a dual steam and water process taking 140 min.

The major fraction of smearable contamination is removed early in the decontamination procedure. The contamination remaining seems to be a film that is removed very slowly with 100 psig steam or 75 psig water. The initial smears on
the WSEP canisters showed a reading of between 8×10^5 and 5×10^6 dpm/100 cm^2.
After decontamination, these canisters showed smear readings between 4×10^3 and
2×10^4 dpm/100 cm^2. Decontamination factors ranged from 139 to 950. Industrial sources indicate that these radiation levels achieved after decontamination
are quite suitable for onsite storage of high-level waste canisters.

In-Cell Canister Storage: Canister storage, filled or empty, within the processing
cell is not expected to be a significant or complex problem. Many proven approaches can be used and have been demonstrated in the nuclear industry. The
method used will be dictated mainly by economics, reliability, simplicity, and minimization of adverse effects on the canister. Since the storage time will be for
only hours or days, many long-term considerations are not applicable such as canister and storage atmosphere compatibility In the WSEP program 33 canisters
filled with radioactive materials have been stored in air and/or water for years
for heat removal with only one minor canister thermowell failure attributed to
stress corrosion cracking.

PNL has developed a preliminary design of a rack for canister storage. It basically consists of a rack constructed from angle or strip 304L stainless steel. If the canister is filled and requires water cooling, a coil may be inserted in the storage position in the rack. This approach is simple and economic. However, the rack must be designed so as to prevent contamination traps which would make it difficult to decontaminate. Storage of the canister in air eliminates contaminating an additional medium such as water, which would eventually have to be treated. Also any possible adverse effects of thermal shocking the canister by placing it in water are eliminated.

Off-Gas Treatment

Multiple steps of off-gas treatment are employed to remove radionuclides and NO_x so that the off-gas may be released to the environment within applicable release guidelines. The effort at PNL is concentrating on characterizing the calciner off-gas to allow selection of optimum effluent treatment equipment. However, future operations will provide data on the performance of various types of effluent treatment equipment.

In the NFS reference flowsheet, calciner off-gas is initially routed through a venturi scrubber-separator and condenser system. This system is designed to remove particulates, nitric acid, water, and volatile ruthenium from the off-gas. The off-gas is also cooled to about 55°C. A negligible amount of NO_x, excluding HNO_3, is assumed removed due to the short residence in the system. The condensed nitric acid containing the particulate and dissolved contaminants is routed to the HLLW blend/feed tank for recycle. Cooled condensate is also continuously recycled to the venturi scrubber-separator for continuous operation. The off-gas is treated later for additional NO_x, ruthenium, particulate, and possibly iodine removal in a production plant.

FLUIDIZED BED CALCINATION

In other work done by Battelle Pacific Northwest Laboratories, calcination of simulated high-level nuclear wastes representative of those to be generated by commercial fuel reprocessors has been tested in a fluidized bed calciner using a new operating technique. An inert material is added to the calciner continuously during operation. The addition of this material improves the operation in many respects: particle size control is not required; a low inventory of fission products is maintained in the bed; coupling to a melter is improved; high sodium-bearing wastes can be calcined; and high feed rates are possible.

This type of operation is used as the drying step prior to mixing the wastes with a frit and melting them to a borosilicate glass form, the most likely form for storage of high-level wastes.

The calciner is heated by in-bed combustion and uses sintered metal filters for primary off-gas cleanup. A simplified atomizing nozzle has been used for introducing feed to the calciner. It consists of concentric tubes and is relatively maintenance free. A wide range of feed chemical compositions has been tested, including wastes with sodium concentrations near 1 M. The process is illustrated in Figure 2.3 and a material balance is given in Table 2.2.

FIGURE 2.3: FLUIDIZED BED CALCINER AND OFF-GAS SYSTEM

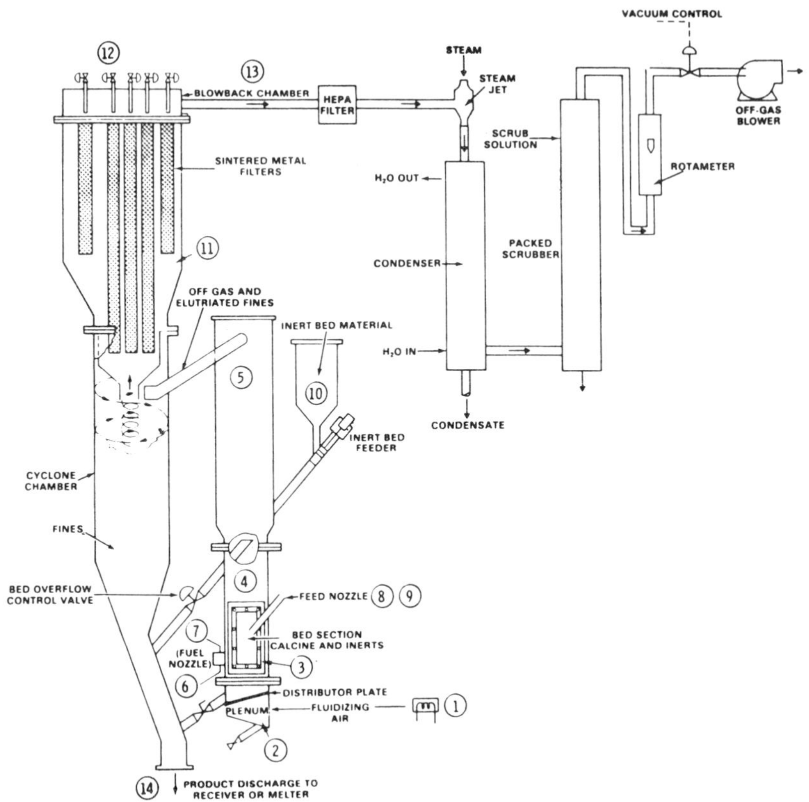

TABLE 2.2: FLUIDIZED BED CALCINER MATERIAL BALANCE

	(1) Fluidizing Air	(2) Plenum	(3) Lower Bed	(4) Upper Bed	(5) Disengaging Space	(6) Fuel	(7) Fuel Atomizing Oxygen	(8) Feed	(9) Atomizing Air	(10) Inert Bed Material	(11) Filter Vessel	(12) Blow Back Air	(13) Filtered Off-Gas	(14) Calcine Produced
Temperature °C	20	360	550	550	475	24	23	25	20		360	100	360	
Pressure psig	50	0.2	-0.2	-0.3	-0.4	80	80	5	90		-0.43	80	-0.9	
Liquid Rate l/hr						4.5		25						
Gas Rate scfm	6.0		6.0				4.4		4.5				37	
Gas Rate acfm	1.46						0.74		0.68				91	
Velocity ft/sec			1.0	6.0	3.0									
Solids Rate kg/hr										3.31				4.47

Source: BNWL-2138

Calcination by Continuous Inert-Bed Method

To operate in the continuous inert-bed manner, the bed section of the vessel is filled with a granular inert material, SiO_2 (quartz sand). This material is cheap, durable, and insensitive to extreme temperature changes. Furthermore, it can be incorporated into accepted glass forms chosen as final products without increasing volume.

A particle size range of about 28 to 65 Tyler mesh is typical. A bed charge of ~15 kg is initially required. Air inflow at a 1 ft/sec superficial velocity is required to fluidize the bed.

Process heat is supplied by the combustion of pure oxygen and kerosene directly in the reaction chamber (bed). Prior to autoignition of the fuel, external heating is needed to heat the bed to ~375°C. A propane heater in the fluidizing air line has been used for rapid heating. When the desired preheat temperature is reached, oxygen and kerosene are metered to the vessel along with nitric acid or feed. Ignition is instantaneous.

Feed solution being calcined either coats the bed, is spray dried, or coats and wears away because of colliding with other particles. The precise mechanism is not quantified. It can be influenced by changing process variables such as temperature, feed composition, atomizing characterisitcs, or jet grinding.

At or near the time feed is introduced, addition of inert bed material is started from the storage hopper to the reaction chamber. The rate of addition and whether it is continuous or semicontinuous depends on the specific run plan. As a lower limit, to maintain proper bed level the rate of addition needs to be equal to the bed material worn down and elutriated. An upper limit would be dependent on the next processing step. For example, when vitrifying, the inert rate would be equivalent to that needed for subsequent glass making. This is roughly equal to the weight of calcine being generated.

The constantly increasing bed level is controlled by periodically opening a valve in the overflow pipe. Constant bed level control could be maintained by continuous overflow through a properly sized pipe; however, for the pilot-plant unit the valve was desirable.

Off-gas passes from the calcining vessel to the adjacent filter vessel. The gas goes through a cyclone section and sintered metal filters before entering the wet cleanup system. The cyclone is being used to reduce the solids loading to the sintered-metal filters. The cyclone is not required but is being tested as part of the equipment development program.

A seal pot connected to the calciner acts as a safety relief device in case of pressurization. The properties of the solid produced depend on the waste feed composition and operating conditions.

Gas Distribution in the Fluidized Bed

The gas velocity in the middle and upper portions of the fluidized bed is much higher than that needed for fluidization. This is due to the additional gas from combustion, atomizing air, and feed evaporation. The degree of elutriation achieved can be designed into the process by the disengaging space configuration and overflow characteristics. With the continuous inert-bed (CIB) process, when a valve is used in the overflow, all gas goes up through the disengaging vessel and to the gas-particulate separator vessel. When the overflow valve is opened, most of the gas will take the path of least resistance through the overflow, sweeping the bed material with it.

Calcination-Vitrification

Calcination and vitrification equipment as envisioned for a commercial reprocessing facility would be operated as a coupled unit. Vitrification could be done using either an in-can melter (15) or continuous ceramic melter (18). This was not always practical in the pilot plant. For example, the in-can melting system, a portable unit, has been coupled with a spray calciner, the fluidized bed calciner, and a wiped-film evaporator. Runs have been made with the in-can melter directly coupled. All gases generated in the canister are filtered through the calciner's sintered metal filters. Glass-forming material (frit) is added from a feeder to the melter proportionately to the CIB product being generated.

The continuous ceramic melter, however, as being tested at PNL, was not designed as a portable unit. Operation with the continuous ceramic melter has been through an indirect coupling or using no coupling at all. Indirect coupling would be simply feeding previously calcined material to the melter while routing the off-gas through the fluidized bed. Coupling to either melter is quite simple and has not been considered a design problem.

Emergency Fluidization and Bed Removal

Work is also being done by Allied Chemicals' Nuclear Services Division at its Idaho Chemical Processing Plant (ICPP) to obtain process and equipment design data for fluidized-bed solidification of the high-level liquid wastes generated during the reprocessing of commercial light water reactor (LWR) fuels. The basic conceptual process design for a commercial fluidized-bed solidification facility is similar to an existing Waste Calcining Facility at the ICPP; process and equipment improvements being designed into the New Waste Calcining Facility are also directly applicable to a commercial facility.

Chemical dissolution of fuel (cladding and meat) at the ICPP and the chop-leach process used and planned for near-term commercial reprocessing plants result in significant differences in the volumetric heat generation rates of the calcined solids. The calcined solids generated at the ICPP have heat generation rates in the range of 5 to 50 Btu/hr-ft^3; calcined solids from the fluidized-bed solidification of LWR first-cycle raffinate will have heat generation rates in the range 4,400 Btu/hr-ft^3 (5 years out of reactor) to 22,000 Btu/hr-ft^3 (1 year out of reactor).

One concern during fluidized-bed solidification of LWR reprocessing wastes is the ability to remove decay heat if the fluidized bed should collapse because of accidental loss of fluidizing air. A collapsed fluidized bed in the calciner in operation at the ICPP has not and could not result in any significant heatup because of the low volumetric heat generation rate. However, collapse of the fluidized bed in a commercial calciner, assuming no remedial action was taken, could result in significantly higher temperatures.

The rate of bed temperature rise in a collapsed commercial calciner bed is sufficiently slow to permit a deliberate response. The heatup rate depends primarily on the decay-heat generation rate and to a lesser extent on the bed diameter, height, and thermal conductivity. With no heat dissipation it would take 0.5 and 3 hr, respectively, with 1- and 5-yr-cooled wastes, for the center of the bed to reach 1300°F, the temperature at which cesium begins to volatilize. Thus, a "hair-trigger" response to a loss-of-fluidization is unnecessary.

Cooling of the calcined solids by redundant fluidizing air systems and removal of the solids from the calciner are proposed for consideration as primary safeguards in the event of a collapsed bed. Fluidizing air at normal velocities (1-4 ft/sec) will maintain the solids at 900°F (the nominal calcination temperature). Fluidizing air provided by the following set of systems would essentially eliminate the possibility of loss of fluidizing air: (1) regular fluidizing air blowers with in-line spares, (2) the plant compressed air system, (3) an emergency compressed or liquid air supply, and (4) emergency fluidizing air blowers powered by gasoline or diesel engines. The design of the fluidizing-air system for a specific plant would depend on the overall plant design and safety analysis, and on specific system reliability requirements.

Redundant systems should be provided for removing the bed if removal is required for maintenance or long-term cooling. Bed removal should be at the discretion of the operators. Candidate bed-removal systems are as follows: (1) normal bed (product) removal system consisting of two or three removal lines each with capability of removing plugs, (2) bed discharge (dump) lines through the fluidizing air distributor plate, and (3) emergency bed removal by a "vacuum" system extending into the bed above.

The consequences of a collapsed bed, even if no actions were taken, would not result in a hazard potential to the public above that in routine plant operation. Process and cell off-gas cleanup systems would still prevent the release of volatile or particulate matter, above acceptable limits, to the atmosphere. A sintered (or partially molten) bed would be an operating problem rather than a safety or environmental hazard. The wall temperature of a partially insulated or uninsulated calciner vessel would be well below the melting temperature of the wall, and decay heat, even for 1-yr old waste, could be dissipated from the vessel wall by natural convection and thermal radiation to the cell off-gas and the cell wall.

Reliable redundant fluidizing air systems for cooling can be designed by application of existing plant engineering and quality assurance procedures. Concepts for bed removal are being evaluated, both theoretically and experimentally.

OTHER SOLIDIFICATION PROCESSES

Phosphate Glass Solidification

The phosphate glass process, originally tested by Brookhaven National Laboratory and further tested at Battelle Northwest (19), consists of an evaporative denitration of the high-level waste after the addition of phosphoric acid, sodium hydroxide, and ferric nitrate. The bottom discharge of the denitrator is melted and collected in a pot for final storage. The overheads from denitration are condensed and reevaporated with the bottom, recycled to the denitrator feed tank, and the overheads fractionated for nitric acid recovery. The aqueous effluent was decontaminated by a factor of 10^7 to 10^8 for the nonvolatiles, and by 10^5 to 10^6 for the radio-ruthenium. Storage of the phosphate glass product was adequate; there was no increase in pressure after the pot cap was welded in place.

Pot Solidification

The pot solidification process, originally tested at Oak Ridge National Laboratory and also further tested at Battelle Northwest (20), is a direct solidification in the storage container used for ultimate disposal. The use of auxiliary evaporator, fractionator, and scrubber equipment provided a 10^9 decontamination factor for the liquid effluent and greater than 10^3 for radio-ruthenium. The sealed containers have undergone testing at the Solids Storage Engineering Test Facility at Hanford.

VITRIFICATION METHODS

There is general agreement on vitrification as the method for final fixation of the HLW from the first generation of power reactor fuel reprocessing plants. Several suitable vitrification processes, differing in significant details, have been developed.

In-Can Melting

The in-can melting process, developed by Battelle at Hanford, has been described in detail on p   . In this process the liquid waste is first calcined in a spray calciner, which converts the waste to a fine, dry predominantly oxide powder. The waste powder and a specially-formulated glass frit, which is added at the base of the calciner, drop together into a stainless steel canister in a furnace. The glass frit waste calcine mixture fuses to form a fairly homogeneous glass. Certain fission products, particularly the noble metals ruthenium, palladium and rhodium do not dissolve in the glass. Instead, they are uniformly distributed as μm-sized crystallites in the glass matrix.

The canister is filled with glass in approximately 24 hours, and replaced with an empty canister. Longitudinal metallic fins may be used inside the canister to conduct heat into the melting area more rapidly during processing and to aid in cooling the interior of the glass during storage. An air tight cap is welded on the filled canister, which is vacuum leak checked before interim storage. Note that in this process the melter becomes the storage vessel, or looking at it another way, a new melter is used for each run.

French Vitrification Process

The French have developed a somewhat similar process which uses a rotary kiln instead of a spray calciner, as shown in Figure 2.4. As in the in-can melting process, the calcine and glass frit are combined as they enter the melter, however the French are using a continuous melter which is drained periodically through a freeze valve. The process installed at Marcoule uses a continuous metallic melter; the process being developed for use with power reactor fuel wastes at LaHague will use a continuous ceramic melter, perhaps heated by direct induction coupling to the molten glass. Continuous metallic melters have a relatively

FIGURE 2.4: FRENCH VITRIFICATION PROCESS

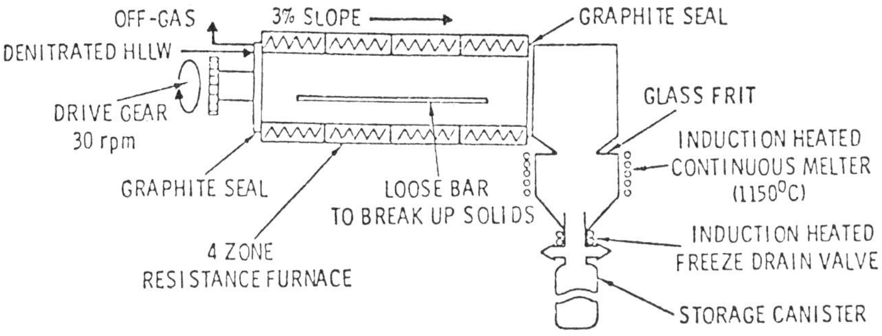

Source: BNWL-SA-5757

short life and have to be replaced every 6-8 weeks. This is one of the incentives
to go to ceramic melters which potentially can last from 2-5 years without replacement. Ceramic melters also permit higher operating temperatures, if desired.

Liquid-Fed Continuous Ceramic Melter

A ceramic melter is being developed at the Pacific Northwest Laboratory which
offers a further simplification, and thus increased reliability, to the vitrification
of HLW. This is the liquid-fed continuous ceramic melter shown in Figure 2.5.
High melting capacity is achieved by conducting an alternating current through
the molten glass. The high melting capacity makes it possible to omit the calcination step used in the previous processes and feed liquid waste directly to the
melter. Thus a relatively complex piece of process equipment, the calciner, is
eliminated completely.

FIGURE 2.5: LIQUID-FED CONTINUOUS MELTER VITRIFICATION
PROCESS

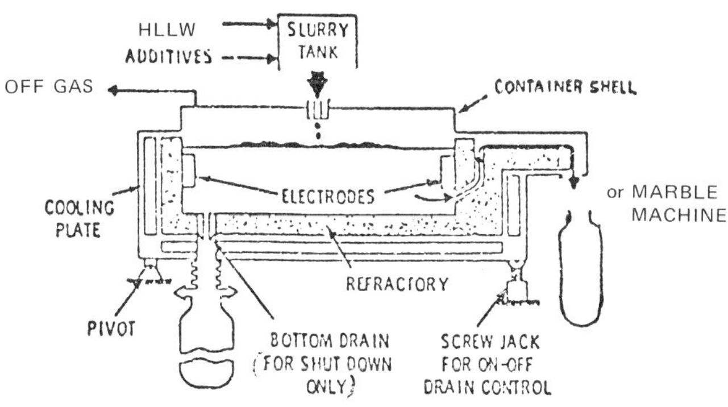

Source: BNWL-SA-5757

The ceramic melter offers the possibility of another potentially desirable process modification. Usually the waste glass will be formed as a single large casting in canisters 1-2 ft in diameter and 10-15 ft long. An alternative is to produce glass shapes such as marbles, which offer flexibility in forming composite final waste forms, using metallic matrices, for example.

Joule-Heated Ceramic Melter

Vitrifying high level waste is basically glass melting. For this reason an investigation was initiated at Battelle Pacific Northwest Laboratories, to see if melter technology from the glass industry could be adapted to vitrifying high level wastes. Ceramic lined all-electric glass melters, which are gaining broader use in the glass industry, showed excellent potential for use in vitrifying high-level waste. All-electric melters are being selected in the glass industry because of high capacity, long life, low volatilization and superior glass quality. Since these characteristics are consistent with the requirements of a high level waste melter, a development program was initiated.

After testing several laboratory scale melters, it was found that startup and restart of a melter with a frozen tank required resolution before it could be used in a radioactive facility. After several techniques were evaluated, the one which provided most flexibility was the sacrificial element startup technique.

To obtain an operating status with a joule-heated ceramic melter it is necessary to heat the glass charge between the electrodes to a temperature at which the glass is conductive. In commercial glass tanks, gas firing is the common startup technique. This approach is not well suited for starting up a radioactive waste melter for two reasons. First, the voluminous noncondensible off-gases might not be compatible with the reprocessing plant. Second, when restarting a frozen tank with submerged electrodes, gas firing might not be successful due to the opaque characteristic of the waste glasses.

To provide a method for initial startup and restart with a frozen tank, the sacrificial element technique was developed. This approach for startup of a joule heated ceramic melter is illustrated in Figure 2.6. Because restart of a frozen tank with submerged electrodes is the most challenging situation, the startup procedure will be explained for this condition.

A set of startup electrodes with a resistance heating element between them is placed in a flange which fits in the cover plate of the melter. This assembly is lowered onto the mating opening on the melter shell and is secured. The vertical startup electrodes are free to translate in the vertical direction. Thus, when the assembly is lowered onto the roof, the electrodes settle to the top of the frozen glass. The startup electrodes are then attached to an independent power supply or in parallel with the main power electrodes. After glass frit is dropped into the cavity and the startup element is buried, the melter is ready for a programmed startup.

Using either a temperature feedback or power control system, energy is dissipated in the sacrificial element. Energy from the sacrificial element heats the surrounding glass. As the power level increases, glass begins to melt adjacent to the heating wire. Further power input results in a molten zone between the startup electrodes. At this condition, the use of the sacrificial element is complete.

FIGURE 2.6: SCHEMATIC OF STARTUP TECHNIQUE

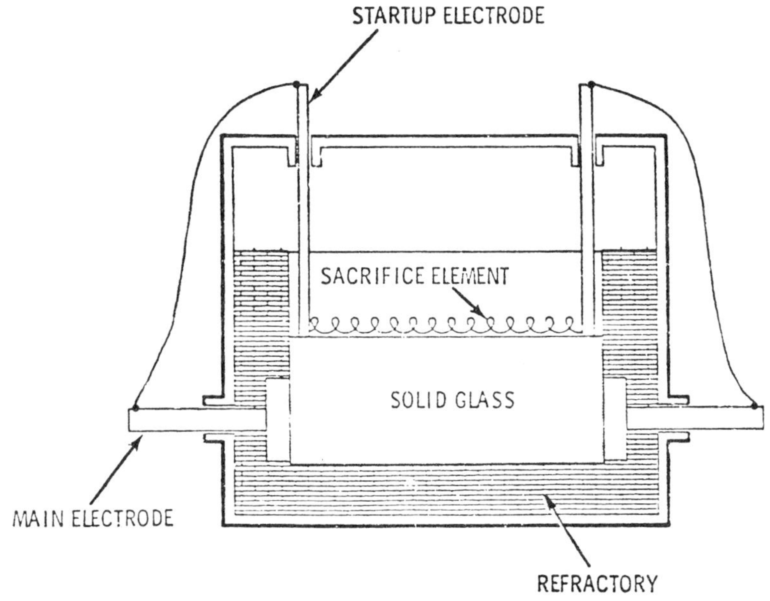

Source: BNWL-SA-5895

The sacrificial element used in all cases has been a common electrical resistance wire. However, other materials like molybdenum disilicide or silicon carbide may be more appropriate. The important characteristic of the sacrificial element is that it is readily dissolved by the glass at high temperatures.

Ultimately the startup electrodes settle down to the elevation of the main electrodes and the molten zone expands to the main electrodes. At this stage, the main electrodes begin to pass current. Finally the main electrodes assume entire control and the startup electrodes can be removed or allowed to dissolve in the molten glass.

This technique has been used for initial startup in two different melters and used to restart one melter with a frozen tank on three separate occasions. Heatup rates as low as 6°C per hour to as high as 300°C have been demonstrated. For a hot cell unit, the initial startup assembly would be an integral part of the melter and no remote operation would be required because both the startup electrodes and sacrificial elements would be consumed during startup.

Restart of a joule heated ceramic melter (JHCM) would require the remote operations outlined above; however, it should be noted that shut-down of the melter would only be necessary in case of an abnormal condition in the melter. Should other equipment require shut down for maintenance, a JHCM would not require shut down because it can operate for extended periods of time at a reduced glass temperature without significant loss of operating life.

Engineering Scale Ceramic Melter: Melter Description — After gaining operating experience and understanding of system operation with small laboratory melters, an engineering scale ceramic melter was designed and constructed. A schematic cutaway view of the assembled melter is given in Figure 2.7. The melting cavity in this unit was 14 inches wide, 30 inches long and 12 inches deep. The glass depth was maintained at 6 inches by the overflow drain pipe.

The drain pipe was located in the overflow cavity which was positioned adjacent to, and centered on, one sidewall of the melting cavity. The overflow cavity was 6 inches wide, 7 inches long and 12 inches deep. The overflow pipe was 2 inches outside diameter and ½ inch inside diameter and was centered in the overflow cavity. The overflow cavity and the melting cavity were powered by independent sets of electrodes. The electrodes in the melting cavity were 2 inches thick, 12 inches wide, 11½ inches high and were suspended from the top of the refractories. The overflow electrode was also 2 inches thick and was suspended from the top of the refractories. The bottom of this electrode was 2 inches above the floor. This allowed molten glass to flow from the melting cavity to the overflow cavity where it could flow out the drain pipe.

FIGURE 2.7: SCHEMATIC OF CERAMIC MELTER

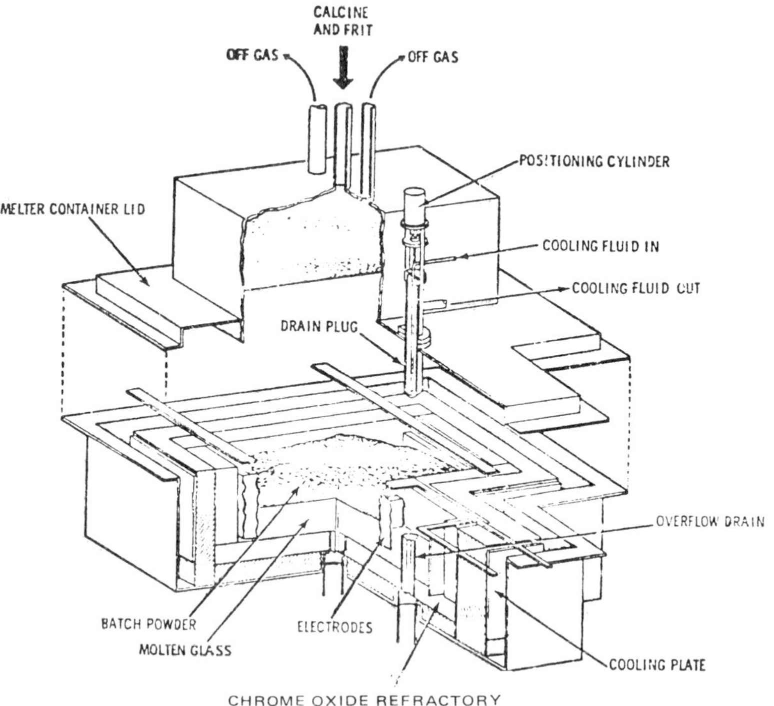

Source: BNWL-SA-5895

The glass contact refractory was Monofrax E which is a fused cast, chrome-spinel material composed primarily of chrome oxide ($\sim$78%). The blocks were all three inches in thickness. These glass contact refractories were backed with zircon bricks also three inches in thickness.

Water cooled stainless steel plates were placed on the outside and bottom of the refractories. The refractories were constrained by compression springs which could be adjusted by positioning screws on the outside of the box. The exterior dimensions of the entire melter were: 4 feet wide, 4½ feet long and 3 feet tall.

In operation, the simulated calcine and frit were fed proportionally into the melter through the centrally located nozzle in the container lid. After the batch was melted and refined in the melting cavity, it flowed under the overflow cavity electrode and out the drain pipe. A fluid cooled plug valve was used to stop glass drainage.

Production Experience While Feeding Calcine and Frit — In preliminary tests the simulated waste calcine and frit were intimately mixed prior to being fed to the melter. After experiencing underpowering in the initial test, smooth rapid melting of the batch was demonstrated. Glass production rates in excess of 45 kg per hour were frequently shown with no problems. Typical operating conditions during these tests were power input of 35 kilowatts at an rms voltage of 160 between the melting cavity electrodes.

After a second solids feeder was purchased and installed, the calcine and frit were fed to the melting cavity through separate lines. Both material streams accumulated in the melter at approximately the same location but little mixing was realized. In spite of this condition the batch did melt into a homogeneous glass in the melter. However, this method of feeding appeared to reduce the capacity of the unit. While melting a calcine with a high content of sodium nitrate, increased capacity was noted.

The unit was started up in mid-January of 1975 and maintained at or above molten glass temperature of 1150°C for 10.8 continuous months. During the nearly 11 months of operation, 20 day-long experiments were completed. Over 3,600 kg (8,000 pounds) of glass were produced by the unit. While feeding powdered calcine and frit, the unit demonstrated a maximum capacity of 60 kg/hr. The average capacity for all batch powder tests was over 45 kg/hr. Thus, this melter has been demonstrated at or above the requirement of a reference 5 MTU/day reprocessing plant.

While feeding the simulated high-level liquid waste (concentrated to 378 l/MTU), the melter has been demonstrated at 29 liters per hour which is equivalent to the needs of a 1.88 MTU/day reprocessing plant.

SOLIDIFIED WASTE FORMS

Waste Glass

An immense number of glass compositions have been examined in screening tests in the twenty years that vitrification of radioactive waste has been studied. Many of the glasses are relatively low-melting borosilicate formulations which fall in the composition ranges given in Table 2.3. The goal is to keep waste loading as high as possible without reducing the quality of the glass.

The complexity and variability in waste composition creates problems. The major waste constituents are the fission products and actinides but there may also be varying amounts of inert sodium, phosphate, iron, chromium, nickel, gadolinium and even fluoride or sulfate. Thus it is not meaningful to talk about a specific waste glass composition unless it is understood to represent a class of formulations, with variations possible for many of the constituents.

Determination of the thermal and radiation effects on a waste glass can require several years, even using accelerated tests. Therefore detailed information on thermal and radiation effects is only being determined on certain generic waste glass compositions, such as the reference glass shown in Table 2.3, which can represent a certain class of formulations.

The reference glass is low-melting, even for a waste glass, as shown by the viscosity, therefore the silica content is at the low end of the range. The B_2O_3 and alkali content are also at the low end of their ranges and are compensated for by a high zinc oxide content, which is the major distinguishing characteristic of this generic glass.

TABLE 2.3: COMPOSITION RANGE OF TYPICAL HLW GLASSES

	Typical Range, (%)	Reference Glass, (%)
SiO_2	27–52	27.3
B_2O_3	9–22	11.1
Alkalies	7–19	8.1
Alkaline earths	0–6	5.9
Al_2O_3	0–1	0
ZnO	0–22	21.3
TiO_2	0–3	0
Waste oxides	20–40	26.3
Temp. for Viscosity of <200 poises	$950°$–$1150°C$	$960°C$

Source: BNWL-SA-5757

Radiation Effects: Although the waste glass will be subjected to high doses of beta, gamma radiation, reaching close to 10^{12} R total, and a neutron flux, it is the alpha particles, and alpha-recoil particles, that have the potential for creating the most radiation damage in the waste glass. The alpha dose in waste glass specimens may be accelerated by substituting relatively short-lived curium-244 for all of the actinide content. A cumulative α dose of almost 2.0×10^{18} per gram has been reached in reference glass specimens that are about 2.5 years old which is equivalent to over 20 centuries storage of actual waste glass.

This equivalence is calculated for waste generated without plutonium recycle, which will be the actual case for the next decade or two. With complete plutonium recycle the waste will contain increased amounts of americium and curium and the alpha dose in 20 centuries will be higher, about 8.0×10^{19} α/gram. It can be seen that the stored energy is approaching an equilibrium, or saturation value, and it is expected that most types of radiation damage will behave similarly, that is, they will approach an equilibrium value which can be defined.

In addition to the radiation-induced changes in stored energy and density, leach rate, physical strength, and helium behavior in the curium-doped waste glasses are being measured. Nothing has been found so far to indicate that glass won't be able to endure the self-radiation to which it will be exposed as a nuclear waste disposal medium, however much more study is needed. The amounts of stored energy shown here, which can be released as heat in thermal excursions, can have only trivial effects. The density of this reference glass is increasing with radiation, that is, it is actually shrinking in volume which should cause no problem.

Thermal Effects: All indications are that the thermal effects on waste glass can be more severe than the radiation effects, however the thermal effects described here will be in the central portions of the glass blocks; a surface "rind" an inch or more in thickness will remain unaffected. Note that there is a certain inversion of thermal and radiation effects. The surface rind will show the largest radiation effects since the higher internal temperatures will tend to anneal out all radiation damage for many years.

The major thermal effect of concern is devitrification, and the associated changes in leach rate which devitrification can cause. In the reference glass the maximum effect occurs in specimens stored at 700°C and is an increase in leachability of about a factor of 10 as shown in Figure 2.8. In specimens stored at 500°C and below no increase in leach rate has occurred.

FIGURE 2.8: LEACHABILITY OF HLW GLASS

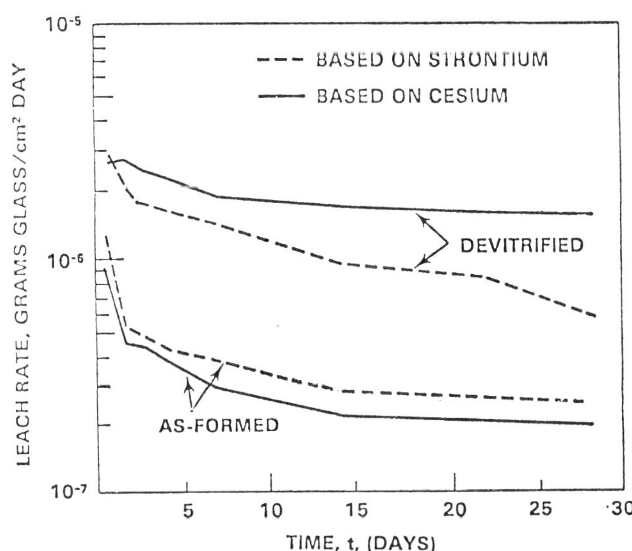

Source: BNWL-SA-5757

Since the walls of the canister will be kept at 400°C or below during storage it is apparent that, as stated earlier, the effects of devitrification will be confined to a relatively small volume of glass in the center of the canister.

The tests shown here were made in specimens held for 2 months. In fact, devitrification had reached an equilibrium at 700°C in just a few days. But at lower temperatures devitrification rates are much slower so specimens of the reference glass are being stored at temperatures between 300° and 600°C for 5 years.

It should be noted also that the leach rates shown in Figure 2.8 were obtained in demineralized water. Of course, except in water basin storage, any water the waste glass may come in contact with won't be demineralized water but a natural water, which may vary from saturated salt brine to rain water. The effect of the type of water upon the leach rate of waste glass is complex and is just beginning to be studied in detail.

Concrete

Although solidification in concrete is a technique usually reserved for low level radioactive wastes, concrete can also be used for the solidification of intermediate- and low heat generating high-level wastes. Work was performed at Brookhaven National Laboratory (BNL) to investigate the properties of Savannah River Plant (SRP) high level waste incorporated into concrete (21).

SRP waste stored in underground tanks is in the form of sludge, supernate and salt cake (formed by evaporation of aged supernate). The sludge is composed principally of $Fe(OH)_3$ and MnO_2 while the salt cake and supernate contain $NaNO_3$, $NaNO_2$, $NaOH$, Na_2CO_3 and $NaAlO_2$. The primary sources of toxicity and heat generation in this waste are ^{137}Cs and [90]Sr. The bulk of the ^{90}Sr in the waste is adsorbed in the sludge while the majority of the ^{137}Cs is found in the supernate.

In a conceptual solidification process, waste would be removed from tank storage by dissolving the salt cake with water and sluicing the sludge and salt-supernate solution from the tank. The sludge would be separated from the liquid fraction by centrifugation and filtration. [137]Cs would be removed from the solution by ion exchange and sorbed onto zeolite. The sludge would be washed to remove its salt content and dried. The forms of waste for a solidification process are then sludge, zeolite, and a salt solution containing relatively little activity. This salt solution could be evaporated to form a salt cake.

Simulated wastes of these types were solidified using Portland type II and high alumina (Lumnite) hydraulic cements. The leachability of these formulations was determined by a modification of the IAEA static leach test method (22). The modification consisted of exposing the entire external surface area of the specimen to the leachant (distilled water) rather than the top surface only. During the first two weeks, the leachant was removed daily for analysis and replaced by fresh leachant.

Subsequently, sampling was done on a weekly basis. Cesium and strontium analyses were performed by atomic adsorption spectrophotometry; a colorimetric method was used for nitrate analysis. Leach rate data were calculated both as a fraction release, F_i, and bulk leach rate, L_B, where

$$(1) \qquad F_i = \frac{a_n}{A_o}$$

and

$$(2) \qquad L_B = F_i \left(\frac{m}{S \cdot t} \right)$$

where a_n = amount of the species of interest leached from the composite during leach period n.

A_o = initial content of the species of interest in the composite.

m = composite mass, g.

S = geometric external surface area of the composite, cm^2.

t = leach time, days.

The bulk leach rate is expressed in units of $g/(cm^2\text{-day})$.

The strontium leachability of sludge-cement composites using high alumina and Portland type II cements is shown in Table 2.4. Increasing the sludge content of the composites results in decreasing strontium bulk leach rates. These leach rates represent the release of strontium from both the sludge and cement.

Since the sludge adsorbs strontium, increasing the sludge content increases the probability that strontium being leached from the composite encounters a sludge particle and is adsorbed. The sludge has a strontium content of 367 ppm while the Portland type II cement contains 720 ppm strontium and the high alumina cement 190 ppm. The leach rates for cement-water pastes are two orders of magnitude higher than for sludge-cement composites suggesting that the sludge also acts to sequester strontium from the cement matrix. As such, the rate at which strontium is leached from the sludge may be lower than suggested by Table 2.4.

The release of cesium from Linde AW-500 zeolite incorporated in cement is a function of the zeolite content of the composite as shown in Table 2.5. A decrease in the cesium bulk leach rate is noted as the zeolite content in the composite is increased.

Based on calculations, the relative amounts of sludge and cesium to be disposed are such that if combined in one composite formulation, there would be 16 parts of sludge to each part of zeolite by weight. A high alumina cement concrete composite formulation was determined using this ratio, incorporating as much sludge as possible while maintaining an adequate composite integrity (1,000 psi compressive strength). This formulation and its leach rate are shown in Table 2.6.

While the strontium bulk leach rate is comparable to that for sludge-cement composites, the cesium leach rate increased almost two orders of magnitude over that for a comparable zeolite-cement composite. The cause of this behavior is not immediately obvious, but apparently represents an interaction between the sludge and the zeolite. This suggests that the sludge and zeolite should be solidified separately.

TABLE 2.4: STRONTIUM LEACH FOR SLUDGE-CEMENT COMPOSITES

Sludge	Cement	Water	Mass, g	Surface Area, cm^2	Leach Time, days	Fraction Release	Bulk Leach Rate $g/(cm^2\text{-day})$
	Formulation, wt %						
. . . . High Alumina Cement							
19.3	56.9	23.8	56.95	55.29	10	$<1.1 \times 10^{-4}$	$<1.1 \times 10^{-5}$
28.2	44.7	27.1	54.10	56.77	10	$<1.1 \times 10^{-4}$	$<1.1 \times 10^{-5}$
37.2	33.5	29.3	54.19	58.48	10	$<1.1 \times 10^{-4}$	$<9.9 \times 10^{-6}$
46.2	22.7	31.1	51.10	56.36	10	$<8.7 \times 10^{-5}$	$<7.9 \times 10^{-6}$
. . . . Portland Type II Cement							
20.0	55.7	24.3	59.75	56.77	10	6.3×10^{-4}	6.7×10^{-5}
30.0	42.6	27.4	54.00	55.99	10	3.5×10^{-4}	3.4×10^{-5}
40.0	29.5	30.5	53.70	56.36	10	$<2.1 \times 10^{-4}$	$<2.0 \times 10^{-5}$
50.0	16.4	33.6	49.90	55.54	10	$<8.3 \times 10^{-5}$	$<7.4 \times 10^{-6}$

TABLE 2.5: CESIUM LEACH FOR HIGH ALUMINA CEMENT-ZEOLITE COMPOSITES

Zeolite	Cement	Water	Mass, g	Surface Area, cm^2	Leach Time, days	Fraction Release	Bulk Leach Rate $g/(cm^2\text{-day})$
	Formulation, wt %						
2.5	79.5	18.0	310.0	168.7	10	7.60×10^{-4}	1.40×10^{-4}
					31	1.36×10^{-3}	8.03×10^{-5}
3.9	76.1	20.0	320.5	167.2	10	6.58×10^{-4}	1.26×10^{-4}
					31	1.16×10^{-3}	7.17×10^{-5}
7.4	72.0	20.6	339.5	178.0	10	5.74×10^{-4}	1.09×10^{-4}
					31	1.05×10^{-3}	6.46×10^{-5}
11.3	66.0	22.7	370.7	187.4	10	4.34×10^{-4}	8.59×10^{-5}
					31	8.58×10^{-4}	5.48×10^{-5}
15.3	59.9	24.8	408.2	202.0	10	3.53×10^{-4}	7.13×10^{-5}
					31	8.18×10^{-4}	5.33×10^{-5}

TABLE 2.6: CESIUM AND STRONTIUM LEACH RATES FOR SLUDGE-ZEOLITE-HIGH ALUMINA CEMENT COMPOSITES

Sludge	Zeolite	Cement	Water	Mass, g	Surface Area, cm^2	Leach Time, days	Fraction Release Cesium	Strontium	Bulk Leach Rate (g/cm^2-day) Cesium	Strontium
		Composition, wt %								
37.0	2.3	28.5	32.5	809.4	328.7	30	4.75×10^{-2}	8.86×10^{-5}	3.90×10^{-3}	7.27×10^{-6}

Strontium content of cement = 367 ppm

Cesium loading of zeolite = 1.0 meq/g

Source: BNL-21571

The low level sodium nitrate waste from the SRP process can also be solidified in concrete. The leach behavior of Portland type II and high alumina cements containing sodium nitrate was investigated (23). The bulk leach rate of $NaNO_3$ in Portland type II cement composites varied from 2.5×10^{-4} to 1.3×10^{-2} g per $(cm^2\text{-day})$ for 5.0 to 30.0 weight percent $NaNO_3$ composites respectively. The leach rate was somewhat higher when high alumina cement was used for solidification. This difference is attributed to a larger average pore size in the high alumina cement concrete.

Advanced Waste Forms

Several other ceramic or composite waste forms and processes for their production are being developed. In most cases the goal is an even safer waste form, although increased processing flexibility and economy are also being sought. Most of them are being considered for second generation plants coming on-stream in the mid 1980s or later.

Supercalcine: Supercalcine is a waste form being developed at Penn State by Dr. Greg McCarthy. He is solution-mixing carefully chosen additives to the liquid waste so that, upon calcination, certain tailor-made, stable, crystalline compounds containing the fission products are formed. This simple procedure fixes individual fission product atoms in a much more chemically stable environment than that of normal oxide calcine. Supercalcine then forms an excellent first step in the formation of composite waste forms such as sintered glass.

Sintered Glass: Sintered glass waste forms are produced by mixing calcine and glass frit and heating with slight pressure in a pseudo hot pressing operation to produce a porous, but tough, consolidated product. Temperatures used are $800°$ to $1000°C$ and pressures range from 10 to 1,000 psi. An advantage of the process is that high-waste loadings, over 50 weight percent, can be achieved.

Glass-Ceramic: Glass-ceramics would not have the devitrification problem described earlier for the waste glasses, being placed in a thermodynamically more stable state during processing. Most of the development of glass-ceramics for HLW solidification is being done in Germany.

Inorganic Ion Exchangers: Probably the first attempt at utilizing ceramics for waste immobilization was by firing clays upon which fission products had been ion exchanged. The waste loading achieved was too low to make the process practical. Recently synthetic alkali titanates have been tested as high capacity ion-exchangers for HLW with some success at Sandia. A problem with this approach is that, due to the chemical complexity of HLW, no one ion exchange medium will hold all of the activity, thus a complex multistep process is required.

Metal Matrices: Metal matrices can be used to produce a composite waste product in which various primary waste forms, such as supercalcine particles, ceramic pellets, or glass beads, or marbles, are embedded in metal. Advantages of this product form are much improved thermal conductivity and impact resistance. The use of metal matrices is being investigated in Germany and Belgium in Europe and at the Idaho National Engineering Laboratory and Pacific Northwest Laboratory in the U.S.

Multibarrier Waste Forms: Perhaps the ultimate waste form is a modification of the metal matrix principle in which the primary waste particles are individually coated with a thin layer of an adherent inert material prior to embedding in metal. Coatings of inert materials such as carbon, silicon carbide, SiO_2 or Al_2O_3 can be applied in a fluidized bed or rotating drum. The multibarrier waste forms offer the potential for applying as much redundant protection, in the form of added barriers, as is desired.

DURABILITY OF STORAGE CONTAINERS FOR SOLIDIFIED WASTES

Most concepts for the disposal of highly radioactive waste involve converting the waste to a solid form like concrete or glass and storing this solid form in metal containers. Two major factors in the final selection of materials for these containers are the compatibility between waste form and container material and the durability of the material at temperatures and stresses expected during service and possible accidents.

According to C.L. Angerman and W.N. Rankin of Savannah River Laboratory, AISI 1020 carbon steel appears to be a better material than other alloys such as *Cor-Ten A*, Type 304L stainless steel, or *Inconel* 600. This choice is based on the results of 10,000 hours of heating tests that showed container compatibility with both concrete and glass waste forms. The selection is also based on (1) analyses of the strengths and (2) oxidation resistance of the alloys under the conditions expected during 100 year storage in air and in various impact and thermal accidents. The thinner wall thickness required for satisfactory performance of the stronger, more-oxidation-resistant alloys is offset by their higher cost per pound.

CLADDING DECONTAMINATION AND DENSIFICATION

Fuel element cladding residues from the chop-leach head end process constitute a low-density, high-surface area metallic waste with a substantial actinide element contamination. The objective of a study conducted at Battelle Pacific Northwest Laboratories was to determine feasibility and methods for removal of long half-life transuranics from the cladding residues and for consolidation of the zirconium and stainless steel residues. The study involved:

(1) Selection and laboratory testing of methods for transuranic removal by (a) aqueous reagents, (b) molten salt reagents and (c) fluxing during melting.
(2) Consolidation of the fuel element hulls by melting and casting Zircaloy, stainless steel, Inconel or alloys of the three.
(3) Testing and evaluating the consolidated fuel hulls for long-term waste management.
(4) Absorption and fixation of tritium into the waste ingots.
(5) Consolidation of transuranics removed.

Review and selection of potential processes for decontamination and melting of chop-leach cladding residues were completed (24), and it was decided that the following process would be developed: (1) a gas phase treatment of cladding hulls by a fluoride to render the corrosion product oxide soluble in dilute aqueous fluoride solutions, (2) oxide removal and light etching in dilute aqueous fluoride solutions (nitric-hydrofluoric acid or oxalic-hydrofluoric acid-hydrogen peroxide solutions), (3) melting of the hulls into an eutectic stainless steel-Zircaloy alloy by induction heating in graphite crucibles with and without a flux or solid adsorbent. If the metal is to be reused for reactor applications, a more conventional cold wall crucible and arc melting of the segregated stainless steel, Inconel, and zirconium alloys would be required. Some results of the investigations are described below.

Actinide Distribution in Chop-Leach Fuel Hulls

Design of surface decontamination and subsequent molten metal decontamination process for fuel hulls requires an accurate knowledge of the distribution, chemical state and type of transuranic contamination of the hull material. An initial sample for analytical determination of these parameters was obtained from a fuel rod (F-2[HQ]) exposed in Saxton Reactor about six years ago. Fuel composition was 6.6% PuO_2 in UO_2. A 0.6 cm piece of F-2(HQ) cladding was lightly surface abraded to remove corrosion product oxide and etched 30 seconds total in two nitric-hydrofluoride acid etch solutions.

These decontaminated pieces were moved to an uncontaminated area and etched 20 seconds more in two etches and briefly leached in concentrated nitric acid for complete surface decontamination. Less than 25 microns were removed from the surfaces in the total treatment sequence. The sample was then dissolved in etch solution in two approximately equal steps.

One sample contained the dissolved outer and inner 0.14 cm surface layer and the other sample the residual 0.28 cm center. Radiation readings with survey instruments after decontamination showed 20 mR at contact and no detectable α. Total α counts and γ scans of the two samples are nearly identical, indicating the completeness of surface decontamination.

The two samples have been analyzed by emission mass spectrometry, α energy analyses and chemical separation of important species for accurate determination of transuranic contamination resulting from uranium and thorium impurity in the original Zircaloy.

Decontamination Process

A small section of the Saxton fuel cladding was treated at 600°C in HF for 30 minutes followed by 1.2 hr in an ammonium oxalate, citrate, fluoride, peroxide solution (25) at 85°C to produce an excellent decontamination and film removal (26). Analyses of the stripping solution and the residual metal have verified that 99.7% of the α activity was removed by the HF-aqueous decontaminant treatment.

The residual 0.3% of α activity compares with the 7 nCi/g found previously for the base metal which indicates the process achieved as much decontamination as possible by surface removal alone. The slow stream of effluent gas from the 600°C HF treatment was routed through two 1 M KOH absorption solutions in 250 ml flasks followed by an activated charcoal absorber. Following treatment of the piece it was found the effluent tube (⅛ in. S/S), the KOH absorption flasks, and the charcoal were all radioactive to approximately the same level. Gamma analyses of the activity showed it to be almost entirely cesium. This indicates the effluent from the HF treatment will have to be treated for cesium activity removal.

The potential for zirconium oxide removal and fuel particle dissolution with molten salts was tested by treating autoclaved (47 mg/dm^2 wt gain) pieces of Zircaloy-2 tubing in a molten salt bath of 37 mol % LiF, 37 mol % NaF and 26 mol % ZrF$_4$ at 660° to 740°C from 1 to 20 min with and without an HF gas sparge. The residual salt was removed from the pieces by treatment in the ammonium oxalate, citrate, fluoride, peroxide solution (25) for 30 to 60 min.

The process apparently produces most of the corrosion in the first 5 min at rates far in excess of those expected (27). It would be difficult to treat hulls in a production line fast enough to reduce corrosion to an acceptable level (<25 μm). The pieces were weighed after removal from the fused salt bath in a vertical position and held to drip in this position for ¼ to ½ min before they cooled below 550°C, the temperature at which this salt solidifies. Even with this draining procedure, the salt dragout would require the purification and retreatment of a significant fraction of the molten salt inventory.

The metal which could be dissolved into the salt would amount to about 49% for the 1.08-cm diam, 0.76-mm wall Zircaloy tubing. The variation in the amount of adherent salt appeared related to the amount of corrosion, which would suggest that freezing of the surface layer of salt occurred as the Zircaloy corroded into it.

One potential product of densification is an alloy of 85% Zircaloy-10% stainless steel-5% Inconel. The preparation, primarily the melting, of this alloy will release absorbed hydrogen and tritium. The approximately 100 ppm of hydrogen evolved could be reabsorbed into a previously cast ingot of the zirconium alloy intended for storage. The absorption rate and equilibrium pressure of hydrogen over the alloy ingot were studied at representative typical low pressures and low concentration.

The absorption temperature was fixed high enough (700°C) that zirconium would dissolve its own oxides and nitrides, which would otherwise impede absorption. Maintaining a large temperature gradient over the length of the absorption specimen to increase the capacity of the ingot for hydrogen while keeping its equilibrium pressure down, produced similar results to the isothermal absorption specimens if the absorbed hydrogen was assumed uniformly distributed.

The hydrogen absorption characteristics of the Zircaloy-stainless steel-Inconel alloy were compared with Zircaloy-2 to evaluate its usefulness as a tritium storage reservoir. In general, the alloy absorbed the hydrogen less rapidly than the Zircaloy-2, but equilibrium pressures of hydrogen over the alloy were similar to those of Zircaloy-2.

Densification Process

A demonstration of Zircaloy-4 clad hull melting by the Inductoslag process was conducted at the U.S. Bureau of Mines, Albany, Oregon. The feed material for this melt was reject PWR Zircaloy-4 fuel clad tubing chopped into 1-in. pieces

and etched by the ammonium oxalate, citrate, fluoride, peroxide bath similar
to that used in the decontamination process. The Inductoslag melting process
uses an induction-heated, split, water-cooled crucible, as described by Clites and
Beal (28). Melting takes place in static, one-third atmospheric pressure helium.
A pool of Zircaloy-4 and CaF_2 is melted on a "starting stub" and subsequent
charge material vibratorily fed into the top of the crucible as the ingot is being
extracted out the bottom of the crucible. Approximately 41 kg (90 lb) of
simulated chop-leached cladding residues were melted into an ingot 10 cm (4 in.)
diameter x ~75 cm (30 in.) long.

Chemical analyses and hardness results from this ingot show that this material
can be fabricated into nuclear components for reuse.

WASTE PARTITIONING AS AN ALTERNATIVE

The partitioning process is usually defined as the separation of nuclear wastes
into an actinide fraction and a second fraction containing all other radionuclides.
Further, the separation is commonly applied to the high-level waste because that
stream contains most of the fission products. The objective of partitioning, however, is to reduce the time for which confinement of at least the major part of
the radioactive waste must be assured.

Therefore, the partitioning process must have a broader scope in order to include
isolation of long-lived fission products such as iodine-129 and technetium-99
and long-lived activation products such as carbon-14. Further, the actinide content of wastes not extensively contaminated by fission products may well exceed
that of high-level waste, and therefore such streams should be included for treatment to separate long-lived radionuclides. The disposition of the collected long-
lived radionuclides, including the actinides, is also an important adjunct to the
rationale for the objective of the partitioning process.

Hence, the steps required for the preparation of the long-lived fraction for subsequent transmutation, extraterrestrial elimination, or other disposal modes become part of the partitioning process. Consequently, reference to the scope of
the partitioning process includes the broad range of separation processes applied
to all types of nuclear wastes; the handling, fabrication, and recycle of actinides
and other nuclides subjected to transmutation; and the elimination of packaged
waste by extraterrestrial transport. Finally, the short-lived radionuclide fraction
obtained by partitioning is to be stored for a modest period such as a few thousand years, but storage is not usually considered to be an integral part of the
partitioning process.

A conference on The Management of Radioactive Waste: Waste Partitioning as
an Alternative was held at the Battelle-Seattle Research Center in Seattle, Wash.
on June 8 through 10, 1976. A number of alternative methods of waste partitioning were presented at the conference and they are described in this section.

Partitioning of Reprocessing Wastes

As part of the Oak Ridge National Laboratory partitioning program, R.E. Leuze
and W.D. Bond of Union Carbide Corporation have prepared detailed material-
balance flow sheets that describe processes for removing actinides to very low

levels in LWR fuel reprocessing wastes. Since the total amount of actinides in reprocessing wastes must be quite small to meet partitioning goals, care must be taken to select processing methods that do not generate large volumes of waste that contain small but significant amounts of actinides (29)(30). Problems encountered with these wastes can be eliminated if the actinides can be recycled into the main process stream. The establishment of effective recycle conditions will require a complete set of material balance flowsheets that indicate the fate of all actinides and show low- and intermediate-level waste stream compositions that are compatible with recycle.

The conceptual processing sequence being investigated for actinide partitioning is outlined in Figure 2.9. It is based on a combination of modified Purex processing and secondary processing of the high-level waste (29). In this concept, the Purex process will be modified so that low- and intermediate-level wastes, all the way through final product purifications, are recycled. A supplementary extraction is assumed to ensure adequate recovery of uranium, neptunium, and possibly plutonium.

It may then be desirable to store the waste for a period of time to allow additional radioactive decay, thus decreasing radiolysis problems during americium-curium recovery. If an appreciable storage time is used, additional plutonium recovery will be required to remove the ^{240}Pu daughter that grows in from ^{244}Cm decay.

FIGURE 2.9: CONCEPTUAL PROCESSING SEQUENCE FOR ACTINIDE PARTITIONING

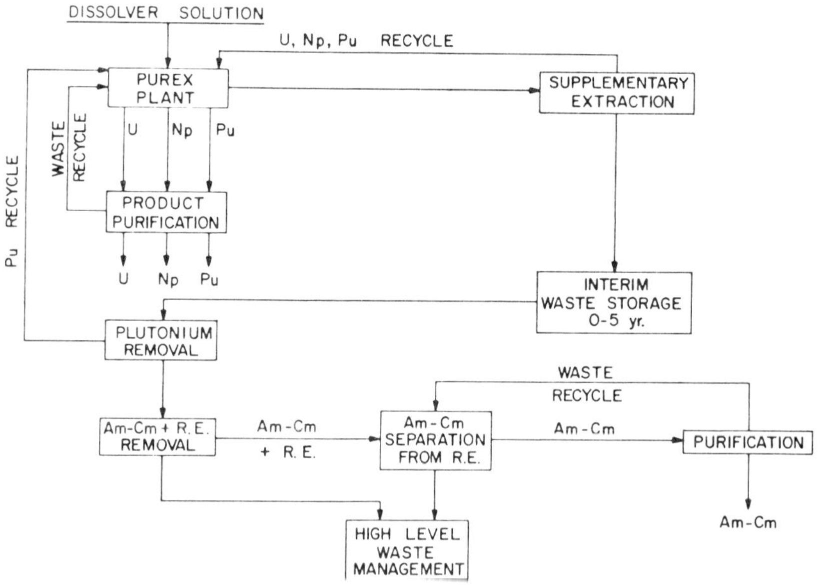

Source: PB-254 737

Process methods that have been used in the past to recover americium and curium
have some rather serious deficiencies when applied to waste partitioning. There
are problems with solids formations for the methods requiring low acidities and
with achieving adequate recovery. Most of the processes use large quantities of
chemical reagents that make low- and intermediate-level waste recycle impossible
and greatly complicate subsequent waste handling (31).

It has been assumed that two distinct processing steps will be used to remove
americium-curium (and transcurium elements when processing actinides that have
been partially "burned" by recycle to reactors) from high-level waste. The purpose of the initial step will be to isolate trivalent actinides (americium, curium,
etc.) plus the chemically similar trivalent lanthanide fission products from the
other waste components, several of which interfere with subsequent processing.
The americium-curium can then be separated from the lanthanides and treated
to obtain the required purity.

Present knowledge is insufficient to determine how completely actinides can
be removed or to adequately characterize and treat wastes that will be generated
during the additional processing required to achieve very high recovery of actinides.
Major technical problems that must be resolved are described below.

Recycle of Low- and Intermediate-Level Wastes in Purex: Conceptual material-
balance flowsheets for a modified Purex process have been prepared (29). These
flowsheets include the recycle of nitric acid as well as most of the aqueous,
actinide-bearing waste streams generated in the partitioning and purification
cycles for uranium, neptunium, and plutonium. Elimination of ferrous sulfamate
as a partitioning reductant is necessary since sulfate would interfere with the recycle of actinide values from partitioning and purification cycle wastes.

Waste streams from solvent purification and from off-gas treatment contain
significant amounts of actinides; however, chemical reagents currently used for
these process steps make the wastes unacceptable for recycle. Experimental work
is needed to define methods that give adequate actinide recoveries to make these
wastes amenable to recycle. The overall concept of extensive recycling in the
Purex process must be thoroughly tested to determine whether it will lead to
accumulations of impurities that interfere with satisfactory overall operations.

Adequate Uranium and Neptunium Recovery by Purex: It seems quite probable
that adequate uranium and neptunium recoveries and nearly adequate plutonium
recovery can be attained with extra extraction stages or with an additional extraction cycle following special adjustment of the high-level waste to enhance extraction. The chief difficulties seem to be with the increased amounts of fission products extracted, which will increase radiation damage to the solvent and may lead
to prohibitive buildup of some fission products in the extraction contactors.

Actinides Sorbed on Solids and "Inextractable" Plutonium: All of the potential
problems with actinides sorbed on solids and with inextractable plutonium have
not yet been defined. Laboratory-scale studies with nonradioactive, synthetic
waste solutions and hot cell studies with 250-g amounts of high-burnup LWR
fuel are in progress to identify areas of difficulty and to develop acceptable solutions. Results of these studies have shown that (1) small amounts of solid residue remain after dissolution of high-burnup fuel; (2) Purex feed solution is metastable, and solids form under certain conditions; (3) considerable amounts of

solids are likely to form when high-level waste solutions are evaporated and stored; and (4) a small amount of plutonium is inextractable.

Thus far, it has not been determined that any of these results offer insurmountable problems, but much more information is needed. Laboratory-scale and hot cell studies should be continued, and attention should be given to the implication of larger-scale processing.

Dissolution of 250-g amounts of LWR fuel that had been irradiated to 31,200 MWd/ton and cooled for two years proceeded quite rapidly in 3 M HNO_3 at about 90°C. The insoluble residue amounted to only about 0.3 weight percent of the fuel and consisted primarily of molybdenum, ruthenium, and technetium with only a few hundredths of a percent of the plutonium and curium. Leaching the residue with 6 M HNO_3 at 90°C decreased the plutonium and curium levels in the residue to well below 0.01%. Although these results indicate no definitive problems, additional information is needed for fuel at other burnup values and for fuel fabricated from mixed PuO_2-UO_2.

When Purex feed (prepared from high-burnup fuel) is stored at elevated temperatures, a highly crystalline solid composed primarily of molybdenum and zirconium at a 2:1 mol ratio is formed. This solid can carry up to 20 mg of plutonium per gram of solid. It has been found that zirconium can hydrolyze at elevated temperatures, even at high acidities, to form colloids or precipitates of hydrous zirconia that also carry plutonium. However, the plutonium can be effectively leached from these solids to very low values with nitric acid. The information available relative to the conditions for formation of these solids also indicates that careful process control will be an effective deterrent to such formation of gross amounts of solids from Purex feed. It appears, however, that there will be some serious problems with formation of these solids during treatment and interim storage of high-level waste.

Test runs have demonstrated that small plutonium losses occur when either highly purified Pu(IV) in 3 M HNO_3 or Purex feed prepared from highly irradiated LWR fuel is extracted with 30% TBP-dodecane. After seven consecutive batch extractions, the plutonium distribution coefficient decreased to ~0.2 and slightly less than 0.01% of the plutonium remained in the aqueous phase. This problem, although apparently not insurmountable, is worrisome and needs additional attention.

Adequate Americium-Curium Removal: Solvent extraction and ion exchange techniques are being considered for both the separation of americium-curium plus lanthanides from high-level waste and for the separation of americium-curium from the lanthanides.

Tributyl phosphate extraction from heavily salted solutions may be a viable method for the recovery of the trivalent elements. If satisfactory methods can be developed for recycling the salting agents, the Purex type of technology might be extended to remove americium-curium and transcurium elements along with the heavy lanthanides at the time of fuel reprocessing. In nitrate-salted systems, the light lanthanides (about 80 mol percent of the lanthanide fission products) can be partitioned from the trivalent actinides and heavy lanthanides (32). Laboratory studies with synthetic wastes have shown that about 95% of the salting agent $Al(NO_3)_3$ can be recovered for recycle by precipitation from 15 M HNO_3

solution. The extra acid required for this step could then be recovered by distillation and reused.

A process based on the use of oxalate precipitation for removal of most of the trivalent elements followed by cation exchange to complete the removal offers several advantages (33): (1) precipitation can be carried out from about 0.5 M HNO_3; (2) the oxalate not only precipitates the trivalent elements but also complexes zirconium so it does not sorb on cation exchange resin; and (3) the composition of the filtrate is satisfactory for subsequent treatment by cation exchange. Most of the process development has been carried out with synthetic waste solutions. In one test made with waste solution prepared from high burnup LWR fuel, the results were very encouraging. The precipitate settled readily, and the supernate contained only about 0.3% of the americium and curium. Cation exchange treatment decreased the americium and curium to $\sim$0.02%.

Although these results indicate that such a method is potentially useful, the use of precipitation and ion exchange methods for highly radioactive solutions on a large scale presents some difficult engineering problems.

Removal of trivalent actinides and lanthanides from high-level waste by extraction with neutral bidentate organophosphorus reagents such as dibutyl-N,N-diethylcarbamylmethylenephosphonate (DBDECMP) or its dihexyl analogue (DHDECMP) appears promising (34)(35). Since the extraction can be made from acid solutions without high concentrations of salting agents, a process based on this technique should not greatly increase the volume of waste or complicate waste management.

Ion exchange chromatography (36)(37) appears to be a possible method for separating the trivalent actinides from the lanthanides both in conceptual flowsheet and laboratory studies. Laboratory studies have emphasized the use of a hydrogen ion barrier for elution rather than a metallic ion barrier such as zinc because this would diminish the addition of nonvolatile solids to the waste streams.

Results of these studies show that a barrier ion composition of 80% H^+-20% Zn^{++} yields good separation, considerably reduces the quantity of zinc that is required, and simplifies the treatment of waste streams from the separation processes. In the Talspeak solvent extraction process (38), which is being considered for this separation, the lanthanides are preferentially extracted into di(2-ethylhexyl)-phosphoric acid (HDEHP) from a feed solution containing carboxylic acids and sodium diethylenetriaminepentaacetate (Na_5DTPA). Both the Talspeak process and ion exchange chromatography use considerable amounts of chemical reagents, and the concepts for low- and intermediate-level waste recycle and final waste management have not been resolved.

Technical Feasibility Studies

The program initiated in FY 1975 at Battelle, Pacific Northwest Laboratories involved, primarily, experimental development studies that resumed and accelerated a previous study to resolve the several major problems in partitioning identified in the feasibility study (30). The program was expanded to include partitioning of solidified waste (e.g., a calcine) such as might be shipped to a central facility for conversion to a storage or disposal form. Two of the experimental studies are described below.

Characterization and Treatment of Solids in High-Level Waste: Finely divided
solids are known to be present in dissolver solution and in other high-activity process streams of fuel-reprocessing and waste treatment facilities. These solids include
cladding fines, residues from the dissolution of the fuel, and precipitates formed
during reprocessing and waste treatment. If small fractions of the plutonium,
americium, curium, or other actinides present in the irradiated fuel are chemically
or physically bound in these solids in unreactive form, they pose a serious problem
to the development of a partitioning process; they limit the degree of removal
of these long-lived elements from the shorter-lived fission products by any process that relies on the actinides being in ionic or soluble molecular form.

Neither the concentrations of the actinides present in such solids nor the characteristics of the solids have been thoroughly investigated. Both questions must
be answered before the technical feasibility of partitioning can be realistically
evaluated or the requirements defined for any solids treatment process used in
conjunction with a partitioning process. Corrective action might include revision
of the initial fuel dissolution flowsheet, secondary treatment of dissolver solution
or high-activity acid waste, or separation of the solids followed by separate treatment or disposal.

The objective of this study was to evaluate the effects of the solids on partitioning by determining the amounts of plutonium, americium, and curium associated
with the solids at various points in the reprocessing of reactor fuels and the
treatment of the wastes, and by measuring the volume, particle size, and other
characteristics of the solids needed in the development of processes for their
treatment and handling. Progress toward the objective was restricted to a single
experiment in which a small quantity of highly irradiated light-water reactor
(LWR) fuel was dissolved and the solid residue analyzed to determine the plutonium, americium, and curium content.

A small quantity of Yankee reactor fuel from an element known to have been
exposed to an average burnup of 39,000 MWd/MT was selected for the experiment. This fuel was considered sufficiently representative of typical LWR fuel
to exemplify the dissolution behavior of the plutonium, americium, and curium.
A part of a fuel rod from this element was cut into sections suitable for dissolution.

The dissolution of the fuel was carried out in a small glass dissolver fitted with
an air-cooled condenser, a dip tube for air sparging, and a thermometer well.
The outlet of the condenser was connected to a trap and a scrubber containing
2 M NaOH to prevent potential loss of fission-product activity to the hot cell
exhaust system.

A quantity of fuel pieces totaling 112 g, including cladding, was placed in the
dissolver which contained 250 ml of 5.5 M HNO_3, heated to 90°C, and held at
that temperature 4 hr. Dissolution of UO_2 in HNO_3 proceeds through the reactions (39):

$$1.5\ UO_2\ +\ 4\ HNO_3\ \longrightarrow\ 1.5\ UO_2(NO_3)_2\ +\ NO\ +\ 2H_2O$$

and

$$UO_2\ +\ 4\ HNO_3\ \longrightarrow\ UO_2(NO_3)_2\ +\ 2\ NO_2\ +\ 2H_2O$$

Both reactions occur, but the first predominates at the acid concentration used here. Dissolution conditions were selected to give a final concentration of 4 to 4.5 M HNO_3 and 1.3 M uranium.

After completion of the dissolution step, the solution was cooled and then filtered through a Teflon filter with a 5 μ pore size. The solids and the cladding hulls were then returned to the dissolver and leached for 4 hours with 250 ml 8 M HNO_3, again filtered, and the filter washed well with 2.0 M HNO_3. The leach solution, washes, and initial dissolver solution were combined, diluted to 1.0 liter, and sampled for chemical analyses. The cladding hulls and the solids were then weighed separately.

The next phase of the experiment was intended to convert the plutonium, americium, and curium in the solids to a soluble form that could be analyzed. This was done by first leaching with a solution of HNO_3-HF followed by fusion of the residue in $NaHSO_4$. The HNO_3-HF leach consisted of heating a weighed portion of the solids in 50 ml 8 M HNO_3-0.05 M HF for 12 hr at 65°C. The residue was filtered and added to 15 g $NaHSO_4$. The $NaHSO_4$ was heated to give a fluid melt, held at temperature for 1.5 hr, cooled, and dissolved in 2.0 M HNO_3. The dissolved $NaHSO_4$ was combined with the HNO_3-HF leachate, diluted to 250.0 ml, and sampled for analyses.

The total weight of the filtered solids obtained from nitric acid dissolution of the fuel was 290 mg and represented 0.32 weight percent of the uranium metal present in the irradiated fuel.

This is in apparent agreement with results obtained at Oak Ridge National Laboratory (ORNL) with a stainless-steel-clad mixed-oxide fuel. Under similar dissolution conditions the ORNL workers observed 0.4 to 0.5 weight percent solids for fuel with exposures of 25,000 to 50,000 MWd/T (40). No attempt was made to identify the constituents of the solids other than the Pu, Am, and Cm, but other workers have reported that the principal elements present are Mo, Ru, Pd, Rh, Zr, Tc, and Sb (40)(41).

The analyses of the clarified dissolver solution and the combined solution obtained from the HNO_3-HF leach and the $NaHSO_4$ fusion are summarized in Table 2.7. The principal alpha activity found in the dissolved solids is plutonium, but the minor amounts of americium and curium indicate a slight tendency for these elements to adhere to the solids.

TABLE 2.7: ANALYSES OF CLEAR DISSOLVER SOLUTION AND SOLUBILIZED SOLIDS

	Content of Dissolver Solution		Content of Solubilized Solids	
	Amount	% of Total in Fuel	Amount	% of Total in Fuel
Pu	1.00 g	99.3	0.0072 g	0.71
^{241}Am	4.06×10^{11} dis/min	99.9	3.5×10^{8} dis/min	0.086
^{244}Cm	3.28×10^{11} dis/min	99.99	4.19×10^{7} dis/min	0.013

Source: PB-254 737

Based on the fractions of plutonium, americium, and curium found in the nitric acid insoluble residues, it is apparent that treatment or elimination of the solids is required if large decontamination factors are to be obtained in a partitioning process. The actinides in the solids, if left in the aqueous high-level wastes, cannot be extracted by conventional processes and will limit the decontamination factors to 1.7, (assuming 0.5% loss of soluble Pu occurs in the Purex process), 1,100 and 8,000 for Pu, Am, and Cm respectively.

Removal of Long-Lived Nuclides from Solidified Waste: The objective of this study was to develop a flowsheet for partitioning starting with a solidified high-level waste, e.g., a calcine, such as might be shipped to a central facility for conversion to a storage or disposal form. The chief aim at the start of the project was to demonstrate a suitable process for getting calcine into a solution that would permit subsequent separation of long-lived nuclides from the waste. Achieving dissolution of the calcine into a nitric acid system appeared to have several advantages, an obvious one being that the resulting solution could probably be used as feed in partitioning flowsheets being developed for other nitric acid waste streams. Thus new technology requirements for processing the solution after dissolution of the calcine could be kept at a minimum. The work for this task was divided into three main areas:

> (1) Laboratory tests with simulated calcine to develop a dissolution process.
>
> (2) Demonstration of the calcine dissolution process with actual calcined high-level waste in a hot cell.
>
> (3) Hot cell demonstration of actinide removal from the feed solution resulting from (2).

The FY 1975 effort was limited to experimental work in area (1).

Simulated calcines were prepared by pot calcination of a low-sodium synthetic high-level waste prepared with the following chemical substitutions: Fe for Ru, Co for Rh, Ni for Pd, Mo for Tc, K for Rb, rare earths for Y + Am + Cm, and U for Pu + Np. This synthetic high-level waste contains solids; i.e., not all of the constituents completely dissolve during preparation of the waste solution.

In the preparation of this synthetic waste, molybdenum is added as the trioxide, MoO_3. The light color of the undissolved solids in the synthetic waste indicated that this compound was not completely dissolved even after agitation of the solution at elevated temperature for 24 hr.

Two simulated calcines were prepared: the first by direct calcination of the synthetic waste, including undissolved solids; the second by calcination of filtered synthetic waste solution. Both types were prepared by evaporating synthetic high-level waste to dryness in a stainless steel beaker and then calcining the solids in the stainless beaker at 900°C in a crucible furnace. The cooled calcine was removed from the beaker, pulverized, and blended to a homogenous mixture.

Acid dissolution, using HNO_3 or HNO_3-HF, and molten salt fusion were used separately and in combination to dissolve the simulated calcines. The acid dissolution tests were performed in covered glass and plastic (for HNO_3-HF solution) beakers utilizing a Teflon-covered magnetic bar for stirring. The molten salt fusions were conducted in nickel crucibles and the melts dissolved in dilute HNO_3.

Initial calcine dissolution tests were performed with simulated calcine prepared from unfiltered synthetic waste solution. The results from acid dissolution tests, presented in Table 2.8, show that digestion with HNO_3-HF is superior to digestion with HNO_3 alone but is not adequate for complete dissolution of the calcine.

Table 2.9 shows results from dissolution tests on the same calcine using molten salt fusions. Caustic-peroxide fusion was the best of the three dissolution methods investigated, yielding 80 to 90% dissolution of the calcine upon contacting the cooled melt with HNO_3.

TABLE 2.8: DISSOLUTION OF SIMULATED CALCINE BY ACID DIGESTION

Acid Test No.	Acid	mℓ Acid/g Calcine	Temp.	Time (hr)	Calcine Dissolved (Wt%)
1	15.7 $\underline{M}$ HNO_3	10	Boiling	4	28
2	Con HCl	10	Boiling	4	75
3	15.7 $\underline{M}$ HNO_3-0.3 $\underline{M}$ HF	10	80°C	5	57
4	15.7 $\underline{M}$ HNO_3-0.03 $\underline{M}$ HF	10	80°C	24	60
5	15.7 $\underline{M}$ HNO_3-0.03 $\underline{M}$ HF	30	95°C	5	70
6	15.7 $\underline{M}$ HNO_3	5	Boiling	2	
	followed by				
	4 $\underline{M}$ HNO_3	30	Boiling	1	35
7	15.7 $\underline{M}$ HNO_3-0.003 $\underline{M}$ HF	5	Boiling	2	
	followed by				
	4 $\underline{M}$ HNO_3-0.01 $\underline{M}$ HF	35	Boiling	1	55
8	4 $\underline{M}$ HNO_3-0.03 $\underline{M}$ HF	25	100°C	1	53

TABLE 2.9: DISSOLUTION OF SIMULATED CALCINE BY MOLTEN SALT FUSION

Fusion Test No.	Fusion Mixture	Temp. (°C)	Wt Ratio Salt/Calcine	Time (hr)	Calcine Dissolved (Wt%)
1	Sodium hydroxide-sodium peroxide	600	10	2	77
2	Sodium hydroxide-sodium peroxide	600-650	4	2	90[a]
3	Sodium hydroxide-sodium peroxide	600-650	4	2	87
4	Alkali carbonate	600	10	2	33
5	Sodium-potassium nitrate	600	8	1-1/4	40

[a]In this test the cooled melt was dissolved in 4 M HNO_3 containing 0.006 M HF.

Source: PB-254 737

Although neither the acid digestion tests nor the molten salt fusions resulted in complete dissolution of this calcine, a combination of the two treatments in series resulted in greater overall dissolution. For example, when a caustic-peroxide fusion was performed on the residual solids from acid test No. 8 (Table 2.8), an overall dissolution of 98% was achieved; when a nitric-hydrofluoric acid digestion was performed on the residual solids from fusion test No. 3 (Table 2.9), an overall dissolution of 99% was achieved.

The simulated calcine used in the above tests was prepared from unfiltered synthetic waste solution. It was surmised that part of the difficulty in dissolving the simulated calcine may result from refractory molybdenum trioxide or other compounds used to prepare the synthetic waste. That is, the partially refractory nature of the simulated calcine may not be produced by the calcination process but may be present with some of the starting materials for the synthetic waste.

For this reason, some simulated calcine was prepared from filtered synthetic waste solution. Nitric-hydrofluoric acid digestion tests on this calcine were more successful than those on calcine prepared with unfiltered synthetic waste. In fact, complete dissolution of the simulated calcine was accomplished by using a two-step acid digestion procedure. The calcine was first digested at 100° to 105°C in 8 M nitric acid for 6 hr. During this treatment 66% of the calcine dissolved. Digestion of the residual solids from this treatment in 8 M HNO_3-0.15 M HF at 100°C for less than 4 hr resulted in complete dissolution of the solids.

Attempts to dissolve the calcine using either an HNO_3-HF digestion only or an HNO_3-HF digestion followed by an HNO_3 digestion did not produce complete dissolution of the calcine. The results indicate that in the HNO_3-HF treatment both the HNO_3 and the HF concentrations are important. Eight molar HNO_3 performed better than either concentrated HNO_3 or 4 M HNO_3 when in combination with HF, and 0.15 M HF gave more complete dissolution than did 0.04 M HF in 8 M HNO_3.

Other studies were also done by BNL on the removal of long-lived nuclides from non-high-level wastes and the development of low-additive flow sheets for removing long-lived nuclides from the fission products.

Partitioning Using Laser Photochemical Methods

The feasibility of photochemically fractionating the actinides in nuclear waste processing has been evaluated on a preliminary basis by T. Gangwer of Brookhaven National Laboratory. The data indicate that there are potentially useful photo-redox reactions. However, there is a serious lack of data on photochemical parameters for the solution conditions that exist in nuclear waste processing. The problem areas relevant to photochemical processing have been identified. The experimental areas that must be investigated in order to further evaluate the photochemistry have been defined. A research and development program is required to determine whether these photochemical reactions can be successfully modified and adapted into a functional actinide fractionating process.

Inorganic Ion Exchange Process for Partitioning

R.L. Schwoebel of Sandia Laboratories proposed two possibilities for partitioning

actinides from the rare earths using the chemistry of the Sandia Solidification
Process, or the titanate process as it is sometimes called (42).

Both of these involve the use of an improved Purex process, described previously
(See Figure 2.9). In this improved Purex process, Pu and Np would be separated
out of the waste stream to acceptably low levels. In the first scheme, the oxalate
process would be used to precipitate Cm, Am, and the rare earths (Figure 2.10).
The high-cross-section rare earths would then be separated from Cm and Am
using a form of the Sandia Solidification Process (SSP) chemistry.

In the second scheme, the SSP chemistry would be applied to separate Cm, Am,
and the rare earths in one step (Figure 2.11). It may be necessary in this case,
however, to separate other species (Rh, Ru, Zr, Cr, Fe, Na) before fission burning can be accomplished. The necessity of this latter step requires further studies
of the detailed neutronics.

FIGURE 2.10: TWO-STEP PROCESS FOR PARTITIONING TRANSURANICS

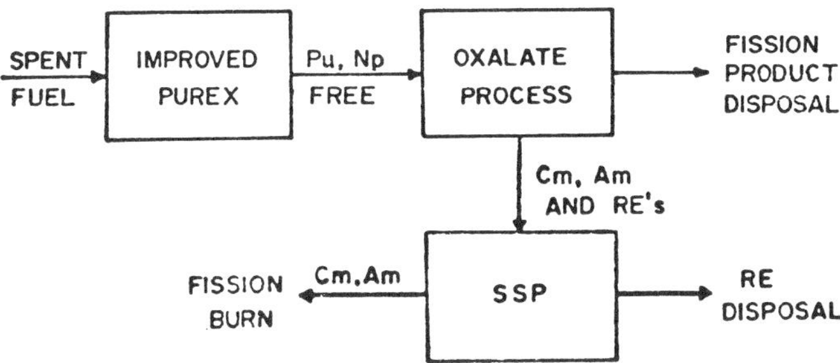

FIGURE 2.11: ONE-STEP PROCESS FOR PARTITIONING TRANSURANICS

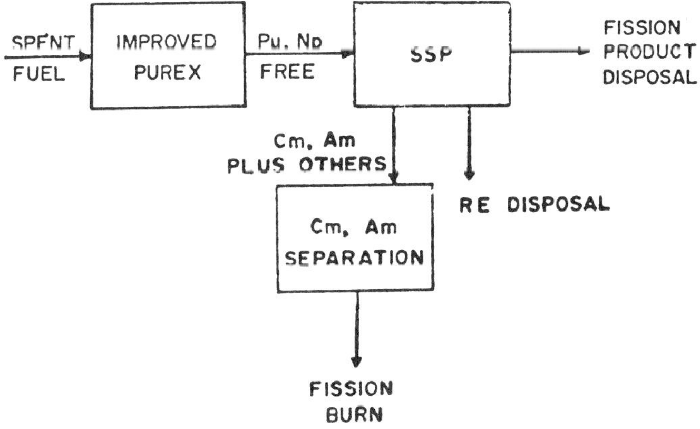

Source: PB-254 737

Chemically Separating Transplutonium Elements from Rare Earth Fission Products

To transmute and destroy the actinides it is necessary to separate them from the waste and then concentrate them for reintroduction into nuclear reactors. Hence, waste partitioning is performed by chemically separating the actinide elements (An) for the process waste streams containing both fission and multiple-neutron capture products.

However, approximately a third of the mass of these fission products resides in the lanthanides (Ln), and this circumstance places a serious obstacle in the path of effectively separating the transplutonium An's. The difficulty arises from the nearly identical chemical properties of the Ln and An series of elements, which results from the filling of the f electron orbits within each series. With few exceptions, the valence of aqueous Ln and transplutonium An ions is tripositive, and further, because of the "lanthanide contraction," the ionic radii of the transplutonium An's overlap those of the major Ln fission products.

Since the valence and ionic radii of these two series of elements are alike, electrostatic bonding, which wholly depends on these parameters, does not provide the discrimination necessary for chemical separation. Thus, the separation of Am and Cm from the much larger quantities of Ln in the waste streams presents the most formidable obstacle to waste partitioning.

To overcome the shortcomings of known methods, C.K. Hulet at Lawrence Livermore Laboratory is investigating a chelation extraction process for the separation of Am(III) and Cm(III) from the trivalent lanthanides in which no aqueous complexants are used. Instead, two successive extractions are performed. For the first, an extraction in which Am and Cm behave like the early lanthanides is used, and for the second, a different extractant in which Am and Cm behave like middle or heavy lanthanides is used.

Numerous studies have shown that Am and Cm behave like Pr or Nd in extractions with di-(2-ethylhexyl)orthophosphoric acid (HDEHP) from dilute mineral acids. The heavier lanthanides are progressively more extractable, with an average separation factor $SF_{z/z-1}$ of 2.4, where the separation factor is defined as the ratio of the distribution coefficients for each species between an organic and an aqueous phase. Hence, Am and Cm can be separated from the heavier lanthanides by extraction with HDEHP.

To separate Am and Cm from the lighter lanthanides, which contain the major lanthanide fission product, several derivatives of 8-hydroxyquinoline (oxine) including 5,7-dichloroxine (HDCO) and 5-nitroxine have been investigated. The results of the experiments indicate a possible 75% recovery of the Am.

Actinide Transmutation by Recycle in LMFBRs

A strategy of actinide burnup in LMFBRs is being investigated by S.L. Beaman and E.A. Aitken of General Electric as a waste management alternative to long-term storage of high-level nuclear waste. This strategy is being evaluated because many of the actinides in the waste from spent-fuel reprocessing have half-lives of thousands of years and an alternative to geological storage may be desired. From a radiological viewpoint, the actinides and their daughters dominate the waste hazard for decay times beyond about 400 years. Actinide burnup in LMFBRs may be an attractive alternative to geological storage because the actinides can be effectively transmuted to fission products that have significantly shorter half-lives. Actinide burnup in LMFBRs rather than LWRs is preferred because the ratio of fission reaction rate to capture reaction rate for the actinides is higher in LMFBRs, and LMFBRs are not so sensitive to the addition of the actinide isotopes.

An actinide target assembly recycle scheme has been evaluated to determine the effects of the actinides on the LMFBR performance, including local power peaking, breeding ratio, and fissile material requirements. Several schemes were evaluated to identify any major problems associated with reprocessing and fabrication of recycle actinide-containing assemblies. The overall efficiency of actinide burnout in LMFBRs was evaluated and equilibrium cycle conditions were determined.

It was concluded that actinide recycle in LMFBRs offers an attractive alternative to long-term storage of the actinides and does not significantly affect the performance of the host LMFBR. Assuming a 0.1% or less actinide loss during reprocessing, a 0.1% loss or less during fabrication, and proper recycling schemes, virtually all of the actinides produced by a fission reactor economy could be transmuted in fast reactors. This approach to actinide waste management is highly dependent upon the success of future reprocessing and fabrication studies.

REFERENCES

(1) J.P. Duckworth, *NFS Waste Solidification Facility—Conceptual Design Manual,* Draft NFS Document, March 8, 1976.

(2) D.E. Larson, *Preliminary Nuclear Fuel Services, Inc. High Level Waste Processing Flowsheet,* Draft NFS Document, March 1976.

(3) W.R. Bond et al, *Waste Solidification Program, Volume 8, Spray Solidification Performance During Final Radioactive Tests in Waste Solidification Engineering Prototypes,* BNWL-1583, Battelle, Pacific Northwest Laboratories, Richland, WA, June 1971.

(4) W.R. Bond et al, *Waste Solidification Program, Volume 6, Spray Solidification Performance During First Radioactive Tests in Waste Solidification Engineering Prototypes,* BNWL-1391, Battelle, Pacific Northwest Laboratories, Richland, WA, August 1970.

(5) W.F. Bonner and A.K. Postma, "Development Spray Calciner and Inconel Melter," *Quarterly Progress Report, Research and Development Activities, Waste Fixation Program, October through December, 1973,* BNWL-1809, Battelle, Pacific Northwest Laboratories, Richland, WA, pp. 2-9, January 1974.

(6) J.D. Kaser, *Evaluation of Atomizing Nozzles for Spray Solidification of Radioactive Waste,* BNWL-1066, Battelle, Pacific Northwest Laboratories, Richland, WA, June 1969.

(7) W.F. Bonner, "Radioactive Waste Management," Report of Visit to England, Germany, and France, July 15, 1975.

(8) H.T. Blair, "Developmental Spray Calciner," *Quarterly Progress Report, Research and Development Activities, Waste Fixation Program, July through September, 1974,* BNWL-1871, Battelle, Pacific Northwest Laboratories, Richland, WA, pp. 15-17, November 1974.

(9) W.F. Bonner, H.T. Blair and L.S. Romero, "Spray Calcination of Nuclear Wastes," Presented at the 78th Annual Meeting of the American Ceramic Society, Cincinnati, Ohio, May 1-6, 1976, BNWL-SA-5764, Battelle, Pacific Northwest Laboratories, Richland, WA, p. 10.

(10) L.S. Romero, "Heated-Wall Spray Calciner," *Quarterly Progress Report, Research and Development Activities, Waste Fixation Program, July through September 1975,* BNWL-1949, Battelle, Pacific Northwest Laboratories, Richland, WA, pp. 13-14, January 1976.

(11) H.T. Blair, "Heated-Wall Spray Calciner," *Quarterly Progress Report, Research and Development Activities, Waste Fixation Program, April through June 1975,* BNWL-1932, Battelle, Pacific Northwest Laboratories, Richland, WA, pp. 14-15, September 1975.

(12) J. D. Kaser and J.D. Moore, "The Development of a Spray Calciner-Melter," *Proceedings of the Symposium on the Solidification and Long-Term Storage of Highly Radioactive Wastes,* February 14-18, 1966, Richland, WA, W. H. Regan, ed., CONF-660208, pp. 326-342.

(13) H.T. Blair and W.A. Ross, "In-Can Melting," *Quarterly Progress Report, Research and Development Activities, Waste Fixation Program, July through September 1975,* BNWL-1949, Battelle, Pacific Northwest Laboratories, Richland, WA, pp. 5-10, January 1976.

(14) J.L. McElroy et al, *Waste Solidification Program Summary Report, Volume II, Evaluation of WSEP High Level Waste Solidification Processes,* BNWL-1667, Battelle, Pacific Northwest Laboratories, Richland, WA, (July 1972).

(15) H.T. Blair, *Vitrification of Nuclear Waste Calcines by In-Can Melting,* BNWL-2061, Battelle, Pacific Northwest Laboratories, Richland, WA, June 1976.

(16) D.H. Siemens, "Summary Report—Waste Decontamination Status and Decontamination Techniques," Battelle, Pacific Northwest Laboratories, Richland, WA, Unpublished Data, June 1975.

(17) D.J. Bradley, "Summary Report—Waste Canister Decontamination and High-Level Waste Contamination Studies," Battelle, Pacific Northwest Laboratories, Richland, WA, Unpublished Data, June 1976.

(18) C.C. Chapman, *Experience with a Joule Heated Ceramic Melter While Converting Simulated High-Level Waste to Glass,* BNWL-2071, Battelle, Pacific Northwest Laboratories, August 1976.

(19) J.L. McElroy, et al, *Phosphate Glass Solidification Performance During Final Radioactive Tests in Waste Solidification Engineering Prototypes* (WSEP), BNWL 1541, January 1971.

(20) J.L. McElroy, et al, *Pot Solidification Performance During Final Radioactive Tests in WSEP,* BNWL 1628, January 1972.

(21) P. Colombo, et al, *Savannah River Laboratory Long-Term Waste Storage Program,* Progress Reports 1-9, July 1973-April 1975.

(22) E.D. Hespe, *At. Energy Rev.,* 9(1), 195, 1971.

(23) B. Manowitz, et al, *Development of Durable Long-Term Radioactive Waste Composite Materials,* Progress Reports 3-6, June 1973-March 1974.

(24) B. Griggs, *Feasibility Studies for Decontamination and Densification of Chop-Leach Cladding Residues,* USAEC Report BNWL-1820, Battelle, Pacific Northwest Laboratories, Richland, WA, July 1974.

(25) A.B. Meservey, "Decontamination and Film Removal," *Decontamination of Nuclear Reactor and Equipment,* J.A. Ayers, Editor, The Ronald Press, New York, NY, pp. 143-155, 1970.

(26) G.J. Dau, Compiler, *Nuclear Waste Management and Transportation Progress Report October through December 1974,* BNWL-1899, p. 35, Battelle, Pacific Northwest Laboratories, Richland, WA, February 1975.

(27) P.D. Miller, E.F. Stephan, W.K. Boyd and W.N. Stiegelmeyer, *The Effect of Heat Treatment and of Salt Composition on the Corrosion of Inor-8 and HyMu-80 During Hydrofluorination,* BMI-X-288, Battelle Memorial Institute, Columbus, OH, May 1964.

(28) P.G. Clites and R.A. Beal, "Inductoslag Melting of Titanium Scrap and Sponge," *Proceedings of Sessions 101st AIME Annual Meeting San Francisco, CA, Feb. 20-24, 1972, Light Metals 1972,* W.C. Rotsell, Editor.

(29) W.D. Bond and R.E. Leuze, *Feasibility Studies of the Partitioning of Commercial High-Level Wastes Generated in Spent Nuclear Fuel Reprocessing: Annual Progress Report for FY-1974,* ORNL-5012, January 1975.

(30) J.W. Bartlett, L.A. Bray, L.L. Burger, R.E. Burns, and J.L. Ryan, *Feasibility Evaluation and R&D Program Plan for Transuranic Partitioning of High-Level Fuel Reprocessing Wastes,* BNWL-1776, Battelle, Pacific Northwest Laboratories, November 1973.

(31) R.E. Leuze and M.H. Lloyd, "Processing Methods for the Recovery of Transplutonium Elements," in *Process Chemistry,* Vol. 4, Pergamon, Elmsford, NY, 1969.

(32) J.M. McKibben, H.P. Holcomb, D.A. Orth, W.E. Prout and W.C. Scotten, *Partitioning of Light Lanthanides from Actinides by Solvent Extraction with TBP,* DP-1361, E.I. du Pont de Nemours & Co., August 1974.

(33) D.E. Ferguson et al, *Chemical Technology Division Annual Progress Report for Period Ending March 31, 1975,* ORNL-5050, pp. 6-11, 30-32, October 1975.

(34) L.D. McIsaac, J.D. Baker and J.W. Tkachyk, *Actinide Removal from ICPP Wastes,* ICP-1080, Allied Chemical Corporation, August 1975.

(35) W.W. Schulz, *Bidentate Organophosphorus Extraction of Americium and Plutonium from Hanford Plutonium Reclamation Facility Waste,* ARH-SA-203, Atlantic Richfield Hanford Company, September 1974.

(36) E.J. Wheelwright and F.P. Roberts, *The Use of Alternating DTPA and NTA Cation-Exchange Flowsheets for the Simultaneous Recovery of and Purification of Pm, Am, and Cm,* BNWL-1072, Battelle, Pacific Northwest Laboratories, July 1968.

(37) J.A. Kelley, *Ion Exchange Process for Separating Americium and Curium from Irradiated Plutonium,* DP-1308, E.I. du Pont de Nemours & Co., November 1972.

(38) B. Weaver and F.A. Kappelmann, *Talspeak: A New Method of Separating Americium and Curium from Lanthanides by Extraction from an Aqueous Solution of Aminopolyacetic Acid Complex with a Monoacidic Phosphate or Phosphonate,* ORNL-3559, August 1964.

(39) M. Shabio and R.G. Robbins, "Kinetics of the Dissolution of Uranium Dioxide in Nitric Acid," *Journal of Applied Chemistry,* Vol. 18, pp. 129-134, 1968.

(40) *Aqueous Fuel Reprocessing Quarterly Report for Period Ending December 31, 1973,* USAEC Report ORNL-TM-4488, Oak Ridge National Laboratory, Oak Ridge, TN, 1964.

(41) J.H. Goode, *Hot-Cell Dissolution of Highly Irradiated 20% PuO_2-80% UO_2 Fast Reactor Specimens,* USAEC Report ORNL-3754, Oak Ridge National Laboratory, Oak Ridge, TN, 1965.

(42) *Sandia Solidification Process,* SAND 76-0105.

NON-HIGH-LEVEL RADIOACTIVE
WASTE TREATMENTS

The following reports were the sources of the material
in this chapter: HEDL-SA-851, HEDL-SA-856, BNL-
21571, BNWL-TR-203, RFP-2471, LA-6252, PB-244 928,
BNWL-1876 and BNWL-1913. See page 359 for a complete bibliography.

VOLUME REDUCTION AND DECONTAMINATION

Non-high-level wastes are generated in every manufacturing and reactor facility
associated with the nuclear fuel cycle. The term non-high-level wastes, as used
here, refers to liquid and solid radioactive wastes from reactors, fuel fabrication
facilities and reprocessing facilities except high-level wastes and fuel cladding
hulls from reprocessing facilities. The radioactivity in non-high-level wastes
consists primarily of short-lived beta-gamma emitters ($<$100 year half-lives),
long-lived alpha emitters (transuranic isotopes with $>$100 year half-lives), or
combinations of both. In general, non-high-level wastes do not contain enough
radioactive decay to pose a heat dissipation requirement as exists for high-level
wastes.

Non-high-level wastes can be treated and packaged to accomplish several purposes: (1) to reduce the volume of waste required for storage, (2) to increase
the margin of safety associated with handling and storing of wastes, (3) to recover transuranic elements, and (4) to reduce the mobility of the wastes during
prolonged storage. The selection of treatments of waste and the selection of
final solidified forms for waste go hand in hand.

The nuclear industry since its inception some 30 years ago has practiced various
methods for treating and handling wastes. In many instances the technology is
not new. It has been and is being used by many other industries with similar
waste problems. For example, municipal refuse is now being sorted and incin
erated by some cities. The only difference for similar applications on nuclear
wastes is the recognized fact that radioactivity must be contained and/or immo-

bilized for storage. Thus, nuclear wastes can be incinerated to reduce the volume, much like city refuse, but the ashes must be stored either in a controlled surface site or in a Federal storage site.

Treatment technology for wastes will be discussed according to types of waste and independently of the radioactivity content. The technology in general is the same for any level of radioactivity. However, a high content of beta-gamma radioactivity requires special shielding for remote operation of equipment.

Non-high-level wastes can be grouped according to the broad generic categories listed below.

> General trash—Combustible and noncombustible trash which includes such materials as cellulosics, plastics, rubbers, HEPA (high-efficiency particulate air), filters, metal and glass items, hand tools, miscellaneous construction and insulation materials, etc.
>
> Discarded equipment—Failed equipment and major items from facility modification or renovation.
>
> Wet wastes—Aqueous solutions and slurries; ion exchange resins; filtering aids and sludges; organic liquids such as lubricating oils, greases, and organic solvents.

The quantities and characteristics of these wastes vary appreciably from facility to facility, even within comparable sections of the fuel cycle, depending largely on the maintenance and operational philosophy used at each site.

Each category of waste can be processed according to a selected treatment to produce an immobile product which can be sent to storage at a controlled surface site or a Federal storage site depending on the radioactivity content in the wastes. Treatment of any primary waste generates additional wastes (secondary wastes) which are similar in characteristics to one of several types of original (primary) waste.

As an example, incineration of combustible wastes produces as part of the operation additional combustible wastes which are also processed. The net result of such a treatment is a significant reduction in the volume for storage even after treating (recycling) secondary wastes. Accordingly, because of the smaller volume, considerable economy in the cost of storage results.

Technologies for General Trash

General trash consists of many materials being discarded from a facility. It contains both combustible and noncombustible materials. Combustible solids consist of a large variety of items such as paper, rags, plastic sheeting, protective clothing, gloves, rubber shoes, wood, filter cartridges, etc., as well as some partially combustible items such as HEPA filters encased in wooden frames. The actual constituents in combustible waste vary depending on the location of operations generating the wastes.

For example, wastes generated outside of a glovebox include primarily cellulosic materials (i.e., paper, cotton clothing and rags, etc.). Wastes generated inside gloveboxes, on the other hand, contain primarily rubber or plastic materials because cellulosic materials are generally excluded from glovebox installations.

Typical compositions used for development purposes are 20 to 50 weight percent cellulosics, 10 to 40 weight percent rubbers and 20 to 40 weight percent plastics (1)(2)(3).

The noncombustible fraction consists typically of (1) metal and glassware, (2) construction and insulation materials (concrete, mortar, etc.), (3) metal-encased HEPA filters, and (4) small discarded equipment, tools, metal filters, and other mechanical devices.

General trash can be treated for several purposes: (1) to segregate combustibles from noncombustibles, (2) to reduce the chemical reactivity of the wastes, (3) to reduce the volume, (4) to recover quantities of fissionable material, and (5) to decontaminate it by removing radioactivity.

Pretreatment (Sorting, Shredding, Classification): General trash almost always consists of both combustible and noncombustible materials, and pretreatment operations are often required prior to primary treatments. Pretreatment of waste, i.e., the physical or chemical processes necessary to prepare waste for primary treatment and/or storage, include operations such as assay, sorting, shredding and classification. In general, the technologies for pretreatment operations are readily available and have been used at ERDA and/or commercial sites.

Sorting — Hand sorting is the simplest method of segregating mixed wastes into constituents which are amenable to treatment by a particular technology. Although this may best be done at point of origin, the wastes are often sorted after collection as part of a waste treatment operation. The categories into which the wastes are sorted will be determined by whatever primary treatment options are selected and used. Hand sorting is used to assure separation of combustible from noncombustible materials; it is also used to identify and remove hazardous materials (e.g., pyrophoric or explosive substances) in the waste treatment area.

Sorting of plutonium contaminated wastes according to the plutonium content has been used for a number of years at ERDA operating sites. For example, at Los Alamos Scientific Laboratory, Los Alamos, New Mexico, wastes are sorted by radioactivity content and separated into combustible and noncombustible fractions (4). At Rocky Flats Plant, near Denver, Colorado, wastes generated by glovebox operations are sorted prior to compaction (5). In these sorting operations protective clothing and radiation shielding are used as appropriate. Sorting of plutonium contaminated waste is accomplished in gloveboxes or in walk-in rooms using pressurized suits (6)(7)(8).

Frequently it is desirable to segregate solid wastes into combustible and noncombustible fractions. Following suitable pretreatment, combustible wastes can be combusted, decontaminated, compacted or packaged. Similarly, noncombustibles can be decontaminated, compacted, melt-cast, dissolved, and/or packaged. Decontamination and melt-casting are discussed under Technologies for Treatment of Discarded Equipment. Dissolution in salts is being developed as an extension of the molten salt combustion technology discussed later under Combustion. Alternatively, noncombustibles are sometimes incorporated directly into cement and other solids for immobilization.

Shredding — Shredding is used on potentially combustible waste materials to produce small pieces for subsequent processing or storage. The principal types of shredding equipment are knife cutters, hammer mills, and variations or combinations thereof.

Commercially available knife cutters used for size reduction are grouped into three broad categories: (1) fly knives mounted tangentially on a rotor to work against stationary bed knives; (2) finger knives mounted axially on a rotor to work against stationary anvils; and (3) finger knives mounted axially on two counter-rotating rotors. Fly knife cutters have the disadvantage of being susceptible to damage from tramp metal in the feed. The knives must be sharpened or replaced periodically.

Shredders of the counter-rotating rotor type are being tested at Hanford Engineering Development Laboratory, Richland, Washington, and at Rocky Flats for application to transuranic contaminated combustible wastes.

Hammer mills consist of pivoted or rigidly mounted hammers on a vertical or horizontal shaft or rotor. The hammers may be rectangular or chisel shaped. Crushing or shredding takes place by impact between the hammers and a breaker plate. These shredders are more effective on brittle and noncombustible wastes. They are less useful for shredding plastics because of the tendency of these materials to bind the hammers.

Classification — Classification by air or inert gases (pneumatic sorting) can be used following shredding to segregate waste materials by density difference. The principal application is in the separation of combustible (low-density) from noncombustible (high-density) wastes.

It is estimated that initial hand sorting should remove about 75% of the tramp metal from the trash fed into the classification system. The wastes are then shredded and introduced into the classifier. The combustible material is carried by the gas stream to a cyclone separator where it is removed and the gas recycled to the classifier. The material of greater density drops counter to the gas stream and is collected in a hopper for periodic removal.

Rocky Flats has a pneumatic classification system under development in support of combustion process demonstration plants. Closed system pneumatic sorting has not been demonstrated for handling nuclear process wastes. However, air classification is used for separating municipal refuse into noncombustibles (glass, metal, etc.) and combustibles (paper, plastics, etc.), the latter then being used as a source of energy (9). Other than radioactivity and the smaller operating scale of the treatment facility, no major differences exist between nuclear fuel cycle wastes and municipal wastes.

Compaction: Compaction is the simplest process for reducing the volume of trash for disposal. Compactors are widely used in the nuclear industry for reducing the volumes of general trash and combustible wastes. The technology is well-established for commercial application. The subject has been reviewed in detail (6)(10). Application specifically for TRU wastes has been discussed by Herald and Luthy (11). In compaction, the waste is compressed inside of a container (such as a 200-liter drum) which then is closed. Some wastes are also amenable to a baling operation in which the wastes are compressed and

then banded so that they cannot expand; the bale can be subsequently packaged
for further disposition.

A typical dry waste compactor consists of a hydraulic system with a ram operating vertically downward, a contoured support plate, frame and safety enclosure with a loading table, vent, and filter system and associated fan, gauges and
controls.

A special application combining shredding and baling is being tested at the Rocky
Flats Plants. The shredder-baler system will shred and compact combustible,
compressible wastes. Combustible wastes which are collected outside the dry-
boxes, are placed in 200-liter drums. These drums are then counted using sodium
iodide detectors for measuring gram quantities of plutonium. All drums containing nondetectable amounts of plutonium are sent to the shredding operation.
The drums are emptied onto a sorting table where tramp materials are removed
to avoid damaging the shredder.

The wastes are then fed by conveyor belt to the shredder where they are reduced
to an assortment of particles averaging 5 mm in size. These particles pass out
of the shredder chute directly into the baler, which is capable of baling particles
as fine as sawdust. Bales are wire-tied before being bagged. The bales measure
50 cm x 60 cm x 120 cm and weigh about 90 to 180 kg. Eight of the bales
will fit in a standard 120 cm x 120 cm x 210 cm fiber-glass-coated plywood box
used for waste shipments and storage.

Combustion Technologies: Combustion treatments under development for radioactive wastes include incineration, pyrolysis, acid digestion and molten salt combustion. incineration involves the burning of combustible materials in air or in
an oxygen-rich atmosphere. Pyrolysis implies burning in an oxygen-deficient atmosphere which provides gasification of part of the waste material. Acid digestion involves oxidation of materials by nitric acid in a concentrated sulfuric acid
media. Molten salt combustion involves air oxidation of combustible materials
in a molten salt environment.

The application of combustion or incineration to the treatment of combustible
solids requires the coupling of several processes or unit operations into a total
waste treatment system which includes feeding, incineration, off-gas treatment
and ash/residue packaging or immobilization. Some waste materials such as
chlorinated rubbers or plastics pose more stringent requirements on the design
because of the generation of soot and because of corrosion by hydrochloric
acid vapor produced during combustion. Special design features are required
to contain the radioactivity and to provide adequate personnel protection. Incineration of combustibles has been widely used as a radioactive waste treatment (6)(12)(13).

The necessary conditions for achieving complete combustion in any incinerator
are: (1) adequate residence time, (2) adequate temperature (to promote complete combustion), (3) turbulence (to promote good mixing), (4) sufficient oxygen. Problem areas in the past have included spalling and warping of construction materials, incomplete combustion (leading to excessive carbon in the ash
and creating problems for the off-gas cleanup system), clogging, fires outside
the furnace chamber (including flashback in the feeder and soot fires in the
off-gas cleanup system), inadequate ash handling and off-gas cleanup systems,

and corrosion (particulated if sulfur or halogen-containing compounds are part of the waste). Experience with radioactive incinerator systems has shown a need for improved performance. Development of improved alternative systems is being actively pursued in the U.S.A. These developments will be available about 1980 for safe and efficient incineration of radioactive combustible wastes on a commercial scale. The following discussion covers the distinguishing features of combustion devices in use or being developed.

Cyclone Incinerator — Older single-chamber incinerators supply excess air in addition to that required for combustion. The use of surplus air results in lower burning temperatures and more fly ash carryover into the combustion gases than is obtained in controlled air incinerators.

At Mound Laboratory, an advanced design of an excess air cyclone incinerator, as shown in Figure 3.1, is being demonstrated in a cold pilot plant at batch rates in excess of 35 kg/hr and with materials spiked with ^{238}Pu ($<$100 nCi/g waste). A distinctive feature of the unit is the combustion of wastes directly in the drum by using the drum as a combustion chamber. The gaseous effluent has shown no detectable radioactivity after filtration through a deep bed fiber filter. Technology should be available in 1978 for commercial plants.

FIGURE 3.1: EXCESS AIR (CYCLONE) INCINERATOR (MOUND LABORATORY)

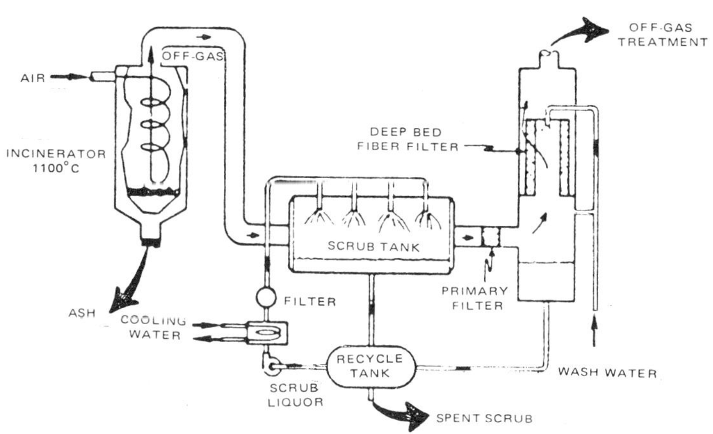

Source: HEDL-SA-851

Controlled Air Incinerator — A controlled air incinerator has been installed at Los Alamos Scientific Laboratory (LASL) (Figure 3.2) as part of a Transuranium Reduction Facility and is undergoing testing on nonradioactive materials. Planned startup of the unit to demonstrate system suitability on a plant operational scale is planned during 1977. Design throughput of the system is 45 kg/hr. Controlled air systems use the concept of multiple-chamber burning to achieve complete combustion of solid waste. Charges of waste enter the first chamber where they

burn in the presence of substoichiometric quantities of air at about 500° to
800°C. The products of partial oxidation and volatilization flow into a second
heated combustion chamber where excess air provides complete combustion at
1000° to 1500°C. This mode of operation produces a nonturbulent combustion
environment in the first chamber and minimizes the entrainment of fly ash.

The most frequently used configuration for controlled air incineration consists
of two stacked, horizontal cylinders. Smaller units do not have a second chamber and use the stack for secondary, or gas-phase, burning. Most designs have
multiple natural gas burners for maintaining temperature control in the chambers.

FIGURE 3.2: CONTROLLED AIR INCINERATOR (LASL)

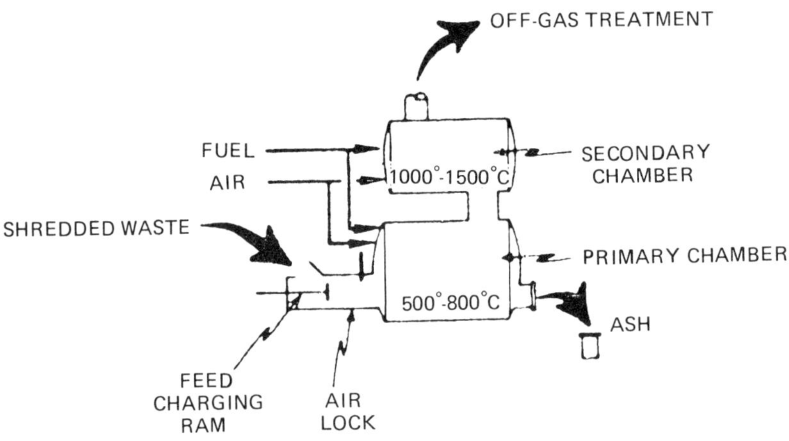

Source: HEDL-SA-851

Pyrolysis-Incinerator — The pyrolysis-incinerator or vertical retort process has
been demonstrated by Pacific Northwest Laboratory (PNL), as shown in Figure
3.3, with simulated, nonradioactive waste at a rate of 15 kg/hr. The design of
a pyrolysis-incinerator system to handle low-level radioactive solid wastes was
begun in March 1974.

In a pyrolysis-incinerator, combustible material is thermally decomposed in an
oxygen-deficient atmosphere to yield a volatile gas which is subsequently combined with excess air and burned in a secondary combustion chamber. A pyrolysis-incinerator depends on accurate control of the amount of air fed to the
burning chambers for efficient operation.

Vortex Incinerator — General Electric at Wilmington, North Carolina has had
a vortex or suspended burning type incinerator in service on uranium fuel fabrication wastes since 1972 as shown in Figure 3.4. Combustible wastes are
shredded and blown tangentially along with excess air into a cylindrical chamber;
centrifugal forces provide turbulence required for complete oxidation and to
assist in separation of the ash from the chamber.

FIGURE 3.3: PYROLYSIS-INCINERATOR (PNL)

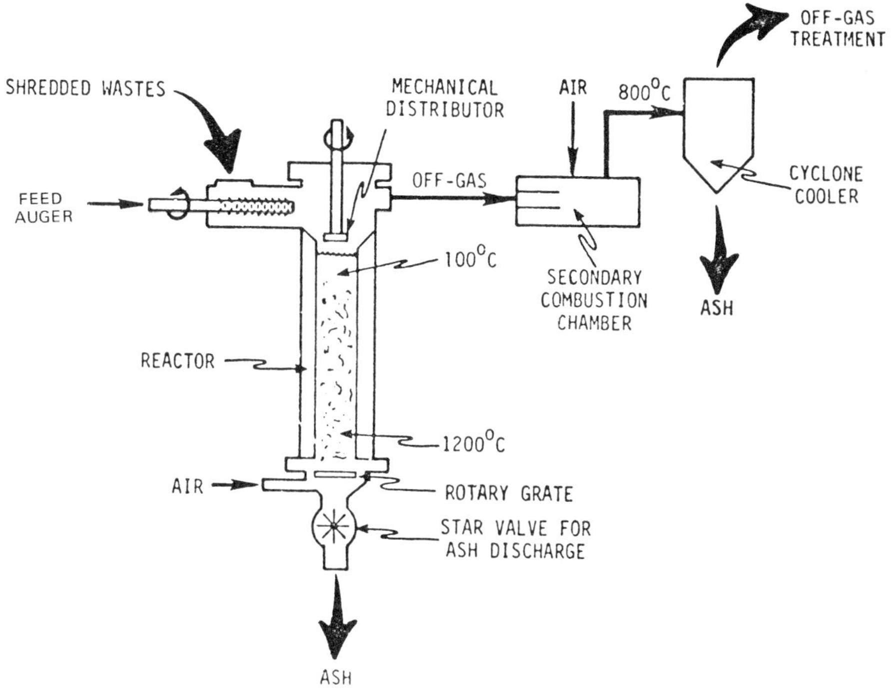

FIGURE 3.4: VORTEX INCINERATOR (GENERAL ELECTRIC COMPANY)

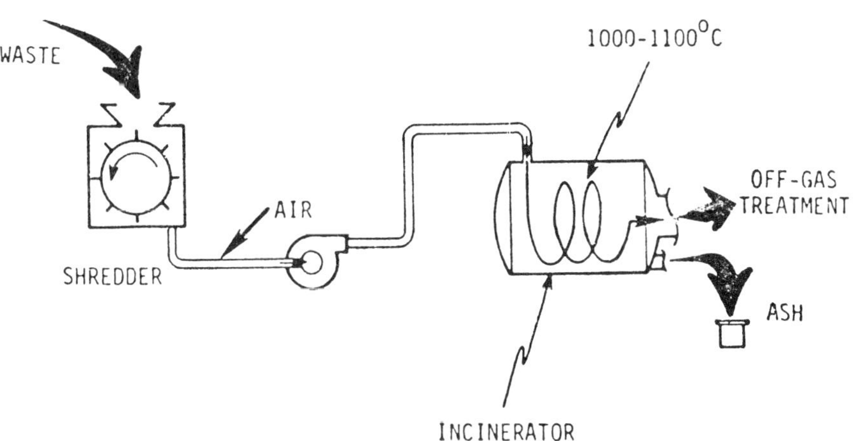

Source: HEDL-SA-851

Primarily designed for high burning rates, this type of incinerator is used in nonradioactive service in the lumber industry to burn bark and scrap. Its primary success has been in applications using only one type of waste material which can be easily shredded or ground to a fine consistency for pneumatic transport.

Fluid Bed Incinerator — Fluidized bed incineration has been demonstrated at Rocky Flats in a cold pilot plant at 9 kg/hr; a pilot plant is now used to burn TRU-non-line-generated solid and liquid wastes. A radioactive demonstration plant has been designed to handle 82 kg/hr (see later section for full description).

In the fluid bed incinerator, shown in Figure 3.5, the shredded wastes are introduced into a vertical, fluidized bed of inert material at 525° to 625°C. The highly agitated fluidized system improves mixing and combustion efficiency. It also provides improved temperature control due to the heat dissipation properties provided by the fluidized inert material. Use of pelletized sodium carbonate as the inert bed material provides additional benefits through neutralization of acidic combustion products, such as HCl, thus permitting use of less expensive construction materials in the incinerator and off-gas treatment system.

FIGURE 3.5: FLUID BED INCINERATOR (ROCKY FLATS)

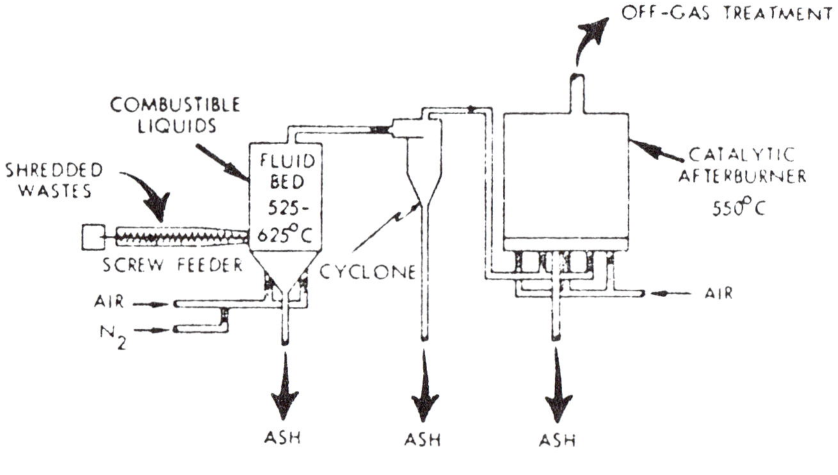

Source: HEDL-SA-851

Rotary Kiln Incinerator — A rotary kiln demonstration plant is now being designed to process ~40 kg/hr at the Rocky Flats Plant (see Figure 3.6). It will be ready for processing plutonium contaminated wastes (>0.02 g Pu/g waste) in 1980. The primary combustion chamber is a tilted horizontal cylinder that rotates on its longitudinal axis to provide a tumbling action for improved combustion efficiency. Combustible solid wastes, HEPA filters, ion exchange resins and liquid wastes from analytical laboratories will be burned or calcined in the diesel-fuel fired rotary kiln at 600° to 800°C. Off-gases pass through an afterburner operated at 1000°C. A rotary kiln has been tested in a cold pilot plant at 2 kg/hr.

FIGURE 3.6: ROTARY KILN INCINERATOR (ROCKY FLATS)

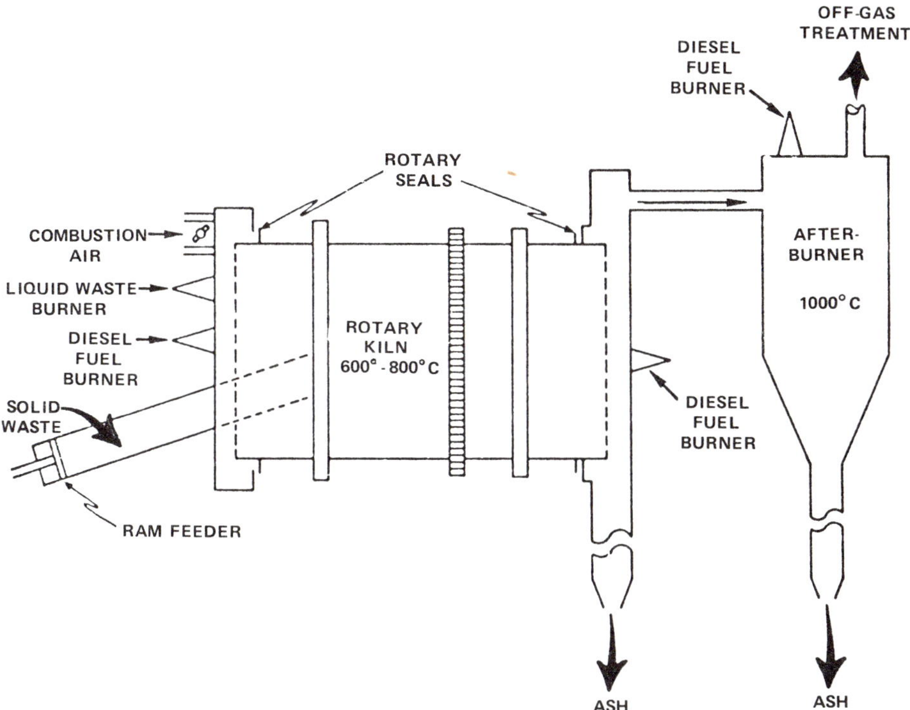

Source: HEDL-SA-851

Agitated Hearth Incinerator — An agitated hearth incinerator capable of processing 70 kg/hr of waste is being fabricated for combustion of low-level (<0.02 g Pu/g waste) contaminated wastes at Rocky Flats (Figure 37). The unit is expected to be ready for operation with contaminated wastes in 1980.

The agitated hearth incinerator uses the motion of the grate to move waste material and constantly expose new unburned material for combustion. Combustibles are charged by a ram feeder onto a circular hearth and moved through the incinerator by rotating rabble arms. The chamber is heated to 600° to 800°C by oil-fired burners. Ash is discharged on a batch basis following completion of combustion. Gases and fly ash are further combusted in an afterburner at ~1000°C. A nonradioactive demonstration model rated at 4 kg/hr has been tested.

Moving Belt Incinerator — A moving belt incinerator used at Hanford by Atlantic Richfield Hanford Company (ARHCO) for plutonium recovery from scrap is shown in Figure 3.8. Combustible wastes are burned at 700° to 800°C by passing them through an electrically heated, horizontal cylindrical furnace on a moving woven wire belt. Volatile components and fly ash are burned in a secondary tubular furnace surmounting the primary furnace. The unit has been operated more than 10 years at 2 kg scrap/hr. Technology is considered available for industry but the design requires some additional known modifications to minimize maintenance problems.

FIGURE 3.7: AGITATED HEARTH INCINERATOR (ROCKY FLATS)

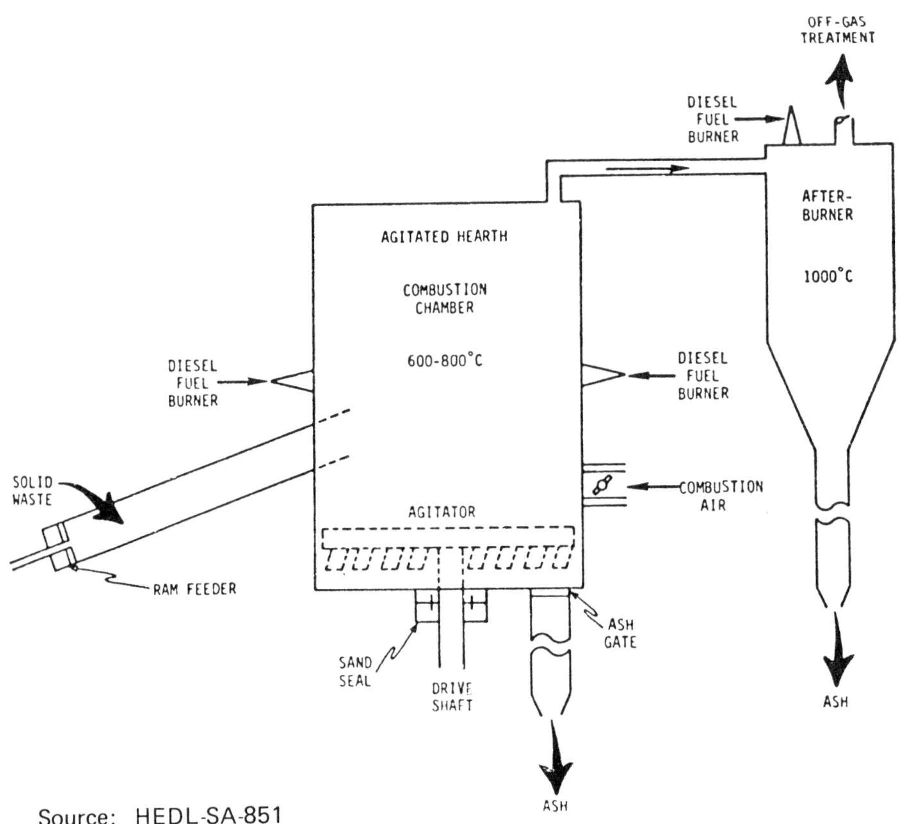

Source: HEDL-SA-851

Acid Digestion — Engineering feasibility of an acid digestion process has been demonstrated by Hanford Engineering Development Laboratory (HEDL) in a 5 kg/hr, nonradioactive test facility shown in Figure 3.9. A 5 kg/hr unit for treating radioactive wastes is being constructed at HEDL and is expected to be ready for processing plutonium-contaminated wastes in 1977. Technology should be available for industry by 1980.

In the acid digestion process, shredded combustible wastes are added to a simmering (250°C) concentrated sulfuric acid solution which carbonizes the waste. Nitric acid is added to the carbonaceous mixture to complete the oxidation to carbon dioxide. A shallow tray is used as the principal design feature to keep the system safe from nuclear criticality while handling quantities of fissionable materials.

FIGURE 3.8: MOVING BELT INCINERATOR (ARHCO)

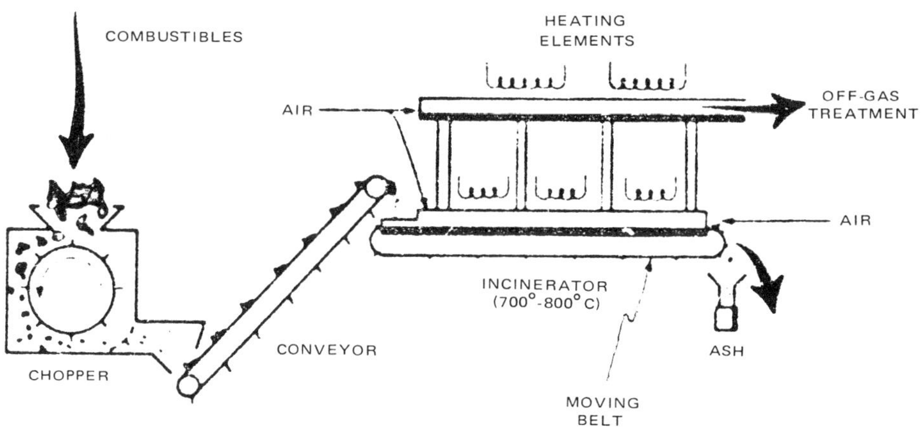

FIGURE 3.9: ACID DIGESTION UNIT

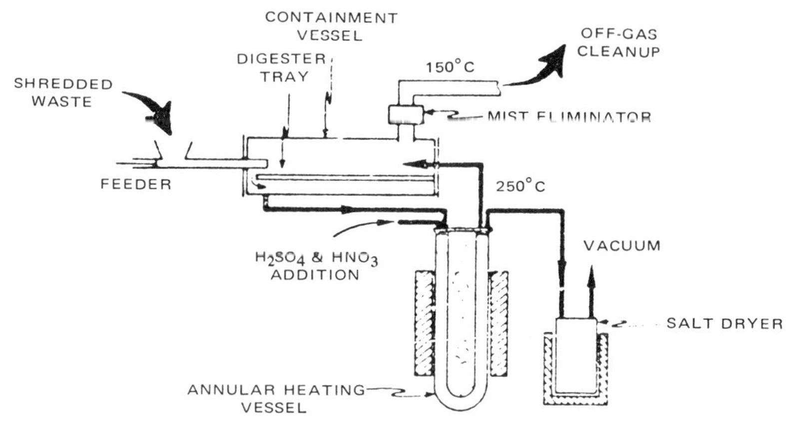

Source: HEDL-SA-851

Solids which accumulate in the system are removed during operation by periodic transfer of a portion of the acid slurry to a separate evaporator pot, from which sulfuric acid is evaporated at 350°C and returned to the digester for reuse. The resulting dry powder is composed primarily of inorganic sulfates and oxides and is thermally stable when heated in air.

Plutonium (including refractory plutonium oxide) is converted to a sulfate precipitate during acid digestion and remains with the process residue. Following
evaporation of sulfuric acid from the residue, the plutonium can be dissolved
from the residue by leaching with dilute nitric acid. The nitric and sulfuric
acids in the off-gas can be recovered and recycled for reuse.

Molten Salt Combustion — The molten salt combustion process, shown in Figure 3.10, has been investigated at Atomics International since 1969 (14)(15).
The process has been tested for the destruction of chlorinated liquid wastes (16),
destruction of toxic and lethal chemicals (17), and destruction of other combustible materials (14). Combustion has been demonstrated on a bench-scale
(0.5 kg/hr) with plutonium and beta-gamma contaminated wastes and on pilot
plant scale (50 kg/hr) with uncontaminated waste.

**FIGURE 3.10: MOLTEN SALT COMBUSTION UNIT (ATOMICS
INTERNATIONAL)**

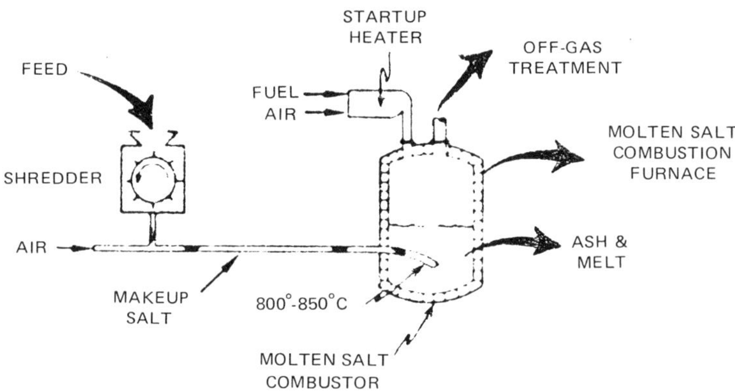

Source: HEDL-SA-851

The molten salt combustion process uses a molten mixture of sodium carbonate
and sodium sulfate salts, contained in an alumina-brick-lined steel furnace, as a
medium for air oxidation of combustible wastes. Shredded combustible waste
materials are fed with air (100% excess) beneath the surface of the molten salt,
which is maintained at 800° to 850°C by the combustion process. Combustion
residues accumulate in the salt.

A portion of salt is removed periodically from the furnace and cast directly
into storage drums. Alternatively, the melt-ash mixture can be processed to
recover sodium carbonate and sodium sulfate for recycle to the combustor, and
to produce an ash and sodium chloride fraction for storage. Plutonium is readily dissolved from the ash (<30 min reaction time) with nitric or hydrochloric
acid. The molten salt process can be used to treat a variety of materials, including dissolution of thin wall metal and glass. Although technically feasible, molten
salt dissolution requires considerable development and demonstration before ap-

plication to the noncombustible waste fractions. Treatment options for noncombustible wastes include size reduction, compaction, melting-casting, dissolution and decontamination. Some of the options differ from those for combustible wastes but are similar to methods used for treatment of discarded equipment.

Technologies for Treatment of Discarded Equipment

When processing equipment used in radioactive service fails or becomes obsolete during the operating life of the nuclear facility, it becomes a waste requiring storage. Discarded equipment normally is flushed and decontaminated prior to its removal from service. Combustible components are removed wherever it is practical to do so.

Reprocessing plant and reactor operations provide the largest and most radioactive pieces of discarded equipment. Typical large items from reprocessing plants consist of dissolvers, columns and concentrators which might be up to 2 meters in diameter and 10 meters in height. Such items can be disassembled at predetermined points to permit removal from their operating positions. Further disassembly may be required for removal from the facility. As an alternative to removal, discarded equipment may be stored temporarily in unused cells or in specially designed storage areas on-site for decay of short-lived radioactivity. Discarded equipment is frequently radioactive enough to require shielding equivalent to one or more meters of concrete.

Discarded equipment at fuel fabrication plants includes such large items as gloveboxes as well as the smaller equipment used in gloveboxes. Direct radiation levels may be low enough to permit direct contact work or work using only light shielding. In facilities for fabricating mixed oxide fuels, however, elaborate containment systems are required to prevent any possible release of plutonium to the environs during decontamination and size reduction operations. Treatment options for discarded equipment include decontamination, mechanical disassembly, and meltdown and casting.

Decontamination: The decontamination (removal of radioactivity deposited on the surface) of radioactive equipment is commonly practiced by nuclear facilities prior to maintenance or storage to reduce the radiation levels and thus minimize personnel exposure during subsequent operations (18)(19). Harsher and more destructive decontamination procedures can be used on equipment destined for storage.

In general, the extent of decontamination which is sought has been held to reducing the radiation to levels which facilitate handling of equipment as a waste. More extensive decontamination can be achieved but each decontamination operation generates additional volumes of waste which must be balanced with the net achievable reduction in contamination levels.

A wide variety of decontaminating agents and procedures are available and the chosen method depends on the nature of the wastes, compatibility with handling systems at the facility, etc. Typical decontamination solutions include alkaline permanganate, mineral acids, detergents, organic acids and chelating compounds. Technology exists to perform almost any desired degree of surface decontamination. However, complete decontamination for unrestricted release is limited to simple items where all surfaces can be monitored to check the effectiveness of the treatment.

Use of high pressure steam (up to 10,000 psi) has been effective as a decontamination procedure. With high radiation levels, remote operation of the spray equipment may be required. If the radiation levels are lower, operators can approach the equipment and manually direct the high pressure spray. For most items requiring treatment, the surface contamination can be removed by steam cleaning or washing in hot water and mild detergent.

Mechanical Disassembly: Mechanical disassembly of discarded equipment becomes necessary if it is too large to move directly to storage. The dismantling serves both to reduce the volume and to simplify handling. Disassembly may also be desirable as a pretreatment for melting, packaging, etc. Technologies for mechanical disassembly operations have been used on a small scale in the U.S.A. at ERDA facilities (19), as well as at foreign facilities (6)(7), and on a larger scale in nuclear facility decommissioning (20)(21)(22).

These operations can employ various types of fixed apparatus (e.g., shearing machine, guillotine) and commonly used tools such as saws, shears, impact tools and cutting torches. Typically, the operations tend to be time consuming and require excessive amounts of labor. However, the work can be done safely and satisfactorily if the need exists.

Meltdown and Casting: Meltdown and casting of discarded equipment, after any required decontamination and/or mechanical disassembly operations are treatment options under development. The resulting ingot represents the minimum volume for the waste and provides a means of fixing the radionuclide contamination.

The radionuclides contaminating the metal equipment are generally on the surface. Melting the metal incorporates most of the radionuclides into the metal matrix where they are immobilized. Some radionuclides such as strontium and plutonium will be preferentially included in the slag covering the melt (23). This presents possibilities for concentrating such radionuclides. The slag from meltdown represents a secondary waste for treatment.

The melting furnaces under consideration are designed similar to small metallurgical furnaces currently used in industry. They will accept pieces of metal with maximum dimensions up to about 30 cm and can operate on either a batch or continuous basis.

Technology for metal scrap meltdown and the production of ingots is well developed in nonnuclear industries; however, the behavior of radionuclides in such processes is not fully understood. Areas which are under development include: definition of radionuclides that will vaporize at the molten metal temperatures; distribution of radionuclides in the final ingot and slag; and development of adequate auxiliary systems (e.g., off-gas treatment). Furnace concepts being considered are: vacuum furnace, induction heating, autogenous melting (e.g., thermite heat source), electric arc, plasma arc and electroslag methods.

Technologies for Treatment of Wet Wastes

Wet wastes consist mainly of aqueous solutions and slurries, evaporator concentrates, spent demineralizer resins, filters and filter sludges, and organic oils and solvents. Evaporative processes are generally used on aqueous wastes with high total solids content. Wastes with low total solids are generally treated by non-evaporative processes to remove contaminants from liquids in order to recycle or discharge the liquid as an effluent. Organic wastes are sent to a combustion process similar to incinerators previously described.

Evaporators, drying and calcination are used as the principal treatment methods for aqueous solutions and slurries with high total solids contents (both dissolved and undissolved). Treatments for wastes with high solids content are well established at reactor sites in the United States. Sources of liquid wastes and treatments normally applied to them are given in Table 3.1. Evaporation is commonly practiced for solutions of high salt content (conductivity >100 mhos/cm at 25°C).

TABLE 3.1: LIQUID WASTES COLLECTED AT LWRs: SOURCES OF WASTE AND TREATMENTS NORMALLY APPLIED

Reactor Type	Liquid Waste	Source	Treatment
BWR	High purity	Equipment drains; low conductivity backwash water	Filtration and ion exchange
BWR	Low purity	Floor drains; water from dewatering of slurry wastes	Filtration and ion exchange (if conductivity below 100 mhos/cm at 25°C) or evaporation
BWR, PWR	Chemical wastes	Laboratory and non-detergent decontamination wastes; ion exchange regenerant solutions	Evaporation
BWR, PWR	Detergent wastes	Laundry wastes; detergent type wastes from decontamination	Filtration and/or reverse osmosis or possibly evaporation
PWR	Miscellaneous wastes	Floor drains, aerated systems and equipment drains, wastes from sampling and primary (boric acid) systems	Evaporation
PWR	Secondary system or steam generator blowdown wastes	Wastes from turbine building, steam generator blowdown (except for ion exchange regeneration)	Filtration and ion exchange

Source: HEDL-SA-851

At reprocessing facilities, the high total solids streams with intermediate levels of radioactivity are called intermediate-level liquid wastes (ILLW). Typical sources of ILLW at a fuel reprocessing plant include scrubber solutions from process vessel off-gas treatment systems, ion exchanger regenerants from spent fuel and/or solidified high-level waste storage basin cleanup systems, cask and plant decontamination solutions, waste solutions from solvent washing systems, and miscellaneous waste solutions from maintenance and laboratory operations. Similarly, aqueous wastes with high solids content arise from ion exchange regeneration, decontamination operations, floor and equipment drainage systems and plant chemistry operations. All of these wastes are evaporated, dried and/or calcined. Similar operations are used for high salt wastes in fuel fabrication plants.

Aqueous solutions containing low concentrations of dissolved solids and low levels of radioactivity (LLLW) are also generated at all nuclear fuel cycle facilities. Treatment options for these wastes include reverse osmosis, flocculation and carrier precipitation, filtration, and ion exchange. Low salt wastes can be decontaminated to a sufficiently high degree that the purified water can be recycled or released to the environs.

At fuel fabrication facilities, LLLW can be generated from washing, etching and rinsing operations, etc., and from solvent extraction system operations if a scrap recovery system is included. The waste streams from a mixed oxide fabrication plant contain only trace quantities of fission products, but may contain significant amounts of transuranic (TRU) contamination.

At reactors, LLLW include high-purity wastes such as filter backwash, filtered phase separator decant liquid, evaporator overheads, some laundry wastes and secondary system wastes at PWR's (steam generator blowdown and turbine building drains), etc.

Organic ion exchange resins used in treatment of LLLW become a combustible waste which is difficult to process in conventional incinerators. Development work to date indicates they could be processed by acid digestion, fluid bed incineration and molten salt combustion techniques described previously. Combustible filters and filtering aids such as cartridge filters and cellulosic filter precoat materials can be incinerated in the usual way. Alternately, resins and filtering aids can be dewatered and packaged with absorbents or incorporated, along with slurry water, into cements for subsequent disposition. Powdered resins are frequently dewatered and packaged in disposable liners for burial or storage.

Typical organic liquid wastes from fuel cycle operations consist of process solvents (tributyl phosphate diluted with a refined hydrocarbon diluent), laboratory solvents, cleaning solvents and lubricants of various kinds.

Waste organic liquids can be burned by spraying into specially adapted oil burners or conventional incinerators used for combustible wastes. Spent solvents from reprocessing plants have received considerable attention and two current approaches to incineration of solvent are described in Figures 3.11 and 3.12. The special nature of spent solvents containing radioactivity and phosphorus pose demanding requirements for off-gas treatment. Although the technology is generally available for incineration of solvents, it has not been fully reduced to operating practice.

Current work at Savannah River Plant (SRP) and at Barnwell Nuclear Fuels Plant (BNFP) is expected to demonstrate such processes. Alternatively, evaporation of small volumes of volatile organic liquids is also a viable decontamination

option. More commonly, because of their small volume, liquids or semiliquids
have been absorbed on vermiculite or similar absorbents for disposal without
further treatment.

FIGURE 3.11: WASTE SOLVENT INCINERATOR UNDER DEVELOPMENT AT SAVANNAH RIVER PLANT (SRP)

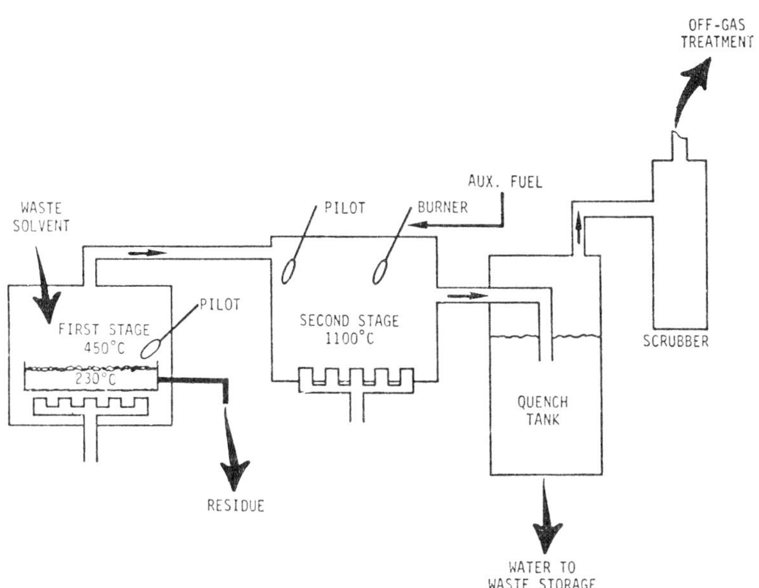

FIGURE 3.12: WASTE SOLVENT BURNER INSTALLED FOR USE AT BARNWELL NUCLEAR FUELS REPROCESSING PLANT (BNFRP)

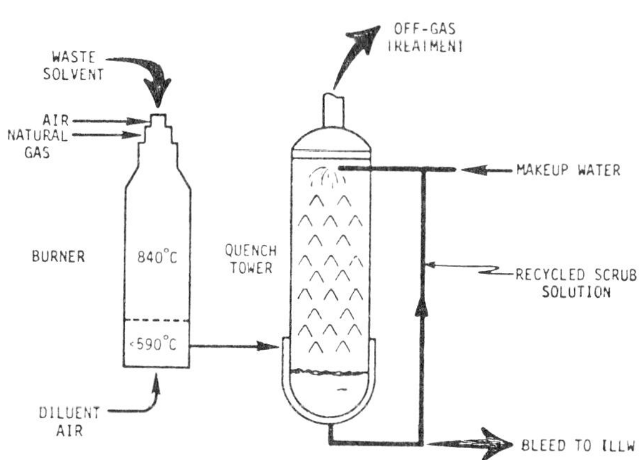

Source: HEDL-SA-851

Evaporation: Evaporation is a process whereby a solution or slurry is concentrated by vaporizing or boiling away the solvent, normally water or mineral acid. The use of evaporation for reducing radioactive waste volumes has been reviewed recently (24)(25). Evaporators coupled with efficient deentrainment devices, provide capability for a high degree of separation for most radioactive materials. The inherently high operating cost of evaporation limits the applications to those liquids which have a high concentration of dissolved solids and require high decontamination factors.

An evaporator consists basically of a device to transfer heat to the solution and a device to separate the vapor and liquid phases. The principal elements involved in evaporator design are heat transfer, vapor-liquid separation, and energy utilization. Careful design of equipment is mandatory for the evaporation of waste liquids with potential for foaming, corrosion, or severe scaling. To resist corrosion and attack from acids, evaporators are usually constructed of stainless steel and are operated at as low a temperature as is practical. The noncorrosive acids may be neutralized prior to evaporation.

Foaming can occur in evaporators due to traces of detergents, or surfactants, but this can be controlled by use of foam breakers (devices inside of evaporator for raising and lowering the temperature of the foam), by operating at low liquid levels, by addition of antifoaming agents, or by spraying water or steam jets on the foam surface. Scale is removed periodically by mechanical scrapers, by acid or alkali washings, and by thermal shock (e.g., as in the bent tube evaporator design) (24). Several variations in evaporation equipment design are available and have been used for many years in the nuclear and nonnuclear industry.

Coil or Pot Type Evaporators — This type is used primarily for batchwise operation; the equipment is simply a vessel in which the liquid to be evaporated is contained and heated, normally by steam flowing through a coil. Vapors leaving the vessel pass through an external vapor-liquid deentrainment device and a water-cooled condenser.

Natural Circulation Evaporators — These are generally heated by condensing steam in shell-and-tube heat exchangers. Both vertical and horizontal tube designs are used, although the vertical design is more common. In the latter, which is shown in Figure 3.13, the steam is condensed in the shell side of the exchanger and liquid circulation is induced in the vertical tubes by vapor formation.

Forced Circulation Evaporators — In this design, liquid is recirculated by a pump through the heating tubes. Circulation rates are high. The heat exchanger is normally a separate unit to facilitate maintenance and replacement.

A vacuum evaporator-crystallizer process has been used for treating liquid wastes at Hanford (26). A forced circulation evaporator concentrates the wastes to produce a slurry with up to 30% crystalline solids by volume. The slurry is returned to an underground storage tank. After a suitable settling time in the storage tank, the concentrated liquor is pumped off leaving a salt cake.

Vapor Compression Evaporators — In this type of evaporator, the vapors are compressed and combined with fresh steam input, thereby making the latent heat of condensation available at a higher temperature and utilizing energy potential of the vapors which are otherwise discarded.

FIGURE 3.13: NATURAL CIRCULATION EVAPORATOR WITH EXTERNAL VERTICAL TUBE HEAT EXCHANGER

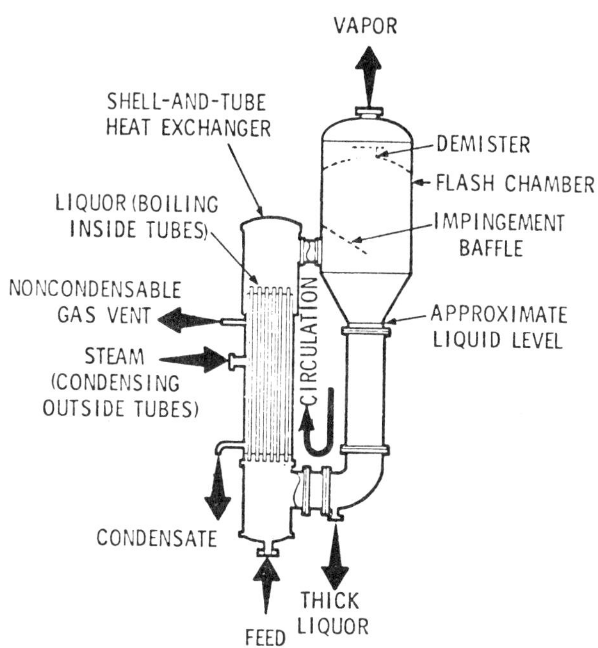

Source: HEDL-SA-851

Multiple Effect Evaporators — These evaporators utilize the vapor from one effect as the heating medium for another effect in which boiling takes place at a lower temperature and pressure. Normally, this type is only used for large-scale application.

Wiped-Film Evaporators — This design permits evaporation of liquids to a much higher concentration of solids than other evaporators. Mechanical energy is used to improve heat transfer efficiency. Liquid waste is fed into a heated cylinder which contains rotating blades or wipers. These reduce the liquid to a film by centrifugal force. The blades also serve to break up foam. A typical design is shown in Figure 3.14. Wiped-film evaporators can also be operated as dryers for waste compositions which are compatible with this mode of operation. Goodlett (27)(28) has discussed use of both vertical and horizontal wiped-film evaporators for reducing Savannah River Plant wastes to salt cake products.

Drying: Drying involves removal of liquid through application of heat to leave a dry solid. Drying has had little application thus far in the disposal of liquid wastes, but increased disposal costs are expected to favor increased use of this technology in nuclear facilities. Various types of dryers have been used or tested to convert radioactive liquid wastes to dry solid residues. The in-drum or in-tank drier consists of a heating element immersed in the liquid to promote evaporation.

FIGURE 3.14: TYPICAL WIPED FILM EVAPORATOR OF HORIZONTAL DESIGN

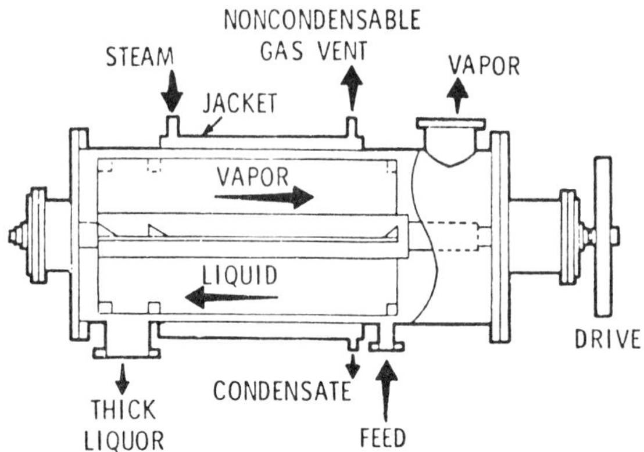

Source: HEDL-SA-851

This method has been used at Hanford in the in-tank solidification process for stored high-salt liquid wastes. In the spray dryer design, the solution is sprayed into a drying chamber and energy vor vaporization is supplied by radiant heating or by heated air. This technology is under development at Rocky Flats for disposal of waste salt solutions. Another alternative is the fluid bed drier, under development for radioactive reactor wastes (radwastes) by Aerojet Energy Conversion Company, an example of which is shown in Figure 3.15. Modifications of all of these units are or could be made available for commercial applications.

The dried product from treatment of aqueous solutions/slurries with high solids content is generally a dry or nearly dry salt cake or powder. In many situations this material is considered suitable for storage. In other situations the material is immobilized further by one of the techniques for cement, bitumen, glass, etc. The technology for drying of salts and sludges is well developed and practiced in the nonnuclear industry, and adaptation for use with nuclear wastes should require straightforward engineering application and design.

Calcination: Calcination, in contrast to drying, is a high temperature process in which liquid wastes are dried and thermally decomposed to form stable, non-fused compounds such as oxides. Most of the calcination process development work with radioactive wastes has been directed toward treatment of high-level wastes, but the technology is also applicable to low- and intermediate level wastes.

In reprocessing facilities, some ILLW streams might be combined with high-level liquid wastes as feed to a calcination process. Options include spray, fluidized bed, pot and rotary kiln calcination. The product form can range from powdery dusts to free-flowing granular material to a porous friable cake. An example of a commercially available fluidized bed calcination system for radwastes is shown in Figure 3.16. This system is specifically intended for treating low and intermediate-level wastes but it can also be used to incinerate combustible wastes, including ion exchange resins.

FIGURE 3.15: FLUID BED DRYING SYSTEM FOR TREATMENT OF
CONCENTRATED LIQUID WASTES
(AEROJET ENERGY CONVERSION COMPANY)

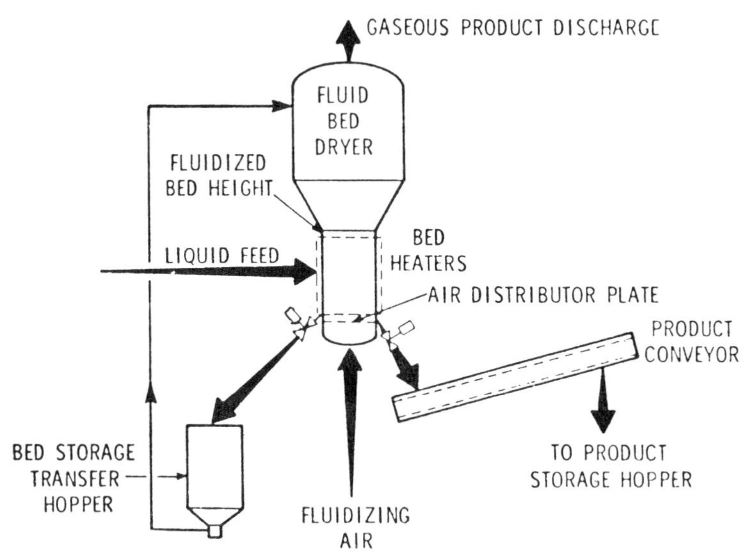

Source: HEDL-SA-851

Reverse Osmosis: Reverse osmosis is a method of purifying waste streams based on the phenomenon known as osmotic pressure. The process involves the separation of solutions of different solute concentrations by a semipermeable membrane. When pressure is applied to more concentrated solution in excess of its osmotic pressure, solvent will flow across the membrane to the less concentrated side. Applications of this concept to radioactive waste cleanup systems have been discussed recently by Kniazewycz (29).

Additional developmental work is needed before wide-spread use of this technology is adopted by the nuclear industry. In terms of the separations achieved, the process is quite similar to evaporation, i.e., a purified liquid stream and a waste concentrate are obtained.

Flocculation: Flocculation and carrier precipitation processes have been used extensively for cleanup of non-high-level liquid wastes commonly as pretreatment to filtration and ion exchange polishing. Chemicals are added to the liquids to produce flocculant precipitates. The radioactivity may be removed by direct precipitation, by absorption on the resultant flocs, or by entrainment in the settling precipitated. The effect is to concentrate the radioactivity into a small volume of insoluble sludge which can be separated by settling or filtering. The solution can be further treated, if needed, by ion exchange. An extensive coverage of the subject is given in reference (30). Flocculation and precipitation methods have been used for treatment of low-level liquid wastes at ERDA facilities and NFS, but other methods are generally preferred by reactor operators.

FIGURE 3.16: FLUIDIZED BED CALCINER SYSTEM FOR TREATMENT OF WASTES CONCENTRATES (ENERGY INCORPORATED)

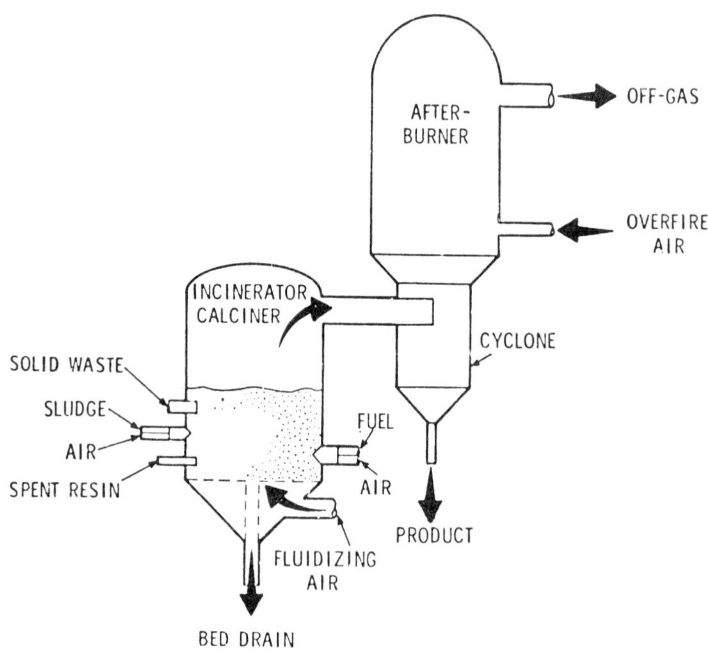

Source: HEDL-SA-851

Use of this technology for LLLW at reprocessing plants is exemplified by the treatment system in place at NFS (Figure 3.17) which is used to treat evaporator and acid fractionator condensates, cooling water, building drains, laundry wastes, and similar LLLW.

Filtration: Filtration is the most commonly used method for the removal of precipitates and particulate matter from liquid waste streams. Other methods used to separate solids from liquids include liquid cyclones (hydroclones), and centrifugation, settling and decantation. Centrifuges are mechanically driven devices of variable design; solids-bearing fluids are fed into a rotating section which imparts rotational motion to the suspension.

Centrifuges have been used to concentrate and dewater waste slurries (including resins) at reactors. A hydroclone is a simple device in which fluid pressure energy is used to create rotational fluid motion in a cyclone thus separating solids from liquids.

FIGURE 3.17: FLOWSHEET FOR TREATMENT OF LOW-LEVEL WASTES (NFS)

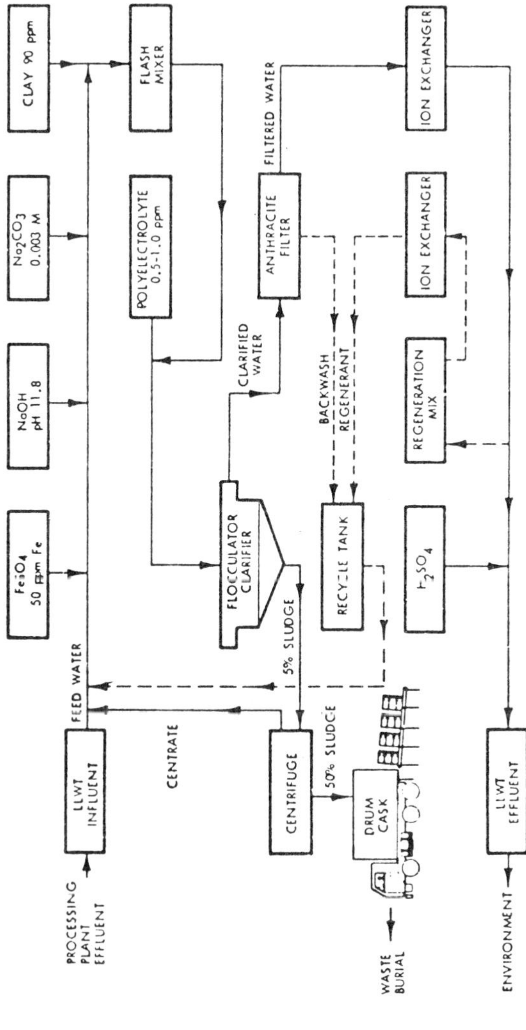

Source: HEDL-SA-851

The porous media used in filtration can be fabricated in a multitude of physical forms. Granular media employed in deep beds for filtering aqueous wastes include sand, diatomaceous earth, anthracite and activated carbon. Combinations of the various filter media are sometimes used. Typical examples of filters with nongranular media are discussed below.

Disposable Filters — Usually constructed of pressed paper, matted fibers or porcelain materials, these filters can be discarded when contaminated.

Metallic Non- or Partially-Cleanable Filters — These filters are constructed of woven wire or sintered metal. Backflushing does not completely clean this type of filter, which is best suited for removal of medium-sized contaminants.

Precoat Filters — These filters are designed as retainers to hold a precoat layer of particulate material which provides the actual filtration. Diatomaceous earth and powdered resins are commonly used as a precoating. The use of powdered ion exchange resins (the Powdex process) also provides decontamination from dissolved ionic impurities. The precoat material is simply backflushed to waste when its capacity is exhausted or the allowable pressure drop is reached.

Metallic Recleanable Filters — Use of this type filter is a relatively new development in the nuclear industry. The filters are designed for reuse following a cleaning operation. The chemically etched disc type is most common.

Gischel (31) has recently reviewed filtering systems that are in use in the nuclear power industry. At boiling water reactors, current practice is generally to use precoat filtering, while at pressurized water reactors disposable cartridge filters are most often used.

Ion Exchange: Ion exchange methods have been widely used for removing dissolved radioactivity from LLLW. The process involves exchange of ions between the liquid and a solid matrix containing ionizable polar groups. Comprehensive reviews have been published concerning commercial applications to the nuclear power industry (32)(33). Both cation and anion exchange resins and zeolites are used. When the exchangers have become fully loaded, they are removed from service and treated as radioactive waste or, alternatively, they may be regenerated by strong acids and strong bases, thus yielding radioactive liquid wastes of high salt content.

Development of a promising new process for decontaminating liquids by use of inorganic ion exchange resins has been reported recently by Sandia Laboratories (34). Natural chromatographic separation of simulated wastes has been demonstrated in passing through the packed bed of a sodium titanate ion exchanger. Actinides show a particularly high affinity for these materials and future interest in waste partitioning could spark an interest in the process for that purpose.

RECOVERY OF TRANSURANICS

Recovery of TRU elements from non-high-level waste and scrap materials has been routinely practiced at many ERDA laboratories and commercial mixed oxide fuel fabrication facilities (35)(36). Plutonium is the major TRU element recovered, and most of the following discussion will be devoted to techniques used for plutonium recovery. Americium-241, a decay product of plutonium-241, can be recovered as a by-product from plutonium recovery operations by solvent extraction and ion exchange methods comparable to those used for plutonium purification. Other TRU elements are seldom present in non-high-level scrap or waste in quantities warranting recovery, although technologies are available and have been used in ERDA fuel reprocessing facilities for the recovery of neptunium, americium, and curium from high-level wastes.

The techniques in use provide recovery of about 90 to 99% of the plutonium, which is usually present in solid wastes as the refractory oxide, PuO_2. The treated wastes are seldom decontaminated to plutonium (or TRU) contamination levels that would permit unrestricted release, although this objective may be approached for surface contaminated metals, glassware, and simple equipment shapes.

An outline of typical plutonium recovery techniques used in ERDA laboratories is shown in Figure 3.18. The most rigorous recovery techniques are those required for plutonium-impregnated materials, such as rags and paper wipes used for surface cleaning.

FIGURE 3.18: TYPICAL PLUTONIUM WASTE RECOVERY OPERATIONS

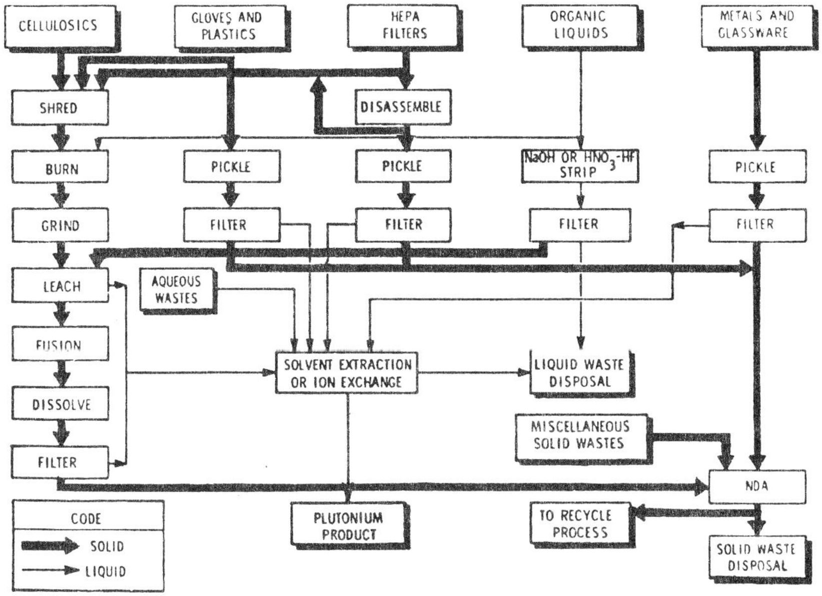

Source: HEDL-SA-856

These are typically burned in plutonium scrap recovery incinerators, and the ashes are ground and leached in a mixture of hot nitric and hydrofluoric acids (HNO_3-HF). Undissolved plutonium in the ashes may be recovered by fusion with a suitable salt, such as 10 to 1 potassium pyrosulfate-sodium fluoride mixture, to produce a nitric acid soluble melt.

Combustion followed by acid leaching can also be used to recover plutonium from shredded rubber gloves, plastics, and the wooden-framed HEPA (High Efficiency Particulate Air) filters used to remove plutonium aerosols from glovebox ventilation exhausts. More often, however, these wastes are leached (pickled) in heated nitric acid or HNO_3-HF mixtures for plutonium recovery.

In addition to conventional heated pickling vats, ERDA laboratories have used ultrasonic energy, hot water ball milling, electropolishing, and steam and water jet sprays to physically remove the plutonium from surface-contaminated metals, glassware, plastics, and rubber gloves. The dislodged plutonium oxide is filtered or centrifuged from the cleaning solution and dissolved by methods similar to those described above.

Plutonium-contaminated oils and solvents can be treated by contacting them with a suitable aqueous solution to transfer the plutonium to the aqueous phase. The oil is decanted or filtered to separate it from the aqueous phase and plutonium residue. Alternatively, the oil may be burned in the combustible waste incinerator.

Once the plutonium is in solution, it can be recovered and purified for recycle by well-established solvent extraction and ion exchange techniques. Aluminum nitrate is added to the feed to complex the fluoride to decrease its interference with plutonium recovery. Tri-n-butyl phosphate (TBP) in an inert diluent is the most commonly used solvent for solvent extraction processes, and strong base anion exchange resins are used for the ion exchange processes.

The Plutonium Reclamation Facility (PRF) at Hanford incorporates many of the previously described plutonium recovery operations (37). Much of the plutonium recovery equipment is located in a canyon-type process cell. Dissolution and solvent extraction equipment is mounted in plug-in modules that allow piping and valve connections to be made through glovebox ports outside the cell. Geometrically favorable process equipment and storage tanks are used to assure criticality safety.

The major features of the solvent extraction recovery process used in the PRF are shown in Figure 3.19. The solvent, CAX, is 20 volume percent TBP in carbon tetrachloride diluent. The feed (CAF) is a mixture of nitric acid solutions from the scrap dissolution operations with aluminum nitrate added to complex the fluoride. A scrub (CAS) of dilute nitric acid is used to remove weakly extracted impurities from the plutonium extract (CAP). Plutonium extracted in the CA column is removed from the solvent in the CC column by stripping with hydroxylamine nitrate reducing solutions (the CCX and CCXF streams), which convert the extracted Pu(IV) to inextractable Pu(III).

The flowsheet modification shown in Figure 3.19 also includes a small solvent flow (CCS) added to the top to reextract uranium impurity from the stripped plutonium product (CCP) (38). After nitric acid addition and a short delay to allow Pu(III) to be reoxidized to Pu(IV), a portion of the plutonium product is returned to the scrub section of the CA column as a reflux stream (CAIS) to increase the plutonium product concentration. Spent solvent (CCW) is treated with HNO_3-HF in the CO column to remove residual plutonium, which is returned to CAF for rework. The solvent is treated further to remove uranium and solvent degradation products before being recycled as CAX and CCS.

FIGURE 3.19: PRF PLUTONIUM RECOVERY SOLVENT EXTRACTION PROCESS

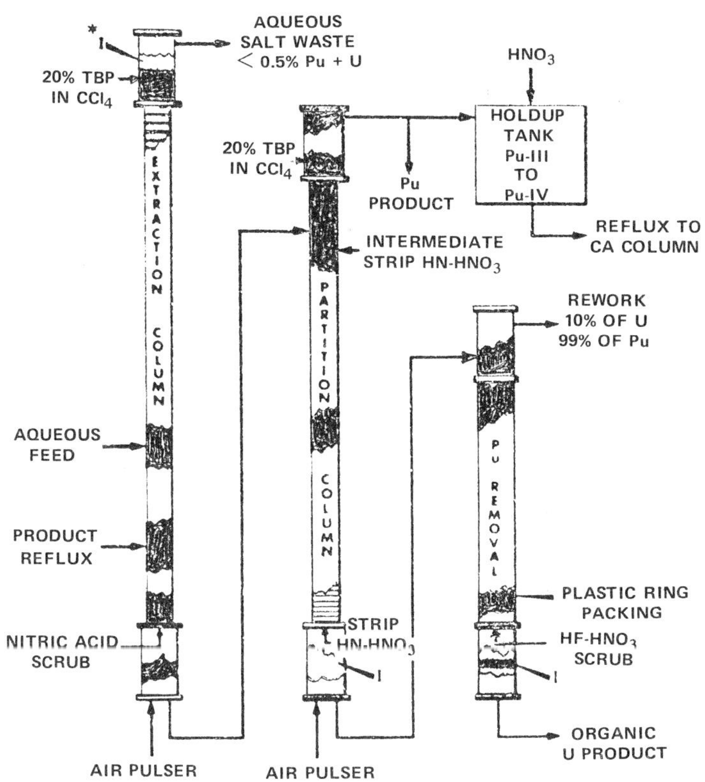

Source: HEDL-SA-856

The PRF also recovers Am-241 from the raffinate (CAW) of the plutonium recovery process (39). The process employs 30 volume percent dibutyl butyl phosphonate in CCl_4 as the solvent to extract both americium and residual plutonium from the high nitrate feed solution, neutralized to a pH of about 1 by addition of NaOH. The americium is selectively stripped from the solvent with dilute nitric acid and concentrated and purified by a cation exchange procedure. Plutonium is removed from the solvent by stripping with a HNO_3-HF solution.

IMMOBILIZATION TECHNOLOGIES

Immobilization, as used in this section, implies a treatment designed to reduce the mobility of as-generated radioactive wastes. In most cases, this is accomplished by incorporating the wastes in a monolithic solid of low dispersibility. There is no official regulation requiring that radioactive wastes be immobilized before disposal; but the Nuclear Regulatory Commission (NRC) now requires that reactors requesting licensing provide a solidification facility for wet wastes, and most of the commercial burial ground sites require that wastes be solidified before burial.

Immobilization technologies can be divided into those that are commercially available and in use and those under development to provide improved product integrity. The major immobilizing agents in use include urea-formaldehyde (U-F) resins, hydraulic (Portland) cement and cement plus additives, and bitumen (asphalt). A list of companies providing radioactive waste solidification systems using these agents is presented in Table 3.2 (40).

TABLE 3.2: U.S. COMPANIES PROVIDING PARTIAL OR COMPLETE RADWASTE SOLIDIFICATION SERVICES AS OF OCTOBER 1975

Company	Location	Solidifying Agent
Aerojet Energy Conversion Co.	Sacramento, CA	*
ANEFCO, Inc.	White Plains, NY	U-F
ATCOR, Inc.	Peekskill, NY	Cement
Chem-Nuclear Systems, Inc.	Bellevue, WA	Cement, U-F
General Electric Co.	San Jose, CA	Cement
Hittman Nuclear & Development Co.	Columbia, MD	Cement, U-F
Protective Packaging, Inc.	Louisville, KY	U-F
Stock Equipment Co.	Cleveland, OH	Cement
United Nuclear Industries, Inc.	Richland, WA	Cement-sodium silicate, U-F
Werner & Pfleiderer Corp.	Waldwick, NJ	Bitumen

*Currently supplies only fluid bed volume reduction system.

Source: HEDL-SA-856

Two other processes used to immobilize waste liquids involve use of absorbents and evaporation to a salt cake. The latter process is not used commercially, but it is being used at Hanford to immobilize sodium salt solutions accumulated during ERDA's plutonium production operations (41). Atomics International (42) also proposes to use the salt media of its molten salt combustion process to encapsulate the ash product as previously discussed by Cooley and Clark. Other immobilization technologies under development include clay fixation, glass formation, and pelletization.

Absorption

Absorbents have been widely used in the nuclear industry to immobilize liquids for transportation and storage. The product is quite leachable and far from being a monolithic product. Therefore, this technique is now largely reserved for oils, solvents, and other organic and inorganic liquids that are difficult to immobilize by other available means.

Widely used absorbents include vermiculite, diatomite, and various dehydrated clays. Typical formulations are one part liquid to two or three parts (by volume) of absorbent. Granular, easily wet absorbents such as vermiculite are mixed with wastes by filling the waste drum with dry absorbent and evenly distributing the liquid over the top of it. Mixer-extruders are available for difficult-to-wet absorbents such as diatomite; the product of the mixer is a thick peanut-butter-like paste.

Salt Matrix

For several years, Hanford has been converting ERDA-generated liquid wastes stored in underground tanks to massive salt cakes by the In-Tank Solidification (ITS) process (44). A flow diagram of the ITS process is shown in Figure 3.20.

FIGURE 3.20: HANFORD'S IN-TANK SOLIDIFICATION PROCESS FOR STORED LIQUID WASTES

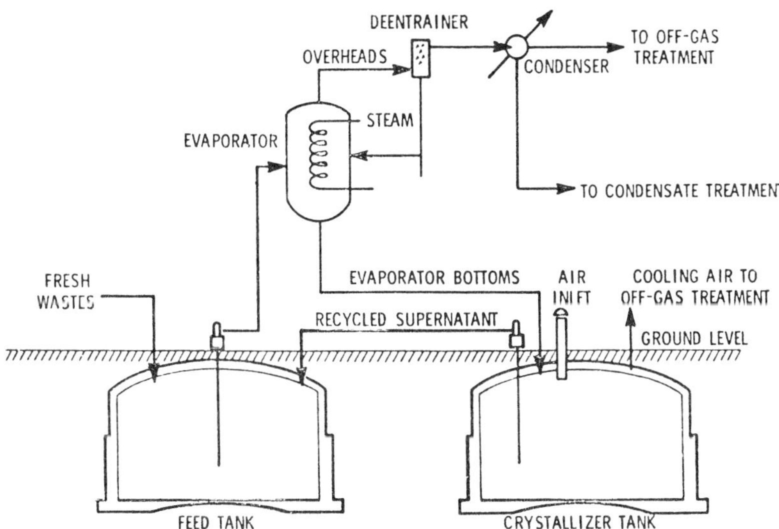

Source: HEDL-SA-856

The waste liquid, typically containing about 1 M NaOH, 2.5 M $NaNO_3$, 0.4 M $NaNO_2$, 0.3 M $NaAlO_2$, and 0.1 M Na_2CO_3, is transferred to an external evaporator for water removal. The concentrated slurry is transferred to the final storage tank for cooling and crystallization. Liquid remaining after crystallization is returned to the evaporator, along with fresh feed, until the desired fill of the final storage tank is obtained. The resulting salt cake is damp and contains about 20 to 30 volume percent concentrated liquor held in the solids by capil-

lary action. About 140,000 m³ of salt cake will be accumulated by the completion of the program in the 1980s. Recent work at both Hanford (43) and Savannah River (44) has shown that wiped film evaporators are capable of reducing
the water content of stored salt wastes sufficiently so that the product solidifies
on cooling. Application of this technology would eliminate the recycle of mother
liquor from the crystallizer-storage tank back to the evaporator and result in a
one-step evaporation process. Wiped-film evaporation technology was discussed
in the preceding section.

Urea-Formaldehyde Resins

Urea-formaldehyde (U-F) resins are widely used as adhesives and grouting materials. They were first investigated in 1971 for solidifying reactor radwastes by
Protective Packaging, Inc., a subsidiary of Nuclear Engineering Co., Inc. Several
firms now market solidification systems based on the use of commercially available U-F adhesives.

The U-F used in radwaste applications is a viscous, water-soluble liquid containing partially polymerized monomethylol and dimethylol ureas. It is mixed with
neutral waste solutions in the proportion of one to three volumes of waste liquid
per volume of U-F (45). Polymerization is initiated by addition of an acid catalyst to lower the pH to about 1 or 2. A concentrated solution of a weak acid
such as sodium bisulfate is normally used to avoid premature gelling that can
occur with an excess of strong acid.

The waste liquid or slurry is mixed with the desired amount of U-F by pumping
the two streams through an in-line static mixer. The catalyst is blended separately as the waste mixture enters the storage drum. The preferred formulations
begin to gel within one or two minutes, and the waste is generally well solidified
within 30 minutes. Curing to a reasonably hard, free-standing solid continues
over a period of several hours, during which time a small amount of free, slightly
acidic water is often released (46). Absorbents, cement, or more U-F liquid can
be added after gelling to absorb the residual water. Water is physically entrapped
in products prepared from aqueous solutions. In an open system, the block can
dry out and become friable. The tendency to dry out is much less severe with
solidified ion exchange resins.

Hydraulic Cement

Immobilization of radwastes by incorporation in hydraulic cements, as typified
by Portland cement, has been practiced for many years (47). The optimum
proportions of cement and waste vary with the type of waste to be immobilized.
For solid waste materials, such as ashes or calcines, the solidified cement products can contain up to about 75 weight percent waste. Solidified products made
with aqueous waste solutions and slurries are limited to about 33 weight percent
waste, including the water fraction of the waste.

A certain minimum proportion of water is required to obtain a workable mix.
Minimum water/cement weight ratios for typical cements are about 0.22 to 0.25.
Some waste solids, such as ferric hydroxide sludges and ion exchange resins,
will absorb large amounts of water; and this excess water must be supplied with
the solid to prevent it from sequestering the required water from the cement
and producing a dry, unworkable mix. Too much water, however, will reduce

the strength and may result in a layer of free water on the surface of the solidified product.

Several additives have been used to improve the setting properties, fission product retention, and/or the volumetric efficiency of cement. These include absorbents and ion exchange materials such as vermiculite and other clays (47)(48), plaster of Paris, and sodium silicate (46). A Portland cement-sodium silicate system has been developed by United Nuclear Industries, Inc. (UNI). Several commercially available radwaste cement systems are similar except for the sodium silicate additive.

The cement mixer used in these continuous flow cement systems is kept small to minimize holdup and to simplify flushing. The UNI system uses a skid mounted Moyno mixing pump which has a total length of about 1.5 m (5 feet) and a cement slurry holdup of about 0.03 m^3 (1 ft^3), including the line to the fill port. Pug mills and similar low-volume mixer-extruders are also used as radwaste-cement mixers (47).

The most distinguishing characteristic of the UNI system is the blending of concentrated sodium silicate solution with the radwaste-cement slurry as it enters the storage container. The sodium silicate additive has three major advantages: it produces a quick set with no free water; it readily solidifies PWR boric acid solutions, which set up poorly with cement alone; and it provides a significant (ca 30%) reduction in the solidified volume, compared with cement without additives (46). Despite the years of experience, solidification with cement is still an art. Each new waste application must be considered individually because of possible interactions between the waste constituents. Some techniques for solidification in concrete are discussed in a later section describing waste handling at Brookhaven National Laboratory.

Bitumen

Bituminization systems for immobilizing both liquid and solid radwastes are used in several foreign countries (47)(49) but experience with bitumen in the United States has been limited to developmental studies. Bitumen, or asphalt, has certain properties that are advantageous for immobilizing low and intermediate level wastes: it is chemically inert; it has good coating properties; and it possesses a certain degree of plasticity.

Perhaps the greatest advantage of bituminization is that the operating temperature of 150° to 250°C evaporates over 99% of the water, resulting in a volume reduction of up to 5-fold for products made with evaporator concentrates compared with conventional cementing techniques. Typical bitumen products contain 40 to 60 weight percent waste solids. Bitumen is also an effective sealant for other waste forms, such as concrete. Several techniques have been developed and used for incorporating waste in bitumen (47). These include a stirred evaporator batch process, a turbulent film evaporator continuous process, and a screw-extruder continuous process. The latter process is available commercially in the U.S. from the Werner & Pfleiderer Corp.

The extruder consists of two or four screws operating in a steel barrel that is divided into several stages heated separately with steam. Preheated bitumen and waste are continuously pumped into the first (cooled) stage where the rotating, self-cleaning screws mix and knead the waste with bitumen and transport the mixture forward through the heated stages to the discharge end. Water is continually evaporated and leaves as steam through large steam domes. The steam, which contains a small amount of entrained radioactivity, is filtered to remove oil and tars, condensed, and returned to the low-level water treatment system.

Bitumen shrinks up to 30% on cooling, and a more optimum fill can be obtained by allowing the receiver to cool after incremental additions. This can be accomplished by placing several receivers on a turntable and adding the desired incremental amount to each in turn until all of the drums are filled. A complete fill may not be desirable for the more radioactive wastes because of radiolytic swelling at accumulated doses of about 10^7 to 10^9 rads (47).

One of the major drawbacks to bitumen is its potential fire hazard, particularly if used to encapsulate oxidants such as nitrates, nitrites, or manganese dioxide. The combustion problem is minimized by using bitumen grades with high flash points ($\geqslant 290°C$). Improved safety can also be obtained by substituting more expensive polyethylene for bitumen. Fires have occurred in bituminization facilities, but they have been readily controlled (49).

Clay Fixation

The previously described salt cake produced by Hanford's ITS program is not considered the most desirable form for long term storage because of the inherent leachability of salt. The Atlantic Richfield Hanford Co. is developing hydrothermal processes which can be used to incorporate the salt in aluminosilicate (clay) minerals (50). In the basic process, concentrated alkaline salt solution is reacted with a powdered clay, such as kaolin or bentonite, at 30° to 100°C to form the insoluble mineral cancrinite, which contains sodium nitrate as part of the mineral cage-like structure. The stoichiometry of the reaction requires two mols of sodium hydroxide for each mol of cancrinite formed. Water, fission products and other minor constituents are trapped within the mineral cages and in the channels enclosed by the cages.

Three versions have been studied: the Lean Clay Process, the Rich Clay Process, and the Clay Calcination Process. In the Lean Clay process, cancrinite crystals are formed by stoichiometric addition of clay to the salt waste. The crystals are washed, mixed with a binder, and formed into massive shapes (e.g., bricks). In the Rich Clay version, the waste solution is mixed with excess clay which serves as a binder for the cancrinite crystals. In the Clay Calcination process, the cancrinite product from either the Lean Clay process or the Rich Clay process is molded into bricks and fired at about 800°C to convert it to nepheline, another durable aluminosilicate mineral. Clay can also be mixed with salt cake and fired directly to form nepheline bricks.

The Rich Clay process has fewer and simpler processing steps than the other versions. The ITS salt wastes (both liquid and solid) are mixed with suitable clay minerals in an extruder mixer to produce a soft, peanut butter-like product which hardens on aging or curing to an adobe-like block. The Rich Clay product is generally softer, more leachable, less cohesive, and has a greater volume (by about 60 to 100%) than the product of the Clay Calcination process which is its main competitor.

Glass

Glass is often considered an ideal form for immobilizing radioactive wastes because of its chemical durability and low leach rates. It also has the advantage of high thermal conductivity, which is important for heat dissipation and temperature control during storage of heat-generating high-level wastes. Glass-making technology is one of the oldest technologies known to mankind, and intensive work has been conducted over the past 20 years to adapt this technology to the incorporation of high-level wastes in glass. A limited amount of work has also been done on the incorporation of non-high-level wastes in glass (51) (52)(53).

In some ways, glass making is simplified for non-high-level wastes because of the much lower radioactivity and because many of the major nonmetallic constituents of such wastes are those used in conventional glass formulations. Ashes and zeolites, for example, are composed of aluminosilicates similar to those used in certain glasses. Sodium salts decompose on heating to provide sodium oxide, a major constituent of soda-lime glasses. It is usually necessary to add glass making additives or fluxes to the waste material to reduce the melting temperature to the desired range of 1000° to 1200°C, to assure the formation of a homogeneous glass, and to provide a sufficiently low viscosity for the melt to flow to the waste container. Borax or commercially available glass frits can be used for this purpose.

Various melter designs can be used. One melter is being developed by Battelle Pacific Northwest Laboratories for converting high-level waste calcine to glass (54). The finely divided waste is fed to a ceramic melter heated by passing an electrical current through a pool of molten glass. Glass can be discharged continuously by overflowing a weir or batch discharged by heating a freeze plug located on a bottom outlet. Resistance heating is used on startup to form the molten glass pool. Samples of glass have been prepared in the laboratory by melting together about 35 weight percent simulated Hanford ITS salt cake with 60 weight percent ground basalt and 5 weight percent boric oxide. The product, resembling obsidian, has a volume about equal to the volume of the salt cake charged (51).

Pelletization

The final waste immobilization process discussed here is pelletization. A variety of pelletization processes have been used in ERDA and commercial fuel fabrication facilities to prepare pellets or compacts of UO_2, PuO_2, and mixed oxides for nuclear fuels and other programs. More recently, several ERDA laboratories have begun investigation of adapting these processes to immobilization of radioactive waste salts, calcines, and ashes (54)-(57).

The two processes being studied most intensively are cold pressing and vacuum hot pressing. The major difference between the two processes is that the cold pressed pellets are sintered to insoluble products by a heat treatment after pressing while the hot pressed product is sintered during pressing. The sintering temperature is about 1000°C with both processes. An optional surface "inertification" step is available in which the pellet surface is sealed with an appropriate ceramic, metal, or epoxy coating.

In the developmental vacuum hot press used by Sandia for hot-pressing ceramic simulated waste forms, vacuums of $\sim 10^{-4}$ Torr are maintained routinely. Sizes range from 1 to 5 cm in diameter and weights from 2 to 80 grams in the pellets produced by hot pressing.

Product Leachability

The leach rate of immobilized waste in water is one of the most important criteria used to evaluate the potential hazards of radioactive waste handling and storage operations. Leach rates are commonly determined by measuring the fraction of a given radioactive constituent that leaches from the solidified waste over a period of time using a specified leach test. The results are expressed as grams of waste solid dissolved per unit area per day. This term is perhaps a little misleading, since certain constituents often leach at a faster rate than the bulk waste solid, e.g., alkali (usually Cs-137 tracer) and alkaline earth elements leach at much higher rates than plutonium from the same type of solidified waste.

The radioactive element of greatest concern in non-high-level wastes is plutonium, since long-lived cesium, strontium, and most other fission products are rejected to the high-level waste during fuel reprocessing. Several waste products, including glasses, bitumen, cement, and cement grouts, have leach rates below 10^{-6} g/cm^2 per day, based on plutonium and rare earth tracer elements. Plutonium leach rates from glasses prepared from non-high-level wastes are comparable to the illustrated leach rates for rare earth elements (50)(53). Cesium and strontium leachability can be reduced significantly by adding certain clays or minerals which have a high affinity for these elements to the solidification formulations (47).

At this time, selection of the immobilization process to be used for each specific radioactive waste requires consideration of both the inherent hazards caused by the radionuclide content and the economic factors involved in solidification, transportation, and storage. The methods in use have generally been satisfactory for the limited types and quantities of non-high-level wastes generated to date. However, the growth of the nuclear industry, the concurrent increase in the amount of wastes for disposal, and the increasing diversity of waste types that will accompany the anticipated recycle of plutonium will dictate that improved immobilization methods be developed that both minimize the waste volume and decrease the mobility of radionuclides in the waste. Methods under development by both commercial firms and ERDA laboratories are aimed at meeting these criteria.

SOLIDIFICATION OF BROOKHAVEN NATIONAL LABORATORY WASTES IN CONCRETE

Brookhaven has no high-level waste from chemical fuel reprocessing operations. The semisolid radioactive material that was generated during the days when the Brookhaven Graphite Research Reactor (BGRR) was fueled by natural uranium is still in storage at the Waste Treatment Facility. There is about 100 cubic meters of wet sludge which when assayed is found to contain approximately 10.5 Ci Pu-239, 1.4 Ci U-238 and 100 Ci of mixed fission products. Plans for the packaging and offsite shipment and storage of this material were initiated during FY 1976 and extend through FY 1980. Preliminary and tentative plans which have been developed to date indicate that the material will be processed and packaged according to existing technologies and then committed to retrievable storage at Oak Ridge.

The largest producer of radioactive waste at BNL was the air-cooled BGRR. Since its shutdown in 1968, the major volume of active wastes originates at the Hot Laboratory complex and the High Flux Beam Reactor (HFBR). Other facilities which contribute contaminated waste include those operated by the Biology, Chemistry, Medical and Applied Science departments. It has been BNL policy to reduce waste volume to a minimum since much of it must be disposed off-site.

Liquid Waste Storage and Treatment

BNL has three 3.8×10^5 liter tanks at the Waste Treatment Facility which serve as temporary storage. Liquid radioactive wastes in tank storage eventually are pumped into a 38,000 liter blending tank where the pH is adjusted to $\sim$7 prior to being fed into an evaporator for volume reduction. Figure 3.21 shows the flow sheet of the Vapor Compression Evaporation Unit at BNL, which has been in operation since 1952 (58).

FIGURE 3.21: VAPOR COMPRESSION EVAPORATION UNIT AT BNL

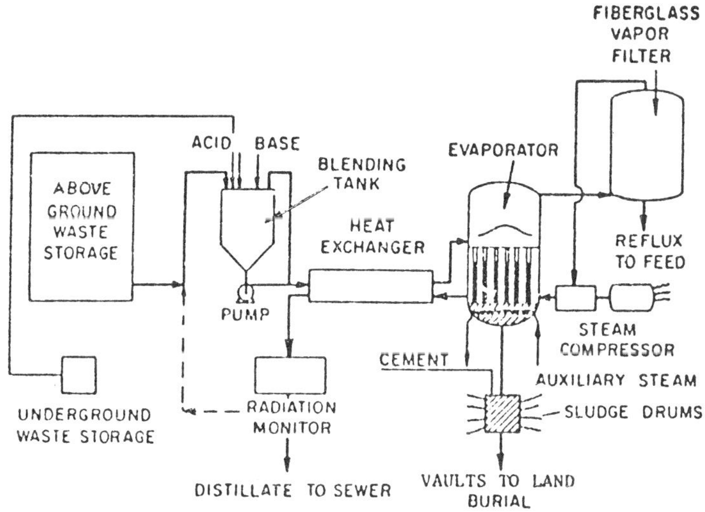

Source: BNL-21571

A concentration factor of about 100 is achieved with an overall decontamination factor of 10^6 to 10^7 which results in a product containing $\sim$30% solids. Before evaporation, typical waste contains 5×10^{-9} Ci/cc consisting mainly of beta and gamma emitters. The slurry produced by the evaporator contains essentially all the activity of the initial waste, except ^{3}H, which is carried over in the distillate. The slurry is transported in portable tanks to a 19,000-liter below-ground interim storage tank at the Waste Management Area for eventual solidification. Storage in this tank generally does not exceed one year.

Solid Waste Treatment

A power baler was installed at the Waste Management Area in 1957 for reducing the volume of solid combustibles. Burning them in an enclosed system has proved uneconomical and difficult, so incineration of solids has not been pursued. Ion exchange columns used in the treatment of liquid wastes are disposed of as solid wastes rather than being regenerated as was formerly done.

Packaging and Solidification for Disposal

Prior to 1959, BNL used 55 gallon drums (nominally 208 liter capacity) as the standard package for sea disposal. Slurry from the evaporator was mixed with cement in a batch-type mixer and poured into drums to harden. Few restrictions were imposed upon these packages other than a minimum density of 5.8 kilograms per liter of capacity in order to guarantee that they sink. Early in 1959, the volume and activity of radioactive waste reached a point where the continued use of 55 gallon drums for disposal became increasingly impractical. Larger packages were required to provide sufficient shielding and at the same time allow for larger payloads.

Three reinforced concrete casks were designed and fabricated to accommodate the various types and levels of radioactive wastes generated at BNL. Pertinent data for the three casks are given in Table 3.3. These casks as designed in 1959 are currently used at BNL.

TABLE 3.3: CONCRETE CASK DATA

Cask No.	Outside Dimensions (m)	Nominal Wall Thickness (cm)	Cask Volume (m^3)	Storage Capacity (m^3)	Cask Weight (kg)
I	1.2 x 1.2 x 2.4	43	3.6	0.18	8.120
II	1.5 x 1.2 x 2.4	30	4.5	1.02	8.390
III	1.5 x 1.5 x 1.8	15	4.3	2.27	4.760

Source: BNL-21571

The Type III cask is generally used for solidification of slurry wastes, baled paper and trash. Type I and II casks having thicker walls are used to reduce surface radiation levels and personnel exposures for more active gamma wastes. The Brookhaven Overpack, a reusable steel protective outer container can be used with these casks to satisfy the requirements of 10CFR71 Appendix B, "hypothetical accident conditions" for the transport of larger than type A quantities of radioactive materials.

Waste slurry from the evaporator is solidified in the 15 cm wall concrete casks by use of a mixture of Portland cement and vermiculite. The vermiculite acts as an absorber and as a cement diluent to reduce the heat evolved by cement hydration during curing. In the solidification procedure, Portland cement and an agricultural grade vermiculite are first dry mixed in a hopper in the proportions of 1 part cement to 3 parts vermiculite by volume. This dry mixture is then transferred into the concrete cask by means of a conveyer belt to the lip level of the cask. The slurry is introduced into the cement-vermiculite mixture through a pipe 1.2 to 1.8 m in length with a 1.3 cm i.d. situated in the center of the cask to a depth of about 2.5 cm from the bottom. Prior to solidification, the slurry is agitated for one day in the slurry tank to obtain a uniform mixture. This mixture is pumped into the cask using a gas operated centrifugal pump. At each pumping, samples are analyzed for total activity and isotopic content. A ball valve throttle is used to regulate the rate of slurry flow. Slurry is pumped into the cask until moisture is uniformly observed on the surface. Subsequently, a steel reinforcing bar grid is placed over the top and a concrete mix of the same formulation used in cask production is poured to seal the cask.

The quantities of materials used to solidify evaporator slurry in a Type II cask are typically: Portland cement, 0.91 m^3; vermiculite, 2.7 m^3; slurry, 1.25 x 10^3 liters. Based on the above proportions, the composite consists of 0.79 cm^3 of waste slurry per gram of mix. The solidified slurry waste has a total activity currently ranging from 0.87 to 2.3 Ci/batch. Solid wastes are also packaged in these concrete casks for disposal. Contaminated machinery and compressed solid combustibles are placed into a cask which is then filled with a concrete mix. Ion exchange resins are normally received at the Waste Management Area in 55 gallon drums which are placed into a cask which is then filled with cement. For small resin volumes or resins in inadequate packaging, the resin may be mixed with the cement used to fill a cask.

Polymer Impregnated Concrete

Since concrete is an open cell structure as a result of the interconnected porosity formed during the cure stage, it is permeable to water and other fluids and therefore exposes a larger surface to leaching than is evidenced by the external geometric surface area. Techniques have been developed at BNL to impregnate this porosity with styrene monomer which is subsequently polymerized in situ (59). This produces a polymer impregnated concrete (PIC) which is essentially impermeable while its strength, durability, and resistance to chemical attack are significantly improved.

Cement mixing and casting may be accomplished by a number of methods. Once the waste has been solidified in concrete, styrene monomer containing a polymerization catalyst is allowed to soak into the casting. After the concrete has been impregnated with monomer, it is heated to induce polymerization. Benzoyl peroxide and AIBN, 2,2'-azobis(2 methylpropionitrile), have been used as polymerization catalysts for styrene monomer at a concentration of approximately 0.5 weight percent in the monomer. Temperatures of 50° to 70°C are used to induce polymerization. Styrene was chosen because of its low monomer viscosity which aids in the impregnation step and its excellent radiation resistance.

Polymer impregnated concrete exhibited bulk leach rates for cesium and strontium at least two orders of magnitude lower than respective cement concrete specimens. The durability of concrete matrix materials are considerably improved as a result of polymer impregnation. Polymer impregnated concretes typically exhibit compressive strengths of 12,000 to 20,000 psi. Failure due to freeze-thaw cycling is eliminated. The resistance of concrete to chemical attack is also greatly improved by polymer impregnation.

EMBEDDING OF WASTES IN THERMOSETTING RESINS

The Grenoble Nuclear Research Center (CENG) of the Atomic Energy Commission has perfected the process of embedding by thermosetting resin. This process, in operation at CENG since 1971, is being developed industrially by the Group for Advanced Atomic Activities (G.A.A.A.). The operational installation for evaporator concentrates at CENG has made it possible to experiment with the process for various categories of wastes produced in lightwater (BWR and PWR) reactors. The principle (60) of the process is to disperse the radioactive residues in a thermosetting resin, e.g., in a polyester resin dissolved in styrene, and to induce polymerization after having obtained a homogeneous mixture. This polymerization is carried out without any input of external heat. Other thermosetting resins can be used.

The process is easy to implement and makes it possible to embed dry residues, ion exchange resins or filtration sludges together with their absorbed water, dry or humid solid wastes, etc. Investigation of this process includes: (1) tests of compatibility between the embedding materials and the wastes having diverse physicochemical properties; (2) tests of quality and safety bearing on the characteristics of the final product obtained after embedding and on the safety conditions implemented in the process; (3) guiding tests designed to provide the data necessary for the design and dimensioning of planned installations. These are supplemented by equipment qualification tests.

This section is concerned primarily with the application of the thermosetting-resin embedding process (61)(62) to light-water reactors (PWR and BWR). These reactors produce low- and medium-activity wastes which are contaminated with fission products or activation products. From the standpoint of the operation of the treatment station, these wastes can be classified into two categories:

(1)   moist wastes that can be transferred by piping
            evaporation concentrates with a base of sulfates (BWR) or borates
               (PWR), detergents
            ion exchange resins in grain or powder form
            filtration sludges

(2)   dry or moist solid wastes
            cartridge filters
            metal castings, glassware, rags, papers, etc.

The first category is the most significant in volume and in activity (see Tables 3.4 and 3.5). The wastes in the first category may require a pretreatment before being embedded.

TABLE 3.4: WASTES PRODUCED BY A BOILING WATER REACTOR

Wastes	Volume, m^3/yr	Activity, Ci/m^3
Ion exchange resins:		
Granular	10	1.25 to 1.25×10^{-1}
Powder (primary loop)	10	8×10^1
Evaporation concentrates	400	1
Filtration sludges	60	5×10^1

Source: BNWL-TR-203

TABLE 3.5: WASTES PRODUCED BY A PRESSURIZED WATER REACTOR

Wastes	Volume, m^3/yr	Activity, Ci/m^3
Ion exchange resins		
Primary loop	15	10^2 to 5×10^3
Auxiliary loops	25	Approx. 10
Miscellaneous effluent		
concentrates	30	1.5
Boron effluent		
concentrates	30	1.5
Detergents	3	1.5×10^1
Effluent concentrates	1	3.5×10^{-3}
Solid wastes		
Filters	7	10^2 to 5×10^3
Miscellaneous		Negligible

Source: BNWL-TR-203

Pretreatments

Pretreatment of Evaporation Concentrates: The concentrates are collected in a
tank where the radioelements are rendered insoluble. These concentrates are
then brought to the dry state, which permits a maximum reduction of the volume to be conditioned and stored. The distillate obtained, which represents 5
to 10% of the primary effluent (before evaporation) is removed from the site
or recycled into the reactor after monitoring its activity. The drying is performed
in a rotary dryer heated by a heat-transfer fluid (steam). The internal part of
the dryer is held at a slight underpressure across a scrubbing column designed
to retain the contaminants possibly entrained by the evaporated water.

This dry product is discharged through a leaktight circuit at a slight under pressure (to avoid any dispersion of activity into the atmosphere) into drums which
are themselves maintained at an underpressure, either directly or via a buffer of
calibrated volume. The dry product obtained has a residual humidity on the
order of 1 to 2%. The drums are then sent to the solidification station.

Pretreatment of Ion Exchange Resins: The resins are collected and stored, together with a portion of their transfer water, in a tank upstream from the Waste
Treatment Station. In this case, the weight reduction generally aimed at in the
process cannot progress to expulsion of the water included in the ion exchange
resins. In fact, experience has shown that their complete drying is an operation
which presents special dangers of inflammability due to their chemical nature
and nuisances due to the decomposition products. Furthermore, it is a lengthy
operation (about 48 hours for 100 liters of resins gorged with water) and one
which does not offer the benefit of substantially reducing their volume.

Therefore, the process has been developed in such a way that the ion exchange
resins, together with their included water, can be embedded by a simple chemical pretreatment carried out in the storage tank itself in about one hour. The
ion exchange resins are then wrung dry, which reduces the weight by lowering
their moisture ratio to about 40 or 60%, depending on whether they are in granular or powdered form. They are then transferred to the embedding station.

Pretreatment of Filtration Sludges: These come from the filtration of liquid effluents on precoat filters. They do not require any chemical pretreatment, but are wrung dry by standard means so that their moisture ratio before embedding is on the order of 40 to 50% by weight.

Pretreatment of Cartridge Filters: If the filtering cartridges are delivered in bulk to the embedding station, it is sufficient to combine them in a consumable basket or support. They are then embedded in this assembly together with their residual moisture.

Miscellaneous Metallic Wastes and Glassware: Brought to the solidification station in bulk, they are placed in a consumable basket before being embedded, like the filtering cartridges.

Embedding of Wastes

The wastes are embedded without supplying any heat, directly in the ultimate storage drums. The embedding technique utilized differs according to the nature of the products to be embedded.

Powdered or Granular Products (Desiccated Concentrates, Ion Exchange Resins, Filtration Sludges): The preaccelerated polymer is introduced into the drum containing the determined quantity of wastes. This mixture is homogenized by a mixer. The duration of this operation to obtain a homogeneous mixture depends on the density of the products, their amenability to moistening, etc., and generally does not exceed 20 minutes. Then the catalyst is introduced and the mixing is continued until reaching a gelation point corresponding to a given viscosity of the mixture, which serves as an indication point to withdraw the mixer. The drum is then removed from the mixing station.

The complete setting of the block (hard block) takes place in about two hours. The polymerization process is accompanied by a variable release of heat, depending on the load. It is possible to adjust the temperature variation by varying the formulation of the mixture.

The embedding installation consists of a drum mixer able to work at several speeds and endowed with an up and down movement. This installation can be confined in a small cell maintained at a slight underpressure in order to decrease the risk of contamination and evacuate the traces of styrene vapors that can be released prior to gelation.

Basket Wastes: A mixture of polymer and a load is mixed with accelerator and catalyst in the storage drum. The drum is then removed from the mixing station. Then the baskets containing the wastes are introduced. The blocks obtained in this way are very hard and can be sawed only with a diamond wheel. The formulation used for the polymerization is adjusted to allow for the transfer time of the drum between the mixing station and the basket introduction station.

Safety of the Process for Operation, Transport and Storage

The safety of the thermosetting resin embedding process is examined for operation of the installation and for the transport and storage (provisional and then long-term) of radioactive wastes conditioned by this process. For operation it is generally admissible to assure the safety of an installation by conceiving of successive barriers between risk sources and the operator or environment. As for transport and storage, the quality of the barrier formed by the conditioning is evaluated with regard to the attacks to which it can be subjected and the consequences that it can have on the environment.

Operational Safety: The risks are those inherent in treated radioactive residues (in particular, risks of irradiation and contamination) and those inherent in the embedding process itself (in particular, risks of chemical origin related to the very nature of the embedding material utilized). It is of interest to examine the contribution made to the Barrier Policy by the equipment and operating method employed in the process. By way of illustration, the case of evaporation concentrates and the case of ion exchange resins will be examined, indicating how barriers are created against the propagation of contamination and irradiation.

Pretreatment of Concentrates and Ion Exchange Resins — The evaporation concentrates are desiccated until a powdery dry product is obtained. The confinement of the contamination is achieved by the system of the dryer and its auxiliary devices (liquid-ring vacuum pump, condenser, condensate tank, etc.) and the collection tank or drum for the dry products.

This system is maintained at an underpressure. This arrangement provides a material barrier of very high efficiency against contamination. No contaminating element is in material contact with the outside medium. The fluid circuits (heating or cooling) are closed circuits. The constitutive water of the concentrates is collected in the tanks and is recycled or rejected after monitoring the radioactivity. The existence of the barrier can be monitored simply by measuring the pressure inside this system (e.g., on the order of 50 mm of Hg in the dryer used at CENG). The efficiency of the barrier can be monitored by measuring the activity of the air in the room.

The irradiation source is formed by the concentrate while drying and the dry product collected in the conditioning drum. The barrier against the danger of irradiation is formed by the biological protection of the cell containing the dryer and drum. However, there are types of dryers in which the amount of concentrates under treatment is quite small (a few liters) so as to require only a protection of small thickness. The dry-product collection drum, integrating all the activity, then constitutes the primary source of irradiation to be placed under protection.

The ion exchange resins present high risks of contamination and irradiation. To avoid the risks of dispersion of contamination, fire or deterioration in the heat which would be involved in their drying, the ion exchange resins are kept continually under water, during their hydraulic transfer and the chemical pretreatment, or moist (adsorbed water remaining after wringing dry); there is a continual material barrier, formed by the tanks and piping, against the spreading of contamination. The risk of irradiation, whose level is generally high, requires a localized lining at the level of the storage tanks, or else the placing of the system under biological protection.

The operation of embedding in the polymer at the mixing station involves: the radioactive risks of irradiation and contamination specific to the treated wastes and the chemical risks related to the nature of the embedding product and reagents (accelerator and catalyst) utilized—risks of fire or explosion, risks of intoxication. To evaluate the magnitude of these risks, it is proper to refer to the characteristics of the embedding resin (flash point, vapor pressure, upper and lower limit of explosivity in air, autoignition temperature, etc.) and of the reagents (catalyst and accelerator) utilized (decomposition temperature, flash point, etc.). These characteristics are given by the suppliers of these products and have been the subject of a complete safety study, confirmed elsewhere by experience.

By way of example, with the mixing and polymerization having taken place at the ambient temperature of 20°C and the vapor pressure of styrene being 4.53 mm Hg, the vapor content in air would be on the order of 0.6% in a static atmosphere, i.e., lower than the lower limit of explosivity (1.1% by volume at 20°C). The simplicity of implementing the process and, in particular, the fact that this operation of mixing and polymerization does not require any provision of heat, makes it easy to create barriers against the risks described.

The operational safety rests essentially on the following arrangements: (1) The mixing station where the mixing of polymer and radioactive residues is performed is in a ventilated enclosure (a dozen renewals per hour is sufficient). It should be noted that as soon as the gelation begins, the styrene participates in the polymerization and is no longer evaporated. (2) The electrical equipment (motors, ventilator, ignition, etc.) is flame-proof. (3) Catalyst and accelerator are introduced in small quantities separately to avoid their interaction.

It should be noted that the dynamic barrier formed by the ventilation and the placing of the working enclosure (cask or cell) at an underpressure enables the risk of spreading radioactive contamination to be contained. Likewise, it should be noted that this risk is decreased appreciably once the polymer is introduced into the drum above the wastes to be embedded.

The other important considerations concerning safety involve the storage conditions for the embedding resin and reagents. These conditions are those used in the plastics industry. There are precise rules to specify the storage conditions in accordance with the quantities in a ventilated room, at a fixed distance from work rooms, etc.

The barrier against the risk of irradiation is created, for example, by the walls of the cell with adequate biological protections in which the mixing station is installed. Depending on the level of irradiation, i.e., the beta and gamma activity of the radioactive wastes, the drum following polymerization must be placed in a conventional biological protection in order to respect the contact dose rate levels for transport of the fabricated blocks.

After embedding, the products obtained are blocks of 100 or 200 liters contained in standard drums. The safety of the conditioning by embedding in thermosetting resin is evaluated by considering the resistance of this barrier to attacking agents in order to prevent the dispersion of radioactive contamination into the environment in the event of a transport accident or in temporary or long-term storage. The attacking agents considered are temperature, fire, water, irradiation, climatic variations, and physicochemical and biological elements in the ground (if buried in the ground).

As for handling and transport, the following characteristics of the final embedded product should be noted: (1) Mechanical strength—Compressive strength is greater than 2 t/cm^2. A 10-meter fall of a 200-liter drum onto a concrete slab was without damage to the block contained in the drum. (2) Fire resistance—To simulate a storage or transport fire, instrumented 20-liter blocks were exposed to gasoline fires above the ignited surface in the hottest zone (700° to 900°C) of the combustion gases. The volume of these blocks is representative of actual blocks. They are of variable composition: polyester plus load of borates, chlorides, sulfates, nitrates or a mixture of those salts, or ion exchange resins.

The repeated experiments showed:

(1)   The blocks containing sulfates, borates, chlorides and ion exchange resins are completely resistant. After 30 minutes of fire, only the surface is deteriorated over a thickness of a few millimeters. The blocks have remained dense and very hard.

(2)   The presence of nitrates in the load does not produce an explosive reaction, even if the load consists solely of those salts (50% of the block by weight). Addition of mineral compounds such as sodium chloride or borate to the embedded load improves the fire resistance.

(3)   With a proportion of nitrate mixed with sodium chloride such that there is 50% polyester, 10% nitrate and 40% chloride, the blocks resist the fire without major damage.

Finally, the safety of long-term storage is judged according to the generally adopted recommendations based on the physical properties of the final product:

(1) Irradiation resistance—At least 10^{10} rads. Note that a dose of 10^9 rads corresponds to the infinite dose absorbed by the embedding polyester, an activity on the order of 200 Ci β, γ of fission and activation products originating from the effluents of a light-water reactor. At an integrated dose of 6×10^9 rads, no mechanical degradation is noted on the specimens subjected to irradiation; the mechanical strength is maintained.

(2) Formation of radiolysis gases—The radiolysis gases (H_2, CH_4, CO_2, etc.) are created during irradiation in quantities small enough that they do not cause any degradation of the specimens subjected to irradiation and containing mineral salts coming from the concentrates. The ongoing irradiation experiments on specimens of polyester embedding different types of residues have as their purpose a precise evaluation of the inflammable gases released in each case, in order to estimate the possible risk created by the storage of a large number of embedded drums in a closed room.

(3) Temperature resistance—No appreciable weight loss was noted for specimens loaded with mineral residues and held at 300°C for several hours. The embedding does not flow.

(4) Resistance to chemical agents—This is verified at actual scale by exposure of embedded specimens exposed to freezing and unfreezing for a long period in a severe climate and in the laboratory in a climatic enclosure by thermal cycling between –20° and +40°C.

(5) Resistance to leaching—The leaching tests carried out by applying the recommendation of AIEA involved radioelements in the form of salts previously rendered insoluble (before desiccation of the concentrates) and of sufficiently long period to be sufficiently attractive from the viewpoint of long-term storage safety in the open air. The results are shown in the following tables (63). The pH of the concentrates between 7 and 11 has no influence on these results.

TABLE 3.6: LEACH TESTING IN DEMINERALIZED WATER*

| Radioelements | Leaching, in cm/day, After | | |
Demineralized Water	200 Days	600 Days	900 Days
Cs-137	1.3×10^{-7}	3×10^{-9}	3×10^{-9}
Co-60	1×10^{-5}	2×10^{-7}	1.7×10^{-7}
Sr-90	3×10^{-6}	2×10^{-6}	1.8×10^{-6}

*According to the AIEA standard

TABLE 3.7: LEACH TESTING IN SOFT WATER AND IN SEA WATER

| Radioelements | Leaching, in cm/day, After | | |
In Soft Water	200 Days	600 Days	750 Days
Cs-137	2.5×10^{-8}	1.2×10^{-8}	4×10^{-9}
Co-60	2.2×10^{-7}	1.8×10^{-7}	1.1×10^{-7}
Sr-90	7×10^{-6}	2.6×10^{-6}	2.8×10^{-7}
In Sea Water			
Cs-137	2×10^{-8}	5×10^{-9}	3.2×10^{-9}
Co-60	2×10^{-8}	1.3×10^{-7}	1.2×10^{-7}
Sr-90	8×10^{-6}	6×10^{-6}	2.8×10^{-6}

Source: BNWL-TR-203

(6) The resistance to direct burial in the ground is examined in order to evaluate the risks of dispersing radioactive contamination after disappearance of the metal drum, in the improbable event that the storage is performed in trenches hollowed out from the soil itself in moist ground. The aerobic and anaerobic conditions representative of burial in the ground were recreated. Standard specimens of pure polyester buried for one year resisted bacterial attack.

However, the homogenization of the mixture of polyester and radioactive residue may cause an outcropping of the embedded load at the surface. Ion exchange phenomena have then been observed in the soil, particularly in moist ground. These phenomena might favor bacterial activity. These experiments, which are being pursued, seem to demonstrate the appropriateness of supplementing the water leaching tests, as recommended by AIEA, by tests allowing for exchange with more complex media such as soils. It is a fact that the very nature of the soil and the moisture ratio have an effect, and that for each site long-duration experiments would have to precede any decision for direct burial in the ground.

The characteristics of the final product were obtained with the commonly utilized polyester. It is found that to solve very special problems (of concentrates with a high content of salts, and which one does not want to desiccate) or to respond to the most severe safety constraints (fire resistance greater than that cited above and, in particular, for embeddings with a high proportion of oxidizing products), the process can make use of base products enabling particularly large amounts of water to be accommodated. It can also use the technique of superembedding to improve the impermeability to ion migration in the soil. Implementation of mixing or the superembedding technique lends itself readily to the use of additives such as fireproof, bactericidal or fungicidal products which can provide additional defense against the attacking agents which might be created by special storage conditions.

Economic Aspects of the Process

This can be evaluated relative to other processes by considering that it makes possible a reduction of the volume and weight of wastes to be stored and simplicity in the auxiliary installations.

Reduction of Volume and Weight: The volume reduction of the wastes (and thus the reduced number of drums to be stored) makes it possible to reduce the size of the temporary and ultimate storage rooms and to decrease the number of transports to the ultimate store. Thus, it has a favorable effect on the costs of construction, transport and storage. This advantage is particularly noteable for evaporation concentrates.

Take as an example 100 m^3 of regeneration solution having a sodium sulfate concentration of 10 g/l (case of BWR's). The primary evaporation leads to the production of 4 m^3 of concentrates with a salinity of 250 g/l, or 1 ton of sodium sulfate. Solidification of these 4 m^3 by the cement corresponds to 40 drums to be stored, since a 200-liter drum can absorb only about 100 liters of concentrate.

The embedding in thermosetting resin of the salts contained in these 4 m^3 of concentrates leads to the production of 7 drums of 200 liters each, the embedding being done with 60% mineral salts. If the embedded activity does not require biological protection for transport, it is of interest to consider the weight reduction: 40 drums of concrete weigh about 22 tons as opposed to 2 tons for the 7 drums of thermosetting resin.

Simplicity of Auxiliary Installations — This affects the economic aspects of the process in the following ways:

(1) The thermosetting resin can be stored either in a tank or in 200-liter drums at ambient temperature. The only restriction, which involves smaller volumes, is the problem of keeping the catalyst at a temperature on the order of 4°C in order to preserve its reactivity in the course of time.

(2) The equipment used for the pretreatment and mixing is current industrial equipment.

(3) The process is implemented at ambient temperature.

(4) The ventilation of the cell does not require any air treatment devices.

(5) Because of the good temperature resistance of the embedding and particularly the absence of any risk of flowing, it is not necessary

to be demanding about the ambient conditions in the storage
places, thereby simplifying their design.

Cost of the Base Material — It is a fact that the cost of a synthetic organic material such as that used in this process is higher than the cost of cement or bitumen, but it can be estimated that the cost is not predominant in the prime cost of the waste treatment, if allowance is made for the favorable effects from the standpoint of finance, reduction of storage rooms and simplicity of auxiliary installations. Be that as it may, this type of embedding uses a high-quality material which provides the conditioning with safety features which are desirable for the long-term storage of low- and medium-activity radioactive wastes.

FLUIDIZED BED INCINERATION AT ROCKY FLATS

A conventional incinerator has been used at the Rocky Flats Plant of Rockwell International for about 15 years to burn combustible scrap containing sufficient plutonium for recovery. The ash is subjected to aqueous processing for removing the contained plutonium. This incinerator has provided a valuable service for the plant; however, it has required frequent maintenance because of equipment corrosion, short refractory life, and mechanical problems primarily in the off-gas scrubbing equipment. Because of the high operating temperature (800° to 1000°C), a refractory type of plutonium oxide is formed. This oxide is difficult to recover from the ash by aqueous processing.

To overcome some of the problems associated with conventional incineration, a program was undertaken to develop a fluidized-bed incineration process. A pilot plant was constructed and has been operated to develop design information and to obtain operating experience. The pilot plant initially was operated with simulated combustible scrap; it is presently burning low-level radioactive waste.

This process utilizes in situ neutralization of the acid gases to minimize equipment corrosion and to eliminate the need for an aqueous scrubbing system. An operating temperature of about 550°C eliminates the need for refractory lined equipment and should produce a more soluble form of plutonium oxide in the ash.

In this process, the waste, which is composed of polyvinyl chloride (PVC), polyethylene, and paper, is chopped in a low speed, cutter-type shredder. The waste is then fed to the incinerator by a constant pitch, tapered screw conveyor at a rate of about 20 pounds per hour. During operation, the shredded waste is introduced into the primary combustion chamber under the surface of a fluidized bed of granular sodium carbonate (Na_2CO_3).

The primary combustion chamber operates at a temperature of approximately 550°C. Hydrogen chloride generated by PVC decomposition reacts with the bed material to form sodium chloride (NaCl). The fluidizing gas is a mixture of nitrogen (N_2) and air. The amount of air used is limited so that pyrolysis of the combustibles occurs and the operating temperature of 550°C is maintained.

Particulates generated in this vessel are then partially removed by a cyclone, and the products of pyrolysis are introduced into the bottom of a catalytic fluidized bed with enough air to ensure complete reaction. Flue gas dust generated in the afterburner at 550°C is fed to a cyclone and a sintered metal filter for removal of the entrained solids. The gas stream then enters an air jet ejector. This ejector provides the motive force for gas flow through the reactors and filtering equipment and also cools the gas stream by dilution. This stream is blended with room air and passes through one stage of high efficiency particulate air (HEPA) filtration before it leaves the incineration room. It then passes through four stages of HEPA filtration before reaching the outside atmosphere.

Process Description

The fluidized-bed incineration process used for combustion of transuranic waste will depart from that of the pilot plant in the method of heat removal and the method of producing the motive force for gas flow through the equipment. In the pilot unit, the heat of combustion was removed by blending a large amount of room air with the flue gas. This becomes impractical as the process is scaled up by a factor of about nine.

Mechanical blowers will be used rather than the air jet ejector that was used in the pilot unit. This will reduce both equipment cost and operating cost. The demonstration plant is being designed for a heat release rate of 1,500,000 kJ/hr (1,500,000 Btu/hr), which should correspond to a capacity of about 82 kg/hr (180 lb/hr). A flow diagram of the process was given in Figure 3.5 and a description of the process is as follows.

Waste will be received in 200-liter (55-gallon) drums. The waste will be introduced from an air lock box (in an air lock room) into the feed preparation area. Large pieces of metal in the waste will be removed by hand sorting to prevent damage to the first shredder (not shown). This metal will be collected and drummed for disposal. Combustible waste from the hand sorting will then be fed to a low-speed, cutter-type shredder for coarse shredding. Associated with the combustibles will be small pieces of metal that were undetected during hand sorting. This tramp metal will be shredded along with the waste.

It is intended that discharged waste from the first shredder be conveyed to an air classifier. This classifier will allow collection of the heavy tramp metal in a glove-ported receiver. Periodically the metal can be bagged out for disposal. Light combustible waste shall then be conveyed in an air stream to a shredder that has narrower and closer spaced blades. The waste will be shredded again and be of a suitable particle size for feeding to the pyrolyzing fluidized bed.

To introduce the waste to the first fluidized bed, a constant pitch screw will be used. This should compact the waste and feed it into the bottom of the fluidized bed. The screw will be variable speed and controlled by the temperature in the catalytic fluidized bed. As the temperature in the catalytic fluidized bed increases, the feed rate of waste to the first fluidized bed will be reduced. As the temperature decreases, the feed rate is increased.

The first fluidized bed will accept the waste and a mixture of air and nitrogen. The temperature of the bed is to be controlled by the amount of air fed to the unit. If more air is required to maintain bed temperature, the amount of nitrogen will be reduced proportionately to maintain a constant flow of fluidizing gas to the unit. As with any fluidized bed, particulate matter will be entrained in the gas stream leaving the unit.

To remove a large portion of particulates, a cyclone is to be installed between the two fluidized beds. This elutriation of solids is the mechanism for removing chlorides neutralized by the sodium carbonate (Na_2CO_3). The cyclone will then pass the combustible gases and a small portion of the solids to the catalytic fluidized bed.

The second fluidized bed shall contain a granular catalyst of chromium oxide on an alumina base. This catalyst will promote oxidation of the pyrolyzed gases at about 550°C. Sufficient air is then added to the bed to provide an excess of oxygen and to ensure uniform and complete fluidization. The combustion products and entrained catalyst are discharged from the afterburner to the dust collection equipment. The gas stream from the catalytic fluidized bed goes through two processing steps to remove the entrained solids. The first step involves a cyclone where 75 to 85% of the solids will be removed.

The second unit is to be a bank of sintered metal filter tubes that will remove remaining dust down to a particle size of one micron. Each set of tubes in this bank shall be sequentially back-blown with air or nitrogen to remove the buildup. Gas stream temperature will be maintained above 400°C by insulation of the dust collection equipment and process lines. As a result, should a process upset occur, the hydrocarbons produced by pyrolysis will not condense and plug the filter tubes. The gas stream could then be cooled for further processing.

The discharge temperature from the filters is calculated to be approximately 500°C. Using a water-cooled heat exchanger, this gas stream will be cooled to about 50°C to prevent damage to the equipment downstream.

The motive force for this system will be supplied by centrifugal blowers. The blowers will be sized to induce enough draft at flowing conditions to return the process pressure back to atmospheric pressure. Room air is to be fed to the inlet of these blowers to control the draft in the process. The pressure sensor for this control loop will be located in the first fluidized bed, and the controller can be set at about 19-cm water column (7 inches) to 25-cm water column (10 inches) below atmospheric pressure.

Blower discharge will be filtered through a set of HEPA filters before leaving the canyon. The process gases and the canyon air then will be ducted to a four-stage HEPA plenum before being discharged to the atmosphere.

Ash Pelletizing

The product from the fluidized-bed incineration process is a dry dust consisting of ash, catalyst fines, sodium carbonate, and sodium chloride. These products are removed from the process by two cyclones and a series of sintered metal filters. Dust from the primary cyclone is primarily ash, sodium carbonate, and sodium chloride. The secondary cyclone, downstream of the afterburner, removes the bulk of catalyst dust generated by the abrasion of bed material. It will also remove a small amount of the dust that passes through the primary cyclone. The flue gas passes through a bank of sintered metal filters that remove essentially all dust that is too fine to be removed by the cyclones.

Dust is an ideal form if the ash has to be processed for recovery of nuclear material; however, it has some disadvantages as a waste product for shipment and storage. Shipment of dust offers potential for rapid dispersion by wind if a shipping container should be breached in an accident. In addition, the fine particle size could increase dispersion should the material be contacted with water. Pressing the dust into a pellet was evaluated as a means of converting the material to a less dispersible form.

Product from the fluidized bed incinerator was used to evaluate pelletization to convert the dust to a bulk mass form. Only 3 to 8% moisture additions were required to produce pellets at pressing pressures of 140 MPa (20,000 psi) to 280 MPa (40,000 psi). Impact strengths varied from 0.13 J (0.09 ft-lb) to 0.7 J (0.5 ft-lb). This process resulted in a product density of about 2,000 kg/m^3, which represents a volume reduction factor of about two. Some components of the mixture are water soluble, however, and a more permanent method of ash immobilization is needed. Sintering was investigated as a means for forming water resistant pellets.

FIXATION OF ASH FROM PYROLYSIS-INCINERATION PROCESS

Along with the development of a pyrolysis-incineration process for the reduction in volume of alpha wastes, Battelle Northwest Laboratories is conducting studies on the conversion to glass of ash residue from the pyrolysis-incineration of simulated waste. A description of the experiments follows.

The feed of composition shown in Table 3.8 was processed by pyrolysis-incineration in laboratory scale studies. Ash obtained in the studies was characterized and converted to a glass. The ash as received was grayish tan, relatively soft material, with particles ranging from about ½ inch down to fluffy fines estimated at less than 100 mesh. The bulk density was determined: as received, 0.211 g/cc; after pulverizing to pass 6 mesh screen, 0.273 g/cc; tap density of pulverized material, 0.409 g/cc.

TABLE 3.8: FEED COMPOSITION

Component	Weight Percent
Rubber tubing	15
Neoprene	15
PVC	20
Tygon	10
Polyethylene	10
Tissue paper	15
Rags	15

Source: BNWL-1913

Composition of the ash as determined by chemical analysis is shown in Table 3.9. It is interesting to note that 4% chloride still remains in the ash. The ash composition appears much like the composition of ash from tissue paper as shown by an emission spectrograph analysis of tissue paper ash. Five glass samples were prepared in which the glass frit:ash ratio was varied systematically from 80:20 to 60:40, typical of the frit:waste ratios being studied in HLW fixation. The zinc borosilicate glass frit was the glass frit that has been studied most extensively for HLW fixation. The first melts were prepared by premixing the glass frit and pulverized ash and melting in platinum crucibles for three hours at 1000°C followed by furnace cooling. The cooled melts ranged from cream color to a mottled tan as the ash content increased. The melts were not true glasses but contained large quantities of crystalline residue; however, they were coherent and nonporous, which is a major goal. Determination of such properties as viscosity will have to wait until more ash sample is available.

TABLE 3.9: CHEMICAL ANALYSIS OF ASH FROM PYROLYSIS-INCINERATION PROCESS

Ash Constituents	Weight Percent
H_2O*	4.6
SiO_2	34.2
TiO_2	16.6
Al_2O_3	17.0
CaO	5.5
$CaCl_2$**	3.6
$CaSO_4$**	1.6
$NaCl$**	1.8
KCl**	1.3
BaO, Cr_2O_3, Fe_2O_3, La_2O_3, MgO, NiO, ZnO	0.02–2.0 each
B_2O_3, Co_3O_4, CuO, MnO_2, MoO_3, SnO_2, ZnO_2	Trace amounts

*Adsorbed water released at 110°C
**Arbitrary assignments of 4.0% Cl and 0.4% S were found by analysis

Source: BNWL-1913

The samples were leach tested by the standard accelerated procedure, the Soxhlet test. Leachability was not a function of ash content, results showed. For all five melts, the leach rate was about a factor of 10 higher than is obtained with the reference HLW glass. The glasses were remelted for three hours at 1200°C and furnace cooled. Results indicated that a considerably reduced leachability was achieved by using the higher melting temperature. Volume reduction factors were calculated using the density of the glasses and taking the density of the initial ash to be 0.4 g/cc. Results are shown in Table 3.10.

TABLE 3.10: ASH:GLASS VOLUME REDUCTION FACTORS

Frit:Ash Ratio	Glass Density, g/cc	Volume Reduction Ash:Glass
80:20	3.0242	1.51
75:25	3.1856	1.44
70:30	3.2318	2.43
65:35	3.3166	2.90
60:40	3.5098	3.51

Source: BNWL-1913

The preliminary experiments on ash fixation indicate that large amounts of typical ash can be incorporated in a monolithic glass form; however, much more development will be required. For instance, it is important that the actual leach rate of plutonium from the glasses be determined before evaluation is carried much further. A basic problem is the chloride content of the ashes. No reference examples of chloride-containing (above fractions of 1%) glasses are known. Thus, the behavior of chloride in glasses will have to be studied in much more depth.

It may be possible to operate the pyrolysis-incineration unit as a slagging gasifier such that, by addition of the glass frit with the waste feed, glass may be obtained directly from the unit. A need for a separate glass making plant could thus be eliminated since the glass plant and incinerator would be combined.

U.S. INCINERATION FACILITIES FOR TREATMENT OF RADIOACTIVE WASTES

The following is a list of incinerators in the U.S. (as of June 1975) which are designed for the treatment of solid wastes contaminated with radionuclides. Included is a description of the types of waste handled.

National Lead Co. of Ohio (NLCO), Fernald, OH

NLCO operates several incinerators to burn various types of low-level wastes contaminated with depleted to low-enrichment (<10% U-235) uranium at their Feed Materials Production Center. The wastes include trash, graphite, and oil.

United Nuclear Corp., Wood River Junction, RI

United Nuclear Corp. has operated an incinerator for wastes slightly contaminated with enriched uranium at their Recovery Operations at Wood River Junction. The unit was installed in 1967, and has a design capacity of 45.5 kg/h (100 lb/h), but is not currently in operation.

Goodyear Atomic Corp., Piketon, OH

In the Uranium Operations Subdivision of Goodyear Atomic Corp. at Piketon, wastes containing small quantities of radioactive materials in the form of UO_2F_2 and uranium oxides are produced. Most of the combustible waste consists of plastic shoe-covers, and varying amounts of paper, cardboard, and rags. To incinerate these wastes, a commercially available incinerator known as the Mark VI radicator was installed in May 1971. This incinerator is run on a batch basis. About 2270 kg/month (5000 lb/month) of contaminated material are incinerated.

Union Carbide Corp., K-25 Plant, Oak Ridge, TN

An incinerator for very low-level radioactive wastes has been in operation at K-25 for ~3 years. It is used to incinerate floor sweepings that contain uranium, so that the uranium can be recovered from the ash. This incinerator operates almost every weekday until, for purposes of criticality control, it must be cooled down and the ashes removed.

Union Carbide Corp., Paducah, KY

In the past, an incinerator with a scrubber has been operated at Paducah to incinerate low-level wastes containing depleted to slightly enriched uranium. The unit had not been used for some time because of corrosion problems and the need to improve scrubber efficiency. Engineering studies have been undertaken. Pending modification to the unit, uranium-contaminated combustibles were stored. Interim storage is not a problem because the generation rate of such material is low.

Union Carbide Corp., Y-12 Plant, Oak Ridge, TN

For >15 years an incinerator and four destructive distillation units have been operated for AEC/ERDA by Union Carbide's Nuclear Division as part of the process to recover nonirradiated, highly enriched uranium (93% U-235) from low-level radioactive wastes at Y-12. All the facilities are currently in regular use, and the distillation units are used ~1 week/month for those instances where material must be kept separate for assay purposes.

The wastes destined for the incinerator and the distillation units are presorted before they are processed. All noncombustibles, plastics, and rubber are removed. The plastics and rubber are sent to the leaching facility for uranium recovery. The incinerated wastes consist mostly of paper, rags, and wood.

Gulf General Atomics (GGA), Nuclear Materials Div., San Diego, CA

GGA in San Diego incinerates combustible wastes contaminated with small quantities of enriched uranium (~93% U-235). The entire incinerator facility is located outside. Wastes are first sorted to remove noncombustibles, plastics, and rubber gloves. The remaining waste is placed loose in 242-liter barrels. Full barrels are sent to the assay area where they are gamma-counted in a barrel counter to determine the U-235 content of each barrel. The barrels are then delivered to the incineration site.

General Electric Nuclear Fuel Plant, Wilmington, NC

General Electric operates a large reactor fuel fabrication plant in Wilmington, NC. Each year this facility produces between (141.5-283) x 10^4 liters (50,000-100,000 ft^3) of combustible waste contaminated with uranium (~2.5 weight percent U-235). An incinerator was installed in 1970 to burn the wastes and to recover some of the uranium.

The wastes are presorted at the generating site before being sent to the incinerator facility. There they are sorted again and noncombustible items are removed. The waste accepted for incineration includes PVC and rubber.

Westinghouse Nuclear Fuel Division, Columbia, SC

The laboratories and fuel fabrication facilities of the Nuclear Fuel Division of Westinghouse produce combustible wastes contaminated with slightly enriched uranium (≤4 weight percent U-235). The wastes consist usually of paper, plastic shoe-covers, gloves, mops, plastic bags, and fiberboard containers. As much as 50% plastic may be present in the waste.

Nuclear Fuel Services, Inc., Erwin, TN

The Erwin facility of Nuclear Fuel Services produces large amounts of combustible waste, some of which is contaminated with uranium of low enrichment and some with highly enriched uranium. An incinerator and calciners are operated at the facility. The incinerator burns wastes containing <1 weight percent of highly enriched uranium; the wastes include filters, sponges, gloves, and paper.

Babcock and Wilcox, Lynchburg, VA

For several years, combustible wastes contaminated with small quantities of U-235 have been incinerated at the fabrication facility of Babcock and Wilcox in Lynchburg. The incinerator is a commercially available model with a throughput capacity of 79.45 kg/h (175 lb/h).

Yankee Atomic Electric Co., Rowe, MA

To treat low-level radioactive combustible wastes generated during operation of the Yankee reactor, an incinerator has been operated since the late 1960s at Yankee's Rowe facility. Combustible cellulosic-type wastes, which include paper, cloth, and shoe coverings, are collected in combustible fiber drums at the generating site. (The drums are reused until they become contaminated.) At the incinerator site, the wastes are unloaded from the drums and put manually into the incinerator.

Kerr-McGee Nuclear Corp., Crescent, OK

During processing activities at the Oklahoma plant of Kerr-McGee, combustible wastes containing uranium enriched to 2 to 3% U-235 are generated. Since July 1972, these combustible wastes have been incinerated so that the uranium may be recovered after further processing of the ash. No PVC is incinerated (very little is used in the uranium plant).

Los Alamos Scientific Laboratory (LASL), Los Alamos, NM

A small incinerator has been successfully operated at LASL for a number of years. The primary purpose of this incinerator is the recovery of plutonium from contaminated wastes. At the same time, there is, of course, a large reduction in volume of the wastes. The first incinerator was installed in 1952 and rebuilt in 1959. The main modifications were (1) attachment of the top of the incinerator to the bottom of the glove box for easier visual inspection and (2) inclusion of a vent system for release of excessive pressure. This modified unit has been used on a batch basis almost continuously. About 8 batches are burned each week, but should the need arise the throughput can be increased to 1 burn about once every 2 hours.

Some presorting and treatment is done at the generating site. Noncombustibles are excluded from the bags destined for the incinerator. In addition, all rags that may have been used to wipe up HNO_3 are supposed to be rinsed in water after each use. If this rinse procedure is followed, wastes should not contain any nitrocellulose.

A new plutonium-processing facility is now being built in Los Alamos. Because the old incinerator has been so successful and because it can achieve a throughput adequate for the new facility, an incinerator system of the same design, size, and materials of construction will be used in the new facility. The only modifications will be in the seal-solution tank and glove-box arrangement. The shape of the seal-solution tank will be slightly changed for purposes of criticality control. The glove boxes will have additional lead shielding so that larger quantities of plutonium can be handled without exposing the operator to excessive levels of radiation. In addition, the air-venting system will be arranged differently and it will be easier to effectively introduce an inert gas into the area in case of fire.

Experimental Program on Incineration: A development program has been initiated to study selected production-level volume- and mass-reduction processes for the treatment of transuranic solid wastes. An initial prototype incinerator is now beginning operation. The feed will include wastes similar to the combustible wastes (including PVC) commonly coming from facilities processing radioactive materials.

Mound Laboratory, Miamisburg, OH

For many years a small incinerator has been used at Mound to incinerate plutonium-contaminated combustible materials so that plutonium can be recovered from the ash. The first facility was identical to the one previously described for plutonium recovery at Los Alamos. Several years ago a new building was constructed at Mound, and a new plutonium-recovery incinerator was built. The basic size and design of the old incinerator were retained, but some modifications were made in the mode of operation and in the off-gas cleanup system.

When wastes are generated at the various stations in the process lines, they are packaged in 4.4-liter (1-gallon) cartons. As at Los Alamos, rags that have been used to wipe up HNO_3 are washed with water and are usually dried before bagging. Once the wastes are packaged, they are moved by conveyor to the gamma-scan laboratory to determine whether they should go to recovery or be discarded. Only discardable wastes are removed from the line. Wastes having recoverable amounts of plutonium are then sorted. Accepted for incineration are polyethylene (such as film), rags, paper, and wood. No glove-box gloves are accepted and neither is PVC.

Rockwell International, Rocky Flats, CO

At the Rocky Flats facility of Rockwell, some of the wastes are contaminated with plutonium. An incinerator for the combustible portion of these wastes was installed in 1959. It is still operating, although modifications have been made several times, particularly to the off-gas cleanup train.

Combustible contaminated wastes are sorted at the generating site and placed into 242-liter drums, each lined with a polyethylene bag. When the drum is about three-quarters full, the bag is sealed. The drums are transported to the assay area where the plutonium in each is determined by gamma-neutron-counting techniques. If a drum contains >0.008 g of plutonium per gram of combustible material, it is sent to the incineration facility; if it contains less plutonium, it is sent for burial. The discard limit soon may drop to 0.0004 g of plutonium per gram of combustible material.

At the incinerator site, the drum lids are removed and the polyethylene bag is placed in the incinerator glove box. The bag is opened, and the contents are again sorted to remove noncombustible or potentially explosive materials. The combustible wastes consist of rags, paper, wood, process filters, plastics, etc. Approximately 30 weight percent of the waste currently incinerated is PVC. Some sludges are also fed to the incinerator in order to dry them.

Two new incinerators will be installed in the plutonium-recovery facility under construction at the Rockwell Rocky Flats plant. One incinerator will burn high-level solid wastes (those containing $>3 \times 10^{-3}$ mg of plutonium per gram of waste) and liquid wastes, whereas the other incinerator will burn low-level solid wastes (containing $<3 \times 10^{-3}$ mg of plutonium per gram of waste). The two incinerators will be near each other in a canyon, but separated by a wall. It will be possible to operate the incinerators from outside the canyon through the use of remotely operated valves and mechanical devices. Any major maintenance will require entry into the canyon, and pressure suits will be required for maintenance personnel.

Before reaching the incinerator, all wastes will have been assayed for plutonium content and segregated into high-level and low-level solid waste. The waste from other generating sites at the plant will be packaged in 242-liter drums and transferred to the incinerator. (HEPA filters will be transported in cardboard boxes.) Wastes generated in the recovery building will be transferred to the incinerator area by a retrieval system and a pneumatic conveyor, as well as by external transfer in a drum to the glove-box system.

Atlantic Richfield Hanford Co. (ARHCO), Richland, WA

A Contaminated Waste Recovery Facility, which includes both incineration and leaching equipment, was built at Hanford in 1961 to recover plutonium from waste materials. The incinerator had not been operating for some time, but was expected to start up again.

The wastes are placed in cartons at the generating sites. Cartons containing >2 g of plutonium are taken to the Recovery Facility, where each carton is introduced into the sorting hood through an air lock and sorted manually. Leachables are placed in the leach pots; combustibles are sent to the incinerator area. The scrap accepted for incineration in the past consisted of PVC, rags, and paper (cartons and towels).

TRITIUM SEPARATION AND FIXATION

Having surveyed the potential radiation problem resulting from tritium production in power reactors, it can be projected that in about two decades the rate of production and the increasing world inventory may dictate controlled storage of tritiated waters. Public opinion may force it before that time.

Since isotope separation may be the only route to controlled storage, the known methods for deuterium separation (heavy water production) have been surveyed and then considered for tritium separation. Because of the extremely low concentration of tritium and the large volume of water, the problem is formidable and with present technologies may be prohibitively expensive. A process with a high separation factor is required.

Of those available, electrolysis and recombination would work but the power costs are too great. Chemical exchange using the hydrogen water reaction fits naturally to the problem and the separation factor is good but to retain an adequate separation factor will require developing a catalyst which will be effective at low temperatures. Solvent extraction is seen as a possible technique and a

few relatively simple experiments would determine its feasibility. The most
novel approach is selective excitation of molecular vibrations. This populates
the higher energy levels of the bond involving the isotopic species in question
and allows it to react at a greatly increased rate. With the availability of tuned
lasers to provide the selective energy input, the concept appears quite feasible.

The following sections describe some of the work being done at Battelle Pacific
Northwest Laboratories on isotopic separation as a means of concentrating tritiated aqueous wastes.

Water Extraction Studies

In 1974 a number of experiments were carried out at BNWL to examine isotopic
separation by extraction of water by organic solvents. Occasionally separation
factors differing from unity were observed, but invariably these were shown to
be the result of experimental errors. It may be instructive to summarize some
of these observations.

Water containing heavy hydrogen isotopes possesses more structure (more and
stronger hydrogen bonds) than does light water. Any separation by solvent extraction would be due to distinctive differences in hydrogen bonding with the
solvent molecules. The latter do exist and are demonstrated by differences in
a wide variety of physical properties, such as: IR spectral shifts and intensities,
NMR shifts, solubility, thermodynamics of solutions, e.g., ΔH and ΔG of solution, volume effects, expansibility, temperature of maximum density, EMF measurements, viscosity, dielectric relaxation, and acoustical absorption. The ranges of
these physical properties are appreciable as various classes of solvents, or solvents
containing specific functional groups, are dissolved.

Studies were carried out with both tritiated water and deuterated water using
ethers, ketones, organic phosphates and phosphonates, and amines. Some of
these, such as the phosphonates and amines form very strong hydrogen bonds
and might be expected to show the maximum effect. A few experiments
were carried out on the extraction of HNO_3 and HNO_2 both of which form very
strong complexes with solvents like tributyl phosphate and ion pair complexes
with amines.

With some of the solvents, dilutions with nonpolar diluents were also tried. Several auxiliary investigations were carried out on the equilibrated organic phase,
such as precipitation of a portion of the water by dilution with CCl_4 and distillation of water from the organic.

There was some indication that dibutyl butyl phosphonate gave a small prefer-
ential extraction of HTO, but the analytical measurements were not precise
enough to obtain a firm value. No solvent showed a separation factor different
from unity by more than 1 or 2%. It is concluded that water isotopic enrichment by solvent extraction is very small. Likewise, addition of solutes to change
the degree of order or structure of the liquid water has a very small effect on
the vapor pressure isotopic ratio.

The small solvent extraction effect is not surprising since the differences in bond
strengths between water molecules and between solvent and water are small.
Thus, although the isotopic effects on the physical properties of one phase are
real and significant, compensation occurs between two phases in equilibrium.

Laser Excitation Experiments

In considering the possible schemes for tritium separation involving laser excitation, initial calculations were directed at multivibrational excitation within the ground electronic state followed by reaction with an added molecular species. There seems little doubt that this can be a highly efficient use of laser energy, and calculations show that combinations exist which permit pumping a molecule to a high vibrational level with a single frequency. Detailed molecular spectra for excited states are not available at the precision necessary to pick out the exact frequencies required and therefore a basic research program is required to obtain them.

Because the detailed molecular spectra for excited states required in the above described scheme are unavailable, initial experiments involved another less demanding spectroscopic scheme to excite HDO to the vibrational 110 level, followed by excitation to the first excited unstable electronic state. Free radicals H + DO + D + HO are expected to be generated and these radicals will be gettered by a free radical scavenger. The source for the vibrational excitation is a Chromatix laser. The schematic of the experiment is shown in Figure 3.22.

FIGURE 3.22: EXPERIMENTAL ARRANGEMENT

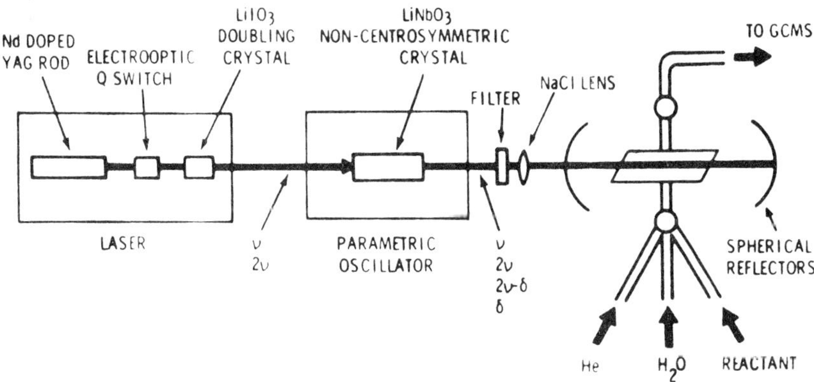

Source: BNWL-1876

The flash lamp around the reaction cell is not shown. Although the tunable laser, which is tuned by changing the temperature of the parametric oscillator, is very low power, calculations indicate that sufficient pumping will be possible to induce reactions sufficient to detect separation. If the experiments involving HDO/H_2O separation were successful or encouraging, separation of HTO would be attempted.

Early experiments indicated that no HD, D_2, or H_2 was formed when HDO vapor (from a 50% D_2O/50% H_2O solution) was bombarded with infrared radiation from a tunable Nd:YAG laser and simultaneous ultraviolet radiation from two xenon flash lamps. It was felt that the UV radiation from the flash lamps was not intense enough to give measurable results at the high frequency required to pump HDO molecules from the 110 excited vibrational level to the dissociation band. The wavelength of the radiation from the laser was changed to excite ground state HDO molecules to the higher energy 011 transition. The lower frequency radiation required from the flash lamps to pump HDO molecules from this excited state to the dissociation band would be in a region of higher flash lamp intensity.

However, experiments using the 011 transition still did not show any HD, D_2, or H_2 formation. An analysis of the initial experimental conditions revealed sizable quantities of oxygen from air in the sample cell. The computer kinetics model developed earlier was used with a new initial oxygen concentration. The result indicated that high concentrations of oxygen in the initial reception mixture would suppress formation of HD, D_2, and H_2.

To avoid this problem, the sample cell was loaded with HDO vapor in a vacuum system. New experiments using this loading technique and only the flash lamps were performed. These did show HD formation and even more H_2 formation. This indicated that the flash lamps do have enough intensity to dissociate H_2O, D_2O, and HDO excited molecules.

Only a few of the lower vibrational excited states populated to any significant extent at room temperature. The relative populations of these states and the fact that HD was formed in detectable quantities imply that the flash lamps currently being used can pump HDO molecules from the 110 vibrational state to the dissociation band. Thus, because of the larger absorbence of HDO at this wavelength ($\sim$2.37 μm) and because the 110 excited state has much of the excitation energy in the O-D stretch, the laser has been set to excite HDO molecules to this 110 vibrational state.

However, the experiments using the vacuum loading technique have revealed a problem. Because some of the lower vibrational states of H_2O, D_2O, and HDO are appreciably populated at 360°K, the enrichment factor could suffer. This "hot band" problem is being considered, and experiments at room temperature were planned to lessen the effect. The kinetics model has been revised for room temperature HDO dissociation and to account for diffusion in the sample cell.

Tritium Fixation

Many different fixation schemes have been proposed. They include hydrate formation, sorption by clays, molecular sieves, and other solids, cements, storage of liquid H_2O in cylinders, organic polymers, metal hydrides and storage of hydrogen gas. Each has its own set of problems which may range from the requirement of high integrity containment for water to high exchange rates with concrete, to limited high temperature stability for organic compounds and metal hydrides.

Among the numerous approaches proposed to limit the spread of tritium into the environs has been the incorporation of tritium into organic polymers (64). Implicit in the formulation of such a polymer is the requirement that the tritium atom be situated in such a manner that exchange with hydrogen (from atmospheric moisture, etc.) be minimized. Thus all polymeric materials in which tritium would be in a labile condition, i.e., alcoholic, carboxylic, ammoniacal groups, etc., must be ruled out. Since hydrogen exchange with carbon-hydrogen skeletal functions, on the other hand, would be minimal, polymers have been sought that would meet this condition. The ideal solution to this problem would be to react tritiated water or tritium gas with a monomeric material which can be readily polymerized. One obvious approach is the reaction of water with calcium carbide to form ditritiated acetylene:

$$CaC_2 + HTO \longrightarrow Ca(OH, T)_2 + TC{\equiv}CT$$

Polymerization reactions of acetylene can then be initiated to fix tritium in a relatively permanent form. Brookhaven National Laboratories have explored this possibility using gamma radiation as the polymerizing technique. Crosslinkages would give a three dimensional saturated product.

The reaction in which acetylene is hydrated using a sulfuric acid solution of mercuric sulfate as the reaction medium has been considered. The resultant product is acetaldehyde, which in the extreme situation could contain as many as four tritium atoms

$$TC{\equiv}CT + T_2O \xrightarrow{HgSO_4} T_3C{-}CT{=}O$$

or normal acetylene could be reacted with tritiated water to give:

$$HC{\equiv}CH + HTO \longrightarrow CH_2T{-}CHO$$

When acetaldehyde is reacted with phenol, a product is formed that is analogous to Bakelite (phenol-formaldehyde polymer).

Initially attempts were made to prepare the polymer using the same procedure as though the reactants were formaldehyde and phenol rather than acetaldehyde and phenol. Although eventually a tough, water-insoluble polymer was formed, considerable time was required for the curing, and significant losses of materials resulted. A modified procedure was adopted which results in a polymer with the desired properties within three or four hours.

It was found that substitution of resorcinol for phenol accelerates the reaction rate significantly. Even so, unless the molar concentration of acetaldehyde relative to the resorcinol is reduced from about 2.2 to 1, the uptake of acetaldehyde becomes slower as the reaction proceeds, until finally a fixed quantity of acetaldehyde continues to reflux at the cold finger. When the reduction of acetaldehyde indicated above was made, the reaction proceeded more completely so that after a reaction time of about an hour only negligible refluxing was observed at the cold finger.

A mol of formaldehyde added to the reaction mixture at this point causes an instantaneous solidification of the polymer. The polymer is cured by heating it to 80°C under a vacuum of about 10 to 15 torrs. The off-gases can be collected in a dry ice-acetone trap and examined by gas chromatography. Acetaldehyde recovered from the cold trap amounts to about 3 to 4%. The product is hard, brittle, and water insoluble.

Tritiated acrylonitrile has also been prepared at BNWL using a Nieuwland catalyst mixture resembling an industrial procedure and it was polymerized under conditions of emulsion polymerization for leach testing in this study. One of

the simplest procedures for incorporating tritium in a polymer is the electrolysis
of tritiated water and the catalytic hydrogenation of polystyrene to form poly-
vinylcyclohexane. The simplicity of the one-step process is particularly attractive
because of the relative ease of containment of the tritium and the absence of
by-products and waste streams. These advantages probably outweigh the apparent
disadvantage of the high-pressure hydrogenation process. The hydrogenation of
polystyrene in the presence of a nickel-aluminum alloy catalyst at 160°C under
60 atm hydrogen pressure has been carried out at BNWL.

Leach Tests: Bakelite was tested by soaking the crude polymer exhibiting
2,874 d g^{-1} min^{-1} in water. An immediate loss of tritium to the wash solution due
to surface contamination of the polymer occurred, but no further increase occurred
on soaking for a longer time. When a sample of the Bakelite was first rinsed,
extended soaking in a rinse solution resulted in no detectable tritium loss to the
rinse solution after six weeks of rinse. Combustion of the polymer and analysis
of water confirmed the tritium level of the polymer.

Similarly, polyacrylonitrile gave no initial loss of tritium due to surface contamination. However, following a rinse no loss of activity to a rinse solution over
an extended period of time (six weeks) was detectable. Hydrogenated polystyrene did not exhibit an initial release and did not undergo detectable leaching
of tritium. The tritium content of polyacrylonitrile and hydrogenated polystyrene was also determined by analysis of water from combustion of the polymers.
Results of the leach tests clearly indicate that all three polymer types adequately
fix tritium.

REFERENCES

(1) C.R. Cooley (comp.), *Commercial Alpha Waste Program Quarterly Progress Report,*
      January to March, 1975, HEDL-TME 75-87, Hanford Engineering Development
      Laboratory, October 1975.

(2) *Transuranic Waste Management Programs,* July 1974 to December 1974, LA-61000-PR,
      Los Alamos Scientific Laboratory, October 1975.

(3) D.L. Seigler, "Incineration Process Fire and Explosion Protection," *Proceedings of the
      Thirteenth AEC Air Cleaning Conference, San Francisco, 12-15, Aug. 1974,* Vol. 2,
      pp. 911-923, March 1975.

(4) R. Mulkin, *Characterization of Transuranic Solid Wastes from a Plutonium Processing
      Facility,* LA-5993, Los Alamos Scientific Laboratory, June 1975.

(5) J.B. Owen, "Management of Plutonium Contaminated Wastes at Rocky Flats," in
      *Management of Plutonium-Contaminated Solid Wastes, Proceedings of the NFA
      Seminar at Marcoule, 1974,* The OECD Nuclear Energy Agency, Paris, 1974, pp. 107-110.

(6) J. Pradel, P. Parsons and E. Malasek, *The Volume Reduction of Low-Activity Solid
      Wastes,* STI/DOC/10/106, International Atomic Energy Agency, Vienna, 1970.

(7) R.H. Burns, G.W. Clare, A.J. Smith and D.A. Dunkason, "The Treatment of Low-Level
      Solid Wastes at the Atomic Energy Research Establishment, Harwell," in *Practices in
      the Treatment of Low- and Intermediate-Level Radioactive Wastes, Proceedings of a
      Symposium at Vienna, December 6-10, 1965,* STI/PUB/116, International Atomic
      Energy Agency, Vienna, 1966, pp. 639-661.

(8) W. Wasling and F. Griffin, "The Treatment of Low-Level Solid Wastes by Compression
      and Baling," in *Practices in the Treatment of Low- and Intermediate-Level Radioactive
      Wastes, Proceedings of a Symposium at Vienna, December 6-10, 1965,* STI/PUB/116,
      International Atomic Energy Agency, Vienna, 1966.

(9) N.W. Snyder, *Chem. Eng.,* 81, No. 22, 65, 1974.

(10) D.M. Anderson et al., *Compaction of Radioactive Solid Waste,* WASH-1167, June 1970.

(11) W.R. Herald and D.F. Luthy, *Compaction of Alpha Contaminated Wastes,* MLM-2282, Mound Laboratory, January 1976.

(12) L. Silverman, "Disposal of Low-Level Combustible Solid Wastes by Incineration," in *Practices in the Treatment of Low- and Intermediate-Level Radioactive Wastes, Proceedings of a Symposium at Vienna, December 6-10, 1965,* STI/PUB/116, International Atomic Energy Agency, Vienna, 1966, pp. 89-115.

(13) W.L. Lennemann, et al., *Incineration of Radioactive Solid Wastes,* WASH-1163, August 1970.

(14) D.E. McKenzie et al., *Disposal of Transuranic Solid Waste Using Atomics International's Molten Salt Combustion Process, II,* AI-ERDA-13151, Atomics International Division, Rockwell International, May 1975.

(15) L.F. Grantham et al., *Disposal of Transuranic Solid Waste Using Atomics International's Molten Salt Combustion Process, II,* AI-ERDA-13169, Atomics International Division, Rockwell International, March 1976.

(16) L.F. Grantham et al., U.S. Patent 3,845,190, 1974.

(17) L.F. Grantham and S.J. Yosim, U.S. Patent 3,899,322, 1975.

(18) J.A. Ayres, *Decontamination of Nuclear Reactors and Equipment,* The Ronald Press, New York, 1970.

(19) M.J. Szulinski, *The Hanford Decontamination Facility,* ARH-SA-181, Atlantic Richfield Hanford Company, June 1974.

(20) W.E. Sande et al., *Decontamination and Decommissioning of Nuclear Facilities—A Literature Search,* BNWL-1917, Battelle Pacific Northwest Laboratories, May 1975.

(21) R.B. Becker, et al., "The Remotely Operated Torch—A Tool for Nuclear Reactor Dismantling," *Conference Proceedings 731105-2, ASME Winter Meeting,* November 11-15, 1973.

(22) G.R. Bainbridge, et al., "Decommissioning of Nuclear Facilities: A Review of Status," *Atomic Energy Review,* Vol. 12, No. 1, pp. 145-160, March 1974.

(23) H.T. Fullam, *High Temperature Methods for Disposal of Contaminated Metal Equipment,* BNWL-B-277, Battelle Pacific Northwest Laboratories, July 1973.

(24) H.W. Godbee, *Use of Evaporation for the Treatment of Liquids in the Nuclear Industry,* ORNL-4790, Oak Ridge National Laboratory, September 1973.

(25) H.W. Godbee and A.H. Kibbey, *Nuclear Safety,* Vol. 16, No. 4, pp. 458-469, 1975.

(26) J.C. Petrie, et al., *Chemical Engineering Progress,* Vol. 72, No. 4, pp. 65-71, April 1976.

(27) C.B. Goodlett, *Concentration of Aqueous Radioactive Waste with Wiped Film Evaporators,* DP-MX-75-17, E.I. du Pont de Nemours & Company, paper presented at AIChE 68th Annual Meeting, Los Angeles, November 16-20, 1975.

(28) C.B. Goodlett, *Chemical Engineering Progress,* Vol. 72, No. 4, pp. 63-64, April 1976.

(29) B.G. Kniazewycz, "The Treatment of Liquid Radwaste by Reverse Osmosis—Conception to Operation," Carolina Power & Light Company, Raleigh, NC, paper presented at ASME-IEEE Joint Power Generation Conference, Portland, OR, September 28-October 1, 1975.

(30) *Chemical Treatment of Radioactive Wastes,* STI/DOC/110/89, International Atomic Energy Agency, Vienna, 1968.

(31) E.H. Gischel, "Radioactive Fluid Filtration Systems," United Engineers & Contractors, Inc., Philadelphia, PA, paper presented at ASME Meeting, New York, NY, Nov. 17-22, 1974.

(32) K.H. Lin, *Use of Ion Exchange for the Treatment of Liquids in Nuclear Power Plants,* ORNL-4792, Oak Ridge National Laboratory, December 1973.

(33) *Operation and Control of Ion Exchange Processes for Treatment of Radioactive Wastes,* Technical Report Series No. 78, International Atomic Energy Agency, Vienna, 1967.

(34) *Chemical and Engineering News,* Vol. 54, pp. 32-33, January 12, 1976.

(35) J.M. Cleveland, *The Chemistry of Plutonium,* pp. 563-684, Gordon and Breach Science Publishers, New York, 1970.

(36) E.L. Christensen and W.J. Maraman, *Plutonium Processing at the Los Alamos Scientific Laboratory,* LA-3542, April 1969.

(37) L.E. Bruns, *Chem. Eng. Progr. Symp. Ser. No. 80,* 63, 156, 1967.

(38) *Proceedings International Solvent Extraction Conference, ISEC 71,* Vol. I, p. 186, Society of Chemical Industry, London, 1971.

(39) W.W. Schulz, *The Chemistry of Americium,* TID-26971, 1976.

(40) L.C. Oyen, *Solid Radwaste Systems,* presented at Radioactive Waste Management for Nuclear Power Reactors, October 20-23, 1975; Short course cosponsored by UCLA, Georgia Institute of Technology, and ASME at Los Angeles, CA, 1975.

(41) J.C. Petrie, R.I. Donovan, R.E. Van der Cook, and W.R. Christensen, *Chem. Eng. Prog.,* 72(4), 65, April 1976.

(42) L.F. Grantham, D.E. McKenzie, R.D. Oldenkamp, and W.L. Richards, *Disposal of Trans-uranic Solid Waste Using Atomics International's Molten Salt Combustion Process, II,* AE-ERDA-13169, March 15, 1976.

(43) R.D. Fox (ed.), *Atlantic Richfield Hanford Company Semiannual Report BB Process Development May 1, 1974 through October 31, 1974,* ARH-ST-118 B, January 1975.

(44) C.B. Goodlett, *Chem. Eng. Prog.,* 7(4), 63, April 1976.

(45) J.H. Leonard and K.A. Gablin, "Leachability Evaluation of Radwaste Solidified with Various Agents," Paper 74-WA/NE-8, presented at ASME Winter Annual Meeting, New York, Nov. 17-22, 1974.

(46) H.W. Heacock and J.W. Riches, "Waste Solidification—Cement or Urea-Formaldehyde," Paper 74-WA/NE-9, presented at ASME Winter Annual Meeting, New York, Nov. 17-22, 1974.

(47) R.H. Burns, *At. Energy Rev.,* 9, 547, 1971.

(48) H.O. Weeren, *An Evaluation of Waste Disposal by Shale Fracturing,* ORNL-TM-5209, Feb. 1976.

(49) G. Meier and W. Bähr, *The Incorporation of Radioactive Wastes into Bitumen, Part 1,* KFK-2104, April 1975.

(50) C.H. Delegard and G.S. Barney, *Fixation of Radioactive Waste by Reaction with Clays: Progress Report,* ARH-ST-124, July 1975.

(51) W.W. Schulz, A.L. Dressen, C.W. Hobbick, H. Babad, *Glass Forms for Immobilization of Hanford Wastes,* ARH-SA-210, April 1975.

(52) M.J. Kupfer, W.W. Schulz, C.W. Hobbick, and J.E. Mendel, *Glass Forms for Alpha Waste Management,* ARH-SA-239, July 1975.

(53) J.A. Kelley, *Evaluation of Glass as a Matrix for Solidification of Savannah River Plant Waste. Radioactive Studies,* DP-1397, Oct. 1975.

(54) J.L. McElroy (ed.), *Quarterly Progress Report Research and Development Activities Waste Fixation Program October through December 1974,* BNWL-1893, February 1975.

(55) J.L. McElroy (ed.), *Quarterly Progress Report Research and Development Activities Waste Fixation Program July through September 1975,* BNWL-1949, January 1976.

(56) R.W. Lynch, R.G. Dosch, B.T. Kenna, J.K. Johnstone, and E.J. Nowak, *The Sandia Solidification Process—A Broad Range Aqueous Waste Solidification Method,* IAEA-SM-207/75, International Symposium on the Management of Radioactive Wastes from the Nuclear Fuel Cycle, Mar. 22-26, 1976, Vienna, Austria.

(57) R. Roy, "Ceramic Science of Nuclear Waste Fixation," presented at the 77th Annual Meeting of the American Ceramic Society, Washington, DC, May 3-8, 1975.

(58) F.P. Cowan and L. Gemmell, "Waste Management Operations at Brookhaven National Laboratory," hearings on Industrial Radioactive Waste Disposal Problems held before the Special Sub-committee on Radiation of the Joint Committee on Atomic Energy, Jan. 28-Feb. 3, 1959.

(59) B. Manowitz, et al., *Development of Durable Long-Term Radioactive Waste Composite Materials,* Progress Reports 7-10, April 1974-March 1975.

(60) P. Pottier, R. Andriot, and D. Cuaz, "Progress in Techniques of Treating and Conditioning Effluents in Research Centers," 4th International Conference on the Peaceful Use of Atomic Energy, Geneva, September 1971.

(61) Y. Sousselier et al., "Conditioning of Wastes from Power Reactors," ANS Winter Meeting, San Francisco, November 1973.

(62) D. Cuaz, A. Limongi, D. Thiery, A. Baer, "Process for Embedding Radioactive Wastes in Thermosetting Resins," *SFRP,* 769, 1974.

(63) "Leach Testing of Immobilized Radioactive Waste Solids—A Proposal for a Standard Method," ed. Hespe, *Atomic Energy Review* 9, no. 1, Vienna, 1971.

(64) L.L. Burger and J.L. Ryan, *The Technology of Tritium Fixation and Storage,* BNWL-1807, 1974.

STORAGE OF RADIOACTIVE WASTES

The following reports were the sources of the material in this chapter: DP-MS-76-39, SAND 76-5229, SAND 75-6093, UCRL-51713, SAND 76-0224, LA-UR-76-1722, TID-27341, NVO-410-38 and PB-260 559. For a complete bibliography, see p 359.

INTERIM STORAGE OF SPENT FUEL ASSEMBLIES

In the uranium fuel cycle initially conceived for light water reactors (LWR), spent fuel was to be discharged from the reactors and then allowed to cool (at the reactor sites) for 5 months. This cooling period would allow short-lived radioactive isotopes to decay and thereby would reduce the heat generation.

After the cooling period, the fuel was to be shipped to a chemical reprocessing plant, which would recover uranium and plutonium. Accordingly, existing storage facilities at LWR sites were designed for only short-term storage of spent fuel. More recently, however, spent fuel has been considered as a possible waste form suitable for interim storage and even ultimate disposal.

Alternatives that have been considered for interim storage of spent fuel include:

> Storage of unpackaged fuel in water-cooled basins or air-cooled vaults when reprocessing is expected within 10 to 20 years.

> Storage of packaged spent fuel in water-cooled basins, air-cooled vaults, concrete surface silos, geologic formations, or near-surface heat sinks when storage for periods longer than 10 to 20 years is expected.

Long-term storage of spent fuel elements would require storage capacity for nearly 500,000 fuel assemblies by the year 2000 (Table 4.1) with up to 50,000 additional assemblies added annually.

TABLE 4.1: PROJECTED FUEL ASSEMBLY DISCHARGES[a]

Year	No. of Fuel Assemblies Annually	Cumulative No. of Fuel Assemblies[b]
1980	5,000	19,000
1985	11,000	57,000
1990	20,000	138,000
1995	33,000	275,000
2000	45,000	476,000

[a]Assumes the reactor schedule of Reference (1)
[b]Includes pre-1980 assemblies

Source: DP-MS-76-39

If the facilities stored 4,800 reference (average) assemblies (1), they would contain 1,500 MTU and would be large relative to existing storage facilities. Two units of this size could be required by 1980; and 100 by the year 2000.

Description of Storage Facilities

Water-Cooled Storage: Most spent fuel today is stored at reactor sites designed to hold from 50 to 150 MT of irradiated fuel (2). The recently designed storage facility at the Barnwell reprocessing plant has a capacity for 400 MT of spent fuel (3). The storage pool at the General Electric Morris Plant is expected to be expanded to 700 MT.

Water basin storage of packaged LWR fuel assemblies provides cooling and radiation shielding and allows visual inspection and direct manipulation of the fuel. The storage basin can also serve as a large heat sink under some accident conditions.

A concept presented in Regulatory Guide 3.24 for a separate facility (i.e., not located at a reactor or fuel reprocessing site) for storage of spent fuel suggests the use of modular basins with each basin having a capacity for about 500 MT of spent fuel (4). The facility is designed to receive, handle, decontaminate and reship spent fuel casks; to remove irradiated fuel from casks; to place the fuel in a storage basin; and to cool and control the quality of the water. Modular construction allows facility expansion with a minimum of additional support facilities and services.

The spent fuel storage basins would be stainless steel lined concrete structures, and if fuel assemblies are to be packaged would presumably be located adjacent to the fuel packaging facility. Fuel assemblies would be transferred to the basin either in water-filled casks or via a transfer canal and would be placed in storage racks within basins similar to those in which fuel assemblies are stored at reactor sites.

The water also is a confinement barrier should radioactivity escape the fuel. Basin water is circulated in closed loop systems through a filter-demineralizer system to maintain water purity and collect radioactive contaminants. Periodically, the system is backwashed and rejuvenated with water and chemicals that are discharged to a liquid waste. A secondary cooling loop to transfer the heat from the basin water to the environment further minimizes the transfer of radioactive materials to the surroundings.

The basin ventilation system maintains the concentrations of radioactive gaseous contaminants at a low level (4)(5). Basins at fuel reprocessing sites can share gas processing systems with the reprocessing plant. Treatment of off-gases to remove tritium and noble gases would be more feasible with water basins than with most other concepts because of the low volume of air vented.

The design for confinement and monitoring of radioactivity for water basin storage of packaged spent fuel would be the same as for storage of unpackaged spent fuel. The fuel container would provide an additional degree of confinement so that continuous filtration of the ventilating air may not be required. If activity escapes, as indicated by air monitoring instrumentation, failed packages would be removed, repaired and overpacked.

Air-Cooled Vaults: Storage of spent fuel in near-surface facilities with either natural- or forced-draft cooling is an alternative for LWR fuel that has been out of the reactor at least 3 to 4 years. In the air-cooled vault storage concept (6) (7), the vaults are reinforced concrete structures, partially buried in the earth, which provide biological shielding, protection from natural phenomena, and secondary confinement of radioactivity.

Concepts for air-cooled storage are shown in Figure 4.1 (8) for fuel from the CANDU (Canada Deuterium Uranium) reactor, and in Figure 4.2 (9) for solidified high-level waste (HLW). These facilities are designed for natural-draft cooling; however, they are similar to forced-draft-cooled vaults in their architectural design, storage arrangement, and flow path of cooling air.

The spent fuel is stored vertically inside sleeves supported by top and bottom tube sheets that form plenums to distribute the inlet air and collect the effluent air. The use of sleeves ensures adequate distribution of cooling air. Fuel is charged into the storage locations by removing the concrete plug from the roof over each sleeve. The plug is replaced after charging.

The decay heat from residual fission products would cause natural convection air flow sufficient for cooling the fuel in a properly designed facility. Such cooling is passive and would require little maintenance and only minimal surveillance.

Because coolant air moves through the vault by natural rather than forced draft, the buoyant force may be too low to pass sufficient quantities of air through high-efficiency filters. In the absence of filters for exhaust air, effective containment of fuel assembly core material depends on the integrity of cladding. Packaging of the fuel assemblies would likely be required to provide adequate confinement. Some release of fission products to the environment could occur if both the fuel assembly cladding and the container fail.

If a high-efficiency filtration system is used for the effluent air to provide a partial barrier to activity spread from failed fuel, the large resistance to air flow by this filter system would require forced-air cooling. Backup fans (excess capacity) and both normal and emergency power may also be required.

FIGURE 4.1: CONCEPT FOR AIR-COOLED VAULT FOR SPENT FUEL

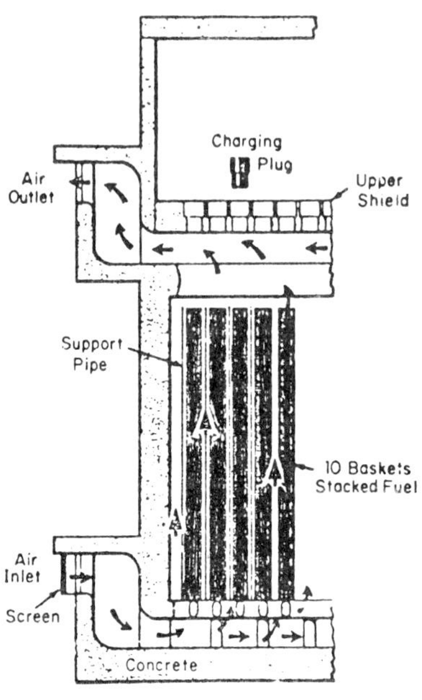

FIGURE 4.2: SURFACE STORAGE FACILITY AIR-COOLED VAULT

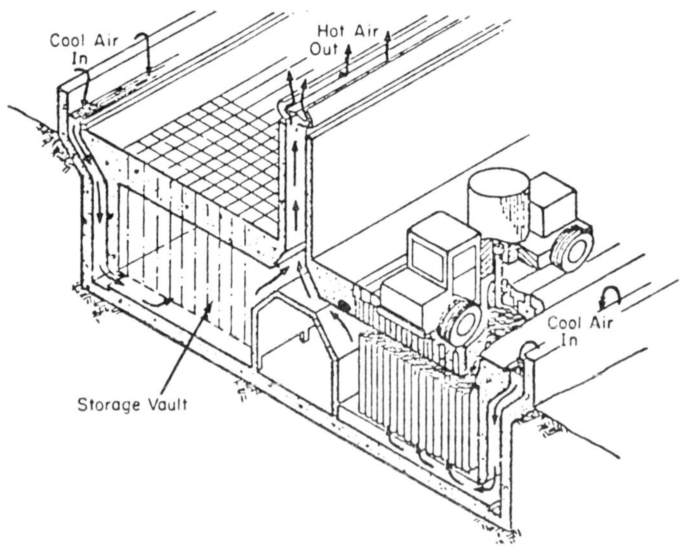

Source: DP-MS-76-39

If the cladding were to fail and subsequently release fission products to the cooling air, prefilters, HEPA (high efficiency particulate air) filters, and carbon beds would remove about 99.9% of any particulates and about 99% of any iodine before the air is released to the atmosphere (10). Noble gases and tritium might be released (11) because treatment of such a large volume of air might be prohibitive. However, atmospheric dispersion of these radionuclides is expected to reduce the offsite concentrations to levels below Federal guidelines. The facility would have radiation monitoring devices to detect and locate radioactivity releases. Leaking assemblies would be located, removed, and packaged to confine the fission products.

Concrete Surface Silo: The concrete surface silo is a column of concrete on the earth's surface with a cavity for a package of spent fuel. Shielding is provided by the thick concrete walls (~1 meter thick). Passive and stable cooling is by conduction through the concrete to the air or by natural draft of air through an annulus between the fuel package and the concrete wall.

Silo designs using conductive cooling have been made for storing heavy-water power reactor (CANDU) fuel (12), but no such designs have been developed for LWR fuel. In the CANDU concept (13), the packaged spent fuel is placed directly in contact with a steel- and lead-lined concrete silo and stored vertically on outdoor concrete pads. Heat is dissipated by conduction through the lead and concrete. This concept has been proposed for relatively low burnup (<10,000 MWD/MTHM) CANDU-type fuel and might not be as practical for storing the higher burnup (av 25,000 MWD/MTHM, max ~33,000 MWD/MTHM) LWR-type fuels.

In the second concept (a modification of that for storing solidified HLW, Figure 4.3 (7), the spent fuel assemblies are packaged in low-carbon steel containers placed in concrete silos and stored vertically on outdoor concrete pads. An annulus between each package and its concrete silo serves as a chimney through which cooling air passes by natural convection.

Radioactivity contained in spent fuel is protected from release by multiple barriers. In the conduction-cooled concept, the fuel cladding, the package, and the sealed steel liner are all barriers. In the convection-cooled concept, the cladding and the package provide confinement. These storage units are designed to withstand credible accident conditions and forces from natural phenomena (4). Loss of confinement would be detected by area survey instruments and by observations of the storage array. Leaking packages would be removed, repaired, overpacked, and returned to a silo.

Near-Surface Storage: Placing spent fuel packages into lined holes in the earth's surface provides storage with passive cooling and stable shielding. In this concept (14), packages containing one or more fuel elements are placed in lined holes and covered with a plug that provides radiation shielding and access. The concept is illustrated by the method used for storage of some CANDU fuels at Chalk River, Canada (Figure 4.4) (14).

The spacing of the holes is determined by criticality requirements and by the heat load that can be dissipated throughout the surrounding medium. Materials such as concrete or tile that are resistant to corrosion attack from the surrounding earth are used for the liner.

FIGURE 4.3: CONCRETE SURFACE SILO, CONVECTION-COOLED

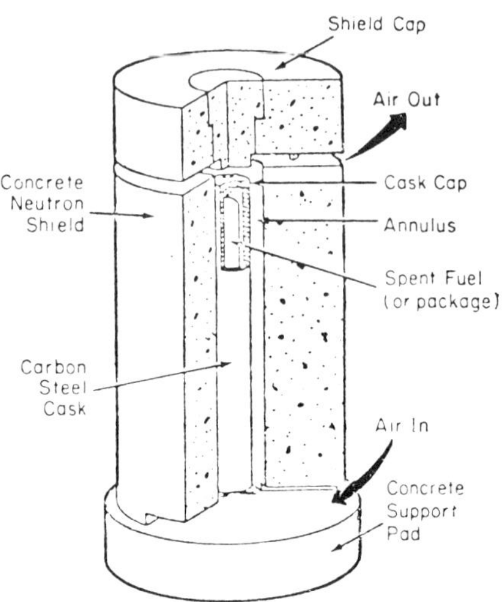

FIGURE 4.4: TYPICAL CONCRETE TILE HOLE FOR SURFACE STORAGE
OF CANDU FUEL

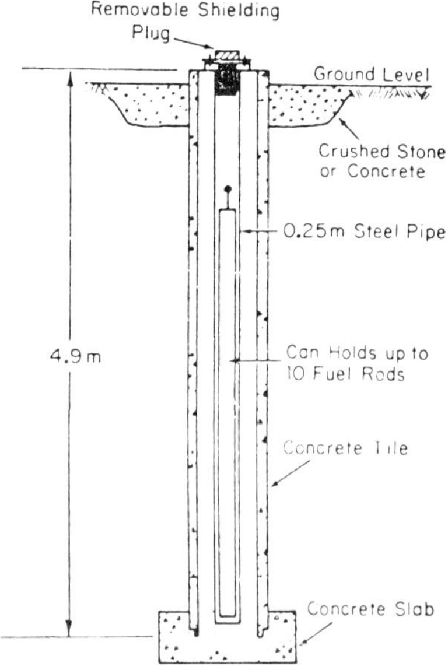

Source: DP-MS-76-39

The water table at the storage site should preferably be sufficiently lower than the bottom of the hole so that during any credible rise in the water table ground water cannot reach the liner. Ground water is easily excluded in arid regions with unsaturated zones 100 to 600 m thick (15). The hole liner and fuel package should keep surface water away from the fuel.

The three confinement barriers for this method of storage are: fuel cladding, fuel package, and the hole liner and shielding plug. Radioactivity monitors throughout the storage site could locate the leak site if a release occurs. Transport of radioactive material to the atmosphere, even if failure occurs, would likely be limited.

Packaging of Spent Fuel

Packaging of spent fuel consists of sealing the fuel assembly inside a container after filling the container with a solid or gaseous heat transfer medium. Spent fuel will probably not be packaged for routine storage until at least 3 to 4 years after the assemblies are discharged to allow decreased heat generation and radioactivity. Routine packaging is expected to occur at central storage facilities rather than at reactor sites because fuel reprocessing is most economical when central processing stations are used (16). Also, shipping cask volumes, and therefore shipping cask weights, can be minimized by shipping unpackaged rather than packaged assemblies.

The simplest container would contain single, intact BWR or PWR spent fuel assemblies, but multiple assembly containers would be more economical. Physical size and/or heat load may restrict the number of assemblies that can be loaded into a single container. A square container with an inside dimension of about 44 cm could hold four intact PWR assemblies or nine intact BWR assemblies. This package would weigh about 3 MT.

An alternative to packaging of the intact spent fuel assembly is dismantling the fuel assembly and storing the fuel elements separate from the nonnuclear materials. The volume of the fuel elements is about 40% of that occupied by the intact assembly.

The container material choice depends on the repository in which the package is to be stored and cooled. Highly corrosion resistant material, such as stainless steel, is required for water storage. Low-carbon steel where moisture can be excluded and aluminum for air-cooled concepts are economic alternatives. Irradiation effects on materials is not a consideration.

When fuel assemblies are enclosed in a container, the heat generated must be transferred from the fuel to the container walls and ultimately to the environment. The additional step of transferring heat from the fuel to the container will cause the fuel assembly to be hotter than during unpackaged storage. This internal temperature rise can be reduced by using a material with high thermal conductivity to fill the space between the fuel rods.

Helium, which is most frequently considered as the filler material for packaged fuel assemblies (8)(12), has a thermal conductivity of 0.17 W/(m-°C) at 100°C. The thermal conductivity of air at the same temperature is 0.035 W/(m-°C). Zinc (melted, poured and solidified), which has been studied as a filler for some

CANDU fuels (8), has a thermal conductivity of about 110 W/(m-°C). Powdered aluminum in a helium atmosphere is a technically simpler filler to use than molten and solidified zinc. The effective thermal conductivity of aluminum powder with 60% packing density and filled with helium equals or exceeds the conductivity of solid zinc.

Some fraction of the spent fuel assemblies presented for packaging will contain fuel elements with breached cladding. A second container or overpack placed over the initial container would restore core confinement to approximately the degree that was provided by the container and the unbreached cladding.

Fuel Storage Costs

Recently, several studies have considered the effects of LWR fuel cycles in which the spent fuel assemblies are stored without reprocessing (the throwaway fuel cycle). In general, studies comparing prompt reprocessing to no reprocessing have agreed that long-term spent fuel storage requires the maximum use of natural resources, separative work, and (perhaps) materials or services while decreasing radiation exposure and potential safeguards threats. Additionally, it is agreed that interim storage costs are only a small fraction of the total fuel cycle costs.

Less agreement has occurred (Table 4.2) over the question of whether or not the value of both the uranium and plutonium recovered by reprocessing is sufficient to offset the costs associated with reprocessing the fuel to recover these materials. The more recent and more detailed studies show cost advantages for recycle of both plutonium and uranium that range from economically advantageous to clearly superior. In all studies, uranium recycle alone is not cost justified, or at best, only marginally justified.

TABLE 4.2: COMPARISON OF RECYCLE COST STUDIES

Source	Date	Conclusion on Reprocessing Cost[*]
GESMO (17)	July, 1974	Minimizes overall costs
Rodger (18)	March, 1975	Exceeds indifference price[**]
ERDA-33 (19)	March, 1975	None[**]
Resnikoff (20)	August, 1975	Not cost justified
Union Carbide (21)	September, 1975	Economically advantageous[**]
SRL (22)	September, 1975	Saves $36 billion by year 2000
AGNS (23)	March, 1976	Clearly superior

[*]In each case, the conclusions are conditional.
[**]These studies present breakeven reprocessing costs as a function of U_3O_8 and separative work costs.

Source: DP-MS-76-39

The cost studies used a variety of fuel cycle models and allocated costs differently between various possible cost centers (shipping, safeguards, fabrication, etc.). Yet general agreement exists on the spent fuel reprocessing costs that exactly offset the value of the uranium and plutonium recovered (breakeven costs) if equal uranium fuel and spent fuel storage costs are assumed.

The situation is different for CANDU reactors, which use natural rather than enriched fuel and, therefore, have only plutonium values for recovery. Recovery of plutonium from CANDU fuel is not currently considered to be cost justified. However, because spent fuel management is not a major cost factor in nuclear power, and because recovery could be economic in the future as the cost of uranium increases, Canada regards its spent fuel as a resource to be stored in retrievable surface storage facilities.

Detailed capital and operating costs for the several spent fuel storage alternatives are not available. Studies of storage of spent CANDU fuel (8)(24) scoped the relative capital and storage costs. Also, a study by ARHCO and Kaiser Engineers (6) for storage of solidified HLW considered the capital and operating costs (until the year 2110) for several concepts of surface storage similar to those considered for spent fuel. These relative costs (Table 4.3) indicated that several concepts may be less expensive than water basin storage at a central site and that the surface silo concept will probably cost more than convection vault concepts.

TABLE 4.3: STORAGE COST RELATIVE TO WATER BASIN STORAGE* AT A CENTRAL SITE

Study	Pool at Reactor	Convection Vault	Surface Silo
AECL (Fuel) (8)(24)	0.6	0.6–0.8	0.8–0.9
ARHCO (HLW) (6)	–	0.9	1.26

*Water basin storage is for unpackaged fuel or waste without overpack.

Source: DP-MS-76-39

Current Fuel Storage Programs

Water basin storage of LWR fuel and in-air storage of experimental HTGR fuel are current practices. A facility for packaging and storing graphite fuel from HTGRs and from the Rover nuclear rocket program has been constructed at the Idaho Chemical Processing Plant (25). The facility is force-draft cooled. Peach Bottom reactor and the Rover nuclear rocket fuels are stored in closed but unsealed packages.

The Canadians routinely use water basins for spent fuel storage, but near-surface storage in an array of tile-lined holes has also been used at CRNL (Figure 4.4). Testing is proceeding (26) on storage in concrete surface silos for an interim spent fuel storage facility to be in service by 1985. This dry storage demonstration seals the fuel in a thick metal container that is surrounded by a 2.4 m diameter concrete shell, ~5 m tall with 0.8 m thick walls. Two electrically heated models were used to test the suitability of the concrete and to demonstrate temperature cycling. An additional cylinder has been loaded with 138 bundles of spent fuel from the WR-1 reactor, and a second was expected to be loaded with spent fuel by mid-1976 (27).

United States programs aimed principally at storing waste from fuel reprocessing can be applicable to spent fuel storage. These include tests by ARHCO of

a concrete surface silo and Sandia studies of salt storage. Most of the available
technology concerning packaging of spent fuels also has been developed for fuels
other than those from LWRs.

Three CANDU concepts involve encapsulation of fuel in mild steel containers
under helium atmospheres (8). In the fourth concept (Figure 4.5), six fuel assemblies are placed in a trefoil-shaped aluminum extrusion with zinc or aluminum as a filler material (8).

FIGURE 4.5: PACKAGING CONCEPT FOR CANDU FUEL

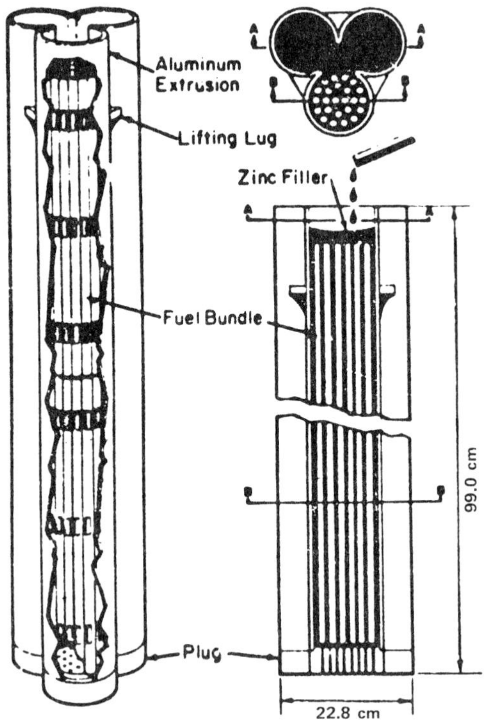

Source: DP-MS-76-39

A concept developed for the FFTF fuel calls for a helium-filled stainless steel
container for each fuel assembly. The assemblies from the Peach Bottom are
packaged in unsealed containers of 0.635 cm thick carbon steel. The containers
are 45.7 cm in diameter and 335 cm long.

Packaged spent fuel assemblies from several different reactors, including Hallam
and Dresden, have been stored underwater at the RBOF Facility (Receiving Basin for Offsite Fuels) at Savannah River for several years.

PILOT PLANT CONCEPT FOR DEEP SALT STORAGE

For nearly twenty years the United States Energy Research and Development Administration (ERDA) and its predecessor, the Atomic Energy Commission (AEC), have pursued a research effort investigating the physical phenomena accompanying the emplacement of radioactive wastes in salt (halite). The program has explored such diverse effects as solid diffusion, thermal and radiation effects, two-phase fluid migration, and mechanical behavior. This research has not revealed any phenomena which would preclude the use of geologic salt formations for the disposal of radioactive waste. Consequently, ERDA has initiated a large-scale demonstration of radioactive waste disposal in a bedded salt formation in the state of New Mexico.

Initial site selection efforts for the pilot plant were conducted for ERDA by the Oak Ridge National Laboratory (ORNL), and an area in the Delaware Basin was selected for further study. In April 1975, ERDA assigned to Sandia Laboratories the responsibility for all aspects of conceptual design, development and implementation of a Radioactive Waste Disposal Pilot Plant (RWDPP) to be located in southeast New Mexico about 50 km east of the city of Carlsbad. This task included continuation of site characterization studies essential to preparation of an Environmental Impact Statement.

In its initial operation as a pilot plant, the RWDPP will accommodate only ERDA-generated wastes, both low-level and intermediate-level waste. The latter category includes the liquid and solvent extraction wastes now stored in tanks at the Savannah River and Hanford facilities. While often referred to as high-level wastes, these wastes are much lower in thermal power and radiation effluent than the solvent extraction wastes from the power reactor fuel cycle. As in the case of commercial wastes, all ERDA wastes will be reduced to an essentially non-leachable solid form before shipment to the repository.

The facility design will be compatible with the storage of commercial high-level wastes, and in situ experiments intended to define the long-term behavior of the wastes in the salt storage environment will begin as soon as progress on underground construction permits. By the time commercial reprocessing plants begin operation, the RWDPP will be able to accept high-level wastes on a "demonstration" scale.

During the pilot plant phase, waste quantities will be limited to those necessary to demonstrate operational safety and to address technical questions which can only be resolved in large scale experiments. Duration of the pilot plant phase required to provide these assurances for low-level wastes is expected to be only a few years. Adequate in situ examination of phenomena associated with high-level waste disposal may require a much longer time (perhaps ten years or more) due to the time required for thermally driven behavior to manifest itself.

An ERDA policy which severely impacts the pilot plant concept is the commitment to provide easy or engineered retrievability of waste throughout the duration of the pilot plant stage. The capability to retrieve waste will be demonstrated at intervals during pilot plant operation. High-level waste presents the most severe retrieval problem because of the thermal and radiation effects on the salt and on the waste package. Most of the high-level waste canisters will be designed and/or emplaced in a manner which prolongs their integrity, thus facilitating their recovery.

Since this mode of emplacement does not provide a complete simulation of the storage environment characteristic of a fully operational repository, a limited but statistically significant number of regular steel canisters will be placed directly into salt. On a still smaller experimental scale, deliberately degraded waste forms will be placed in direct contact with the salt in order to accelerate the observation of any migration of radionuclides that may occur. These latter emplacement modes imply a retrieval method based on overcoring the waste; in effect, recovering a salt canister containing the waste.

After successful completion of the pilot plant demonstration, the RWDPP will be converted to full scale repository operation and will be licensed by the Nuclear Regulatory Commission (NRC). To facilitate this conversion, applicable NRC standards will be met in the initial pilot plant design. In the fully operational repository phase, waste recovery will still be possible, but difficult and expensive.

Acceptance criteria related to waste form and packaging are under discussion with prospective producers of the various waste types. Present RWDPP concepts require that all wastes be solid, noncombustible, and stabilized in some fashion to retard dispersal and leaching should the container be breached. Low-level waste will undoubtedly be incinerated, and it will probably be stabilized in concrete or bitumen. Two ERDA containers (2.4 m^3 fiber glass coated plywood boxes and 200 liter steel drums) will serve as prototypes for commercial low-level waste packages.

It is expected that most of the intermediate-level waste will be packaged in cylindrical steel canisters. The Zircaloy cladding hulls from light water reactor fuel elements constitute a special category of intermediate-level waste. Uncompacted, they require an inordinately large storage volume in the repository, but volume reduction is difficult and expensive because of the pyrophoric nature of Zircaloy. Both mechanical compaction and smelting are under consideration; volume reduction factors for the two processes are six and nine, respectively.

The liquid high-level waste will be calcined to a dry powder and stabilized in a low leachability solid such as zinc borosilicate glass. Solidified high-level waste will be packaged in stainless steel canisters, about 30 to 40 cm in diameter and 3 to 4.5 m in length.

The age and thermal power density of high-level waste is very important to the design of the underground storage facilities. Age, measured in terms of the time since the spent fuel was discharged from a reactor, determines the effective half-life of the waste. This, in turn, determines the total energy deposited in the salt over a given interval of time and thereby influences the macroscopic behavior of the salt bed. On a mesoscopic scale, decrepitation phenomena establish the maximum permissible temperature ($\sim$250°C) in the salt. That temperature, for a given canister configuration, depends on the thermal power density of the waste. During the first five years or so of repository operation, all the high-level waste will be at least ten years old. For the ten year old waste in the form of zinc borosilicate glass, the thermal power density will be the order of 25 W/l and the acceptable areal power density will be about 370 kW/ha.

Of the several waste categories for which the repository is being designed, only the low-level waste is amenable to handling by contact methods; all other waste types require shielding and/or remote handling. It is logical, therefore, to divide the repository into two facilities: a cold facility for the low-level waste and a hot facility for all other wastes. The present design concept develops both the low-level waste and intermediate- and high-level waste repositories in the same salt horizon at a depth of about 800 m. The two-storage areas will be served by a common man-and-materials shaft but will be separated by air locks. The two areas will have independent ventilation systems. Separate waste transport shafts, communicating with completely isolated surface facilities, will be required.

Because of its heat generation, the high-level waste will ultimately require more storage area than other waste types. If all the waste from ERDA facilities and from the domestic power reactor fuel cycle is stored in the RWDPP, the storage areas required for the low-, intermediate- and high-level wastes generated through the year 2000 are estimated to be 100, 90 and 270 ha, respectively.

Should additional storage be required for any reason, the less pure salt at higher horizons can be developed to accept low-level waste without any problem. While these units are less suited to a high-level waste pilot plant, operational experience and experiments could develop the knowledge needed to allow future utilization of these less pure salts for high-level waste disposal.

The low-level and hot surface facilities will accommodate waste arriving by both rail and motor truck. Temporary surface storage will be provided for limited quantities of waste to allow for surges in shipment and for operational downtime. Provision for decontaminating shipping casks and waste containers and facilities for overpacking ruptured waste packages will be necessary. Except for the treatment of site-generated decontamination wastes and air filters, no other waste processing will be carried out at the RWDPP.

DEEP ROCK DISPOSAL

An especially attractive concept for disposal of radioactive, toxic nuclear power reactor wastes is the deep rock disposal concept, in which the wastes are buried in a hole drilled into deep, dry, geologically inactive crustal rock. Since the volume of waste is relatively small, drilling costs add minimally to the cost of power production, and no further maintenance of the site after sealing of the hole should be necessary.

However, because the wastes remain radioactive and toxic up to 10^5 years after their production, the most careful possible analysis of their mobility is required before implementing this concept. In order to minimize this mobility, the deep rock disposal proposal in one version envisions burying waste sufficiently young that the thermal energy of radioactive decay produces melting of the waste matrix and some of the surrounding rock, thereby producing, on cooling and re-solidification, a glassy cocoon containing the material. An additional benefit of the heating is that any ground water present in the surrounding rock (or hydrated water driven off the minerals by heating) should be kept from contact with the hot waste material, thereby reducing the likelihood of a waterborne path back to the biosphere.

The entire spectrum of mechanical and chemical interactions between the heated materials near the borehole is only now beginning to be outlined (28)(29)(30).

The deep rock disposal scheme is feasible when proper attention is paid to the thermal expansion of the waste.

DEEP UNDERGROUND MELT PROCESS

Description of Concept

DUMP, for Deep Underground Melt Process, was proposed in 1972 for the management of high level nuclear waste (31). The concept (as depicted in Figure 4.6) called for direct placement of liquid radioactive waste from fuel reprocessing operations deep underground into rubble-filled void spaces (called chimneys) formed by the underground nuclear detonations. Capacities and properties indicated on Figure 4.6 are for a typical 5 tons per day fuel reprocessing plant.

FIGURE 4.6: WASTE DISPOSAL IN A NUCLEAR CHIMNEY

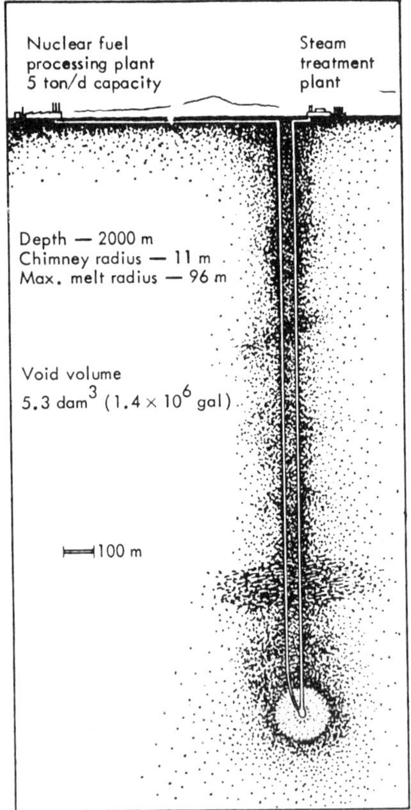

Source: UCRL-51713

Figure 4.7 depicts the time sequence of this waste disposal process. During the waste addition phase, decay heat maintains self-boiling conditions within the cavity; the vapor effluent is condensed at the surface and either recycled through the plant, or directly reinjected for cooling the cavity. Following waste addition, the cavity is allowed to boil dry and the access holes are sealed. Subsequently, the decay heat melts the waste plus surrounding rock. The molten rock dissolves the waste and the molten system continues to grow, reaching its maximum radius in times typically estimated at tens to hundreds of years. When the rate of conductive heat loss from the system exceeds that of the radioactive decay heat output, the system begins to cool and refreeze.

FIGURE 4.7: SEQUENCE OF EVENTS IN THE DUMP PROCESS CHIMNEY STARTING WITH LIQUID WASTE ADDITION

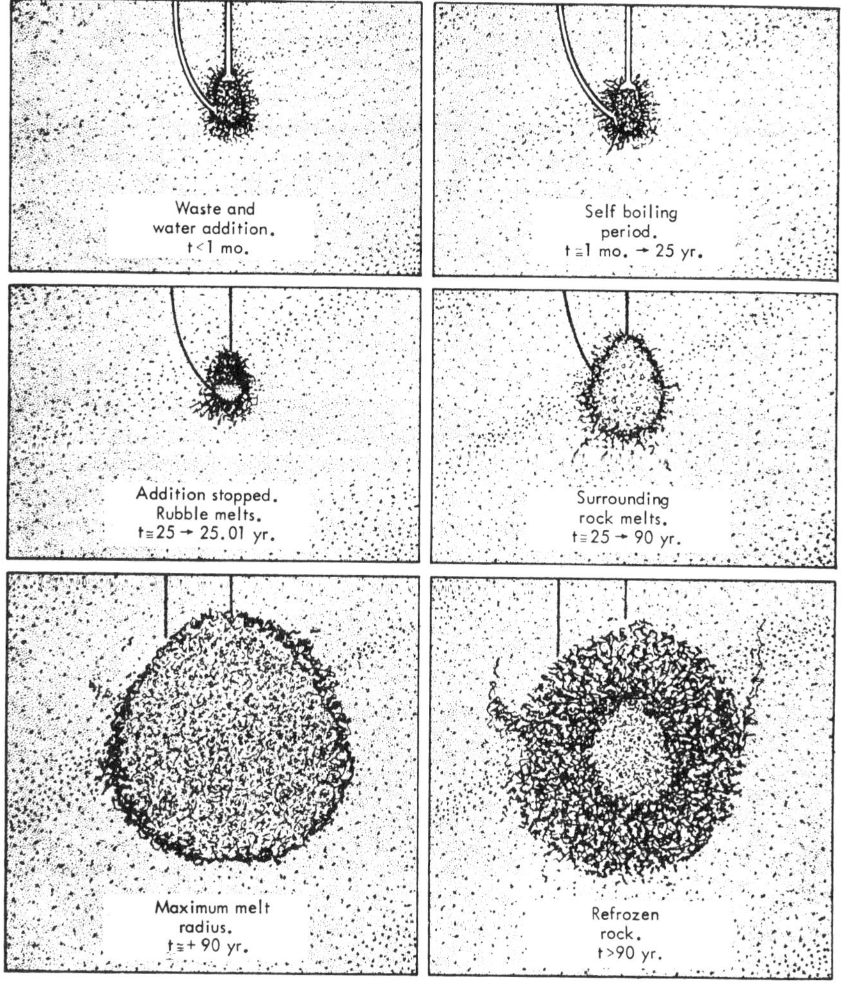

Source: UCRL-51713

Total resolidification would require millenia. By that time all fission products
will have essentially decayed away, and the only remaining radioactive components will be the actinides. The final result would be the encapsulation of this
radioactive material in an insoluble rock matrix deep underground.

At a thousand years following the initial emplacement the relative hazard of this
resolidified rock, based on its actinide content, would be less than 1% of that of
naturally occurring pitchblende as determined by the method discussed by
Claiborne (32).

This concept for waste disposal and its modifications have been discussed and
analyzed by several authors (33)-(37). One such modification is the direct placement of waste, either in liquid or presolidifed form, in underground cavities excavated by conventional means (38) as depicted in Figure 4.8.

FIGURE 4.8: LIQUID WASTE EMPLACEMENT IN A MINED CAVITY

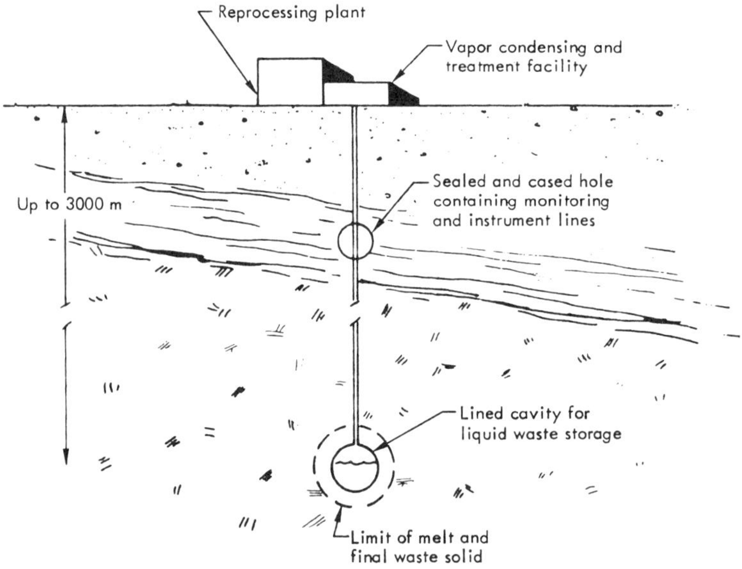

Source: UCRL-51713

Another modification involves using refractory capsules which contain the waste
and which, by melting the rock immediately beneath them, embed themselves
deep within the earth (39)(40). A vexing problem with this latter approach is
the selection of a container material that can maintain its integrity in a molten
rock system.

In all these concepts the heat from radioactive decay melts rock in situ deep underground and an insoluble rock matrix eventually encapsulates the waste. The
acronym DUMP has been selected as a generic designation for these radioactive
waste disposal methods.

Advantages of Process

Potential advantages of DUMP include:

(a) Elimination or reduction of many processing operations and their attendant risks. These operations include interim storage, chemical separation, and solidification.

(b) Potential for safe disposal of all levels of liquid waste, including intermediate and low level.

(c) Elimination of radioactive waste transportation and its attendant risks.

(d) Permanent binding of the radioactive waste in a rock matrix deep underground, assuring its permanent elimination from the biosphere.

(e) Provision of a secure disposal method, assuring permanent safety from effects of disasters such as floods, sabotage or warfare.

Evaluations of the concept performed by various groups (41)(42)(43) indicate that it shows great promise as a safe, efficient, and economic solution to the waste management problem.

Problem Areas

Possible Policy Conflicts: In assessing the DUMP concept, certain problem areas have been identified. These include a possible conflict with AEC policy, which requires retrievability of wastes and storage at a single centralized facility (44). An inherent weakness in this policy is that where retrievability is possible, the risk of unauthorized retrieval becomes apparent. In this regard, retrievability may be equated with vulnerability. It is believed however, that if a safe and effective ultimate disposal method can be proven, the requirement for retrievability can be eliminated.

Since the DUMP concept requires permanent disposal at the site where the waste is generated [the fuel reprocessing plant (FRP)], the requirement for one centralized storage facility could not be met unless one FRP could suffice for the entire country. However, the additional risk incurred by increasing the number of disposal sites should be more than offset by the elimination of the need for transportation of the hazardous waste. The DUMP concept would be admirably suited for application at Nuclear Power Parks, since it would allow for siting a safe, permanent waste disposal facility within the site boundary. The DUMP concept would also be applicable to the permanent disposal of wastes accumulated at Retrievable Surface Storage Facilities (RSSF).

Misconceptions: A second category of problems is based upon certain misconceptions of the DUMP concept. First, it must be emphasized that a nuclear explosive is not necessary to the concept as some appear to believe (45). Although nuclear chimneys are the most economical and expeditious method for forming the required void space deep underground, the economics of the concept are such that, relative to alternative waste management concepts, the added costs of conventional cavity construction by mining would be trivial.

The political and psychological disadvantages of cavity construction using nuclear explosives could outweigh the advantages and unduly delay the development of a potentially valuable technology.

A second misconception is an expressed concern regarding potential ground water contamination during the liquid addition phase, and also at later times. The question of possible ground water contamination can be evaluated for both the liquid waste addition phase and the molten rock phase of the DUMP process.

During the liquid addition phase, the interior of the cavity is maintained at essentially atmospheric pressure. At a depth of 1.5 km, the formation pressure would be about 15 MPa hydrostatic. Therefore, any flow would, of necessity, be from the surrounding formation into the cavity. In applications where presolidified waste is added, e.g., disposal of waste accumulated at an RSSF, the waste could remain sealed in containers during the brief addition phase, thereby precluding water contamination problems.

The possibility of ground water contamination during the rock melting phase of the DUMP process warrants consideration in some detail. Consider the system of molten rock plus radioactive waste resulting from a terminated DUMP operation. The core of the system will have a temperature at or above the melting point of the rock. Any excess heat will go into heat of fusion to melt more rock, the process thereby maintaining essentially uniform temperature within the molten core. Surrounding this core will be a series of decreasing isotherms ranging from the melting point of rock, through the boiling point of water, and eventually reaching the ambient temperature of the native rock. Somewhere in this continuum there will necessarily exist an isotherm at which liquid water will flash to vapor. This isotherm will serve as a thermal barrier separating liquid water from the core and its radioactive contents. This thermal barrier prevents water intrusion and minimizes the possibility of water contamination during the rock melting phase.

A third misconception relates to the possible migration of the molten rock mass. It has not been possible to identify any physical mechanism that would cause such migration, since conductive heat flow is nondirectional. Assuming that there is such a mechanism, however, it is readily shown that the translational movement associated with a nonspherical growth of the molten rock is very limited.

The higher surface-to-volume ratio associated with a nonspherical shape increases conductive heat transfer from the molten body and thereby decreases the ultimate volume of molten rock. Thus, even if directional growth were to occur, the increased rate of conductive heat loss would preclude the possibility of movement over any significant distance. Any process of growth or movement would necessarily be self-limiting.

Technical Problems: A third category of problems associated with DUMP is the one termed "real" or technical problems. These include:

> Engineering difficulties
>
> Fate of volatile radionuclides
>
> Geological alteration due to phase
> change of a large volume of rock

Convective phenomena (shape)

Consequences of seismic activity during
addition phase.

Chemical compatibility of waste and
medium

Perhaps the most difficult of these would be the engineering problems dealing
with waste handling, logistics, and materials. DUMP conductivity phenomenol-
ogy is the subject of ongoing research at Sandia Laboratories (46).

Cost-Benefit Analyses

To gain insight into the relative desirability of a number of high-level waste disposal concepts, generic cost-benefit evaluations of these concepts have been made.
Figure 4.9 gives a categorization of options. These are taken largely from the
paper of Kubo and Rose (45).

Three basic approaches to the treatment and disposal of high level nuclear waste
are indicated. The first (pathway 1) seems to be favored by the AEC's Division
of Waste Management and Transportation, although the ultimate method of disposal of solidified waste has not been chosen. The second approach (pathway 2)
basically involves partition of the waste into a short half-life component, i.e., fission products, and a long half-life component, i.e., the actinides. Such a separation is economic only when the ultimate disposal mode selected is too expensive
to be applied to both components of high-level waste. The third approach (pathway 3) envisages minimal or no treatment of the high-level waste prior to disposal.

Only those disposal modes involving geologic disposal will be discussed, inasmuch
as the other paths and disposal modes (those with shaded backgrounds) are
clearly uneconomic or have significant potential uncertainties associated with
them.

The first approach is amenable to a number of ultimate disposal modes, including
either melting or nonmelting scenarios subsequent to the introduction of the
stored waste into wellbores, conventionally mined cavities or cavities created by
explosives. Also, a number of geologic media could be used for such disposal,
including evaporites, granite, or shale. It can be seen that the "melting" disposal
in rock is one variation of the DUMP concept.

The third approach is, of course, the simplest and least expensive of the DUMP
variations since it involves the least treatment and handling of the high-level
waste.   As previously discussed, this concept involves the direct disposal of liquid waste in an underground cavity, followed by its "self-solidification."

Approaches one and three have been analyzed for both their inherent safety and
for their economic costs (1974 dollars); the results of these analyses are given in
Tables 4.4 and 4.5. The assumptions involved in the preparation of Table 4.4
are outlined in Appendix A of UCRL-51713. The figures presented in Table 4.5
were derived from data given in References (31) and (42).

FIGURE 4.9: WASTE MANAGEMENT OPTIONS

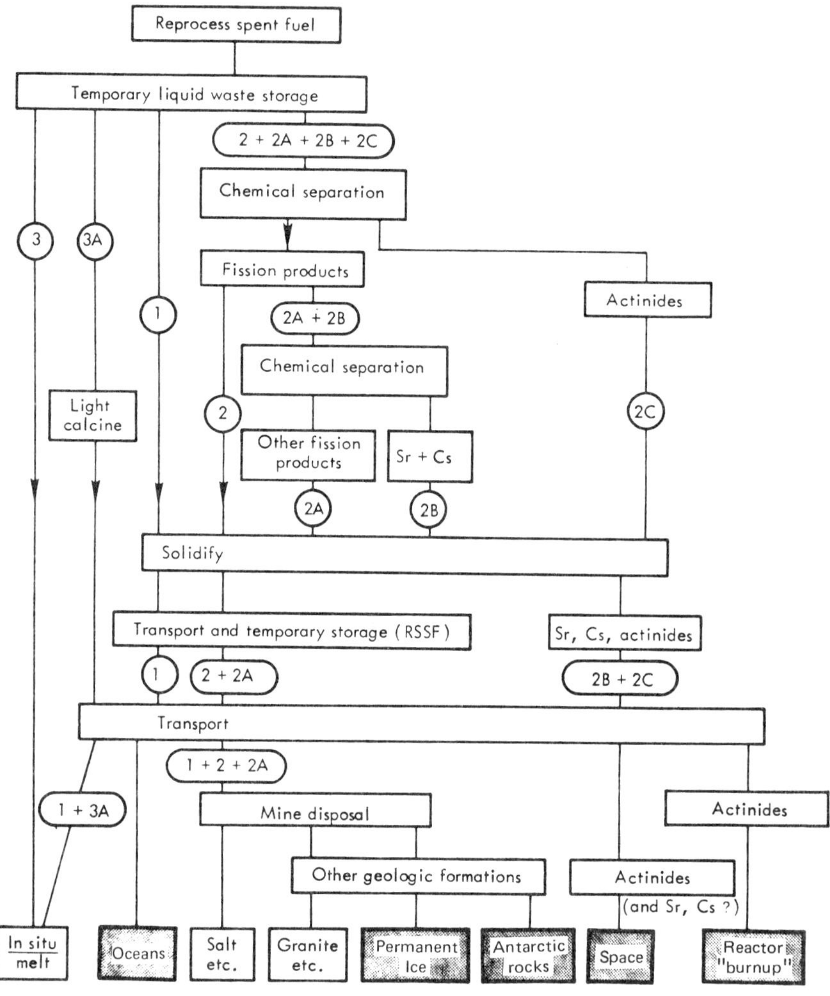

Source: UCRL-51713

TABLE 4.4: 100-YEAR DOSE COMMITMENT FROM NUCLEAR POWER WASTE MANAGEMENT, man-rem/10^3 GWe-y

Operation	Solidification + engineered geologic storage	Solidification + long term RSSF	Solidification + RSSF + conv. geol. disp.	Solidification + RSSF + DUMP	Solidification + DUMP	Direct DUMP
Interim liquid storage	5 000	5 000	5 000	5 000	5 000	5 000*
Solidification	2 000	2 000	2 000	2 000	2 000	—
Interim solid storage	3 000	3 000	3 000	3 000	3 000	—
Transport to RSSF	—	3 000	3 000	3 000	—	—
RSSF	—	10 000	5 000	5 000	—	—
Transport to ultimate storage site	2 000	—	2 000	2 000	2 000	—
Ultimate storage	500	—	500	500	500	500
Total	12 500	23 000	20 500	20 500	12 500	5 500

*Liquid addition phase.

TABLE 4.5: SUMMARY COST COMPARISON OF VARIOUS WASTE MANAGEMENT CONCEPTS

Process	Mill/kWeh	10^6 \$/10^3 GWe-y
Solidification + eng. geologic storage	0.05	440
Solidification + long term RSSF	0.04	350
Solidification – RSSF + eng. geol. disp.	0.07	610
Solidification – RSSF + DUMP	0.06	550
Solidification + DUMP	0.03	230
Direct DUMP	0.01	90

Source: UCRL-51713

Although the direct DUMP sequence appears to have less adverse health impact than the alternative concepts, the overall differences appear to be of no great significance. On the other hand, the economic savings realized by the implementation of the DUMP process are substantial, i.e., up to 80%. By the year 2000, these savings could amount to several hundred million dollars annually on a national scale.

SEABED DISPOSAL OF HIGH-LEVEL WASTES

A workshop was held under U.S. auspices at the Marine Biological Laboratory facilities, Woods Hole, Massachusetts on February 16 through February 20, 1976. International and interdisciplinary participation allowed a fruitful discussion of all technical and scientific aspects of the concept of seabed and subseabed disposal of HLW. The conclusions reached by the various subgroups dealing with disposal canister research, the water column, undersea geology and biology are given below.

The Disposal Canister

Waste Form: When spent fuel from nuclear reactors is reprocessed, it is dissolved and the uranium and plutonium are recovered by solvent extraction techniques. The first solvent extraction cycle gives rise to an aqueous raffinate containing most (over 99%) of the fission products, and is commonly known as high-level waste.

The high-level waste contains a large number of different fission products and some transuranic elements, together with small amounts of unrecovered uranium and plutonium. The initial quantities of these elements present are governed by the irradiation history of the fuel. Approximately 1 kg of fission products is created for each 1,000 MW days (thermal).

The high-level waste can also contain materials from other sources such as: burnable poisons or alloying elements incorporated in the original fuel; and corrosion and solvent degradation products coming from reprocessing operations.

Various solidification processes are being developed to immobilize the ions in high-level waste with the prime object of making the wastes easier to store (see previous chapters). On the basis of current technology, the best method seems to be to incorporate the radionuclides into a leach-resistant glass lattice of high radiation and thermal stability and relatively good thermal conductivity.

The high-level waste is vitrified with glass-making constituents such as silica and borax at temperatures of 1000° to 1200°C. Typically, the vitrified waste contains up to 25% of fission and waste-product oxides and smaller quantities of transuranics. Increasing the proportion of waste incorporated into the glass eventually results in a poorer glass. The molten vitrified high-level waste is either cast or poured in a metallic container, and after it has cooled sufficiently, it can be moved to either a water-cooled or air-cooled engineered store. Cooling is necessary to remove the radioactive decay heat that comes predominantly from the shorter-lived fission products. Depending on the size and shape of the container and also on the method of cooling, it is possible to vary the quantity of fission product decay heat that can be accepted into the vitrified high-level waste without excessive temperatures being reached, e.g., sufficient to melt or devitrify the glass in the center of the container. Countries planning to introduce industrial scale high-level waste solidification plants during the next decade would be expected to start with waste that has been cooling longest so that, initially, heat will not be a major problem.

Much work has been done to characterize vitrified high-level waste for surface storage situations and more is planned. Measured leach rates in fresh water range from 10^{-4} to $10^{-9}g/cm^2$ per day (typical density 2.7 g/ml) but are very specific to the experimental technique employed, the radionuclides involved, and the composition of the water.

It is known that under certain conditions re.atively soluble phases can separate from the glass and because in time glass might lose its monolithic structure and become fragmented, this has to be allowed for in assessing rates of possible release of activity. Vitrification is regarded as an irreversible process step, apart from heat annealing or the possibility of remelting to transfer the material into a new mold. Work has been done on using calcination as a solidification process for high-level waste because much lower temperatures (about 600°C) are required. However, calcine is not generally regarded as a suitable form for storage or disposal because of its dusty nature, low thermal conductivity, and high solubility. Other solidification forms, such as ceramic, may be developed in the future.

Container: In considering the feasibility of metallurgical schemes for encapsulating the conditioned high-level waste in materials which will exhibit a long life, say 10^3 years, in deep seawater, one must either extrapolate the most promising behavior from relatively short-term corrosion tests on some of the more recent metals and alloys, or one must look at the metallurgy of artifacts that have been recovered from marine archaeological digs after seawater immersion for up to several thousand years under various conditions. Ancient materials which have been good, or in some cases virtually unaffected, after thousands of years in relatively shallow seawater are precious metals, lead and some copper alloys. Although waste forms buried in the deep oceans will suffer heat and radiation effects, the archaeological evidence is regarded as promising. New materials which appear to be excellent in marine environments include titanium. For evidence of deep-water corrosion on intermediate time scales, some submarine cables have been in existence for more than 100 years.

Canister materials can range from carbon steel to heat-resisting, water-corrosion-resisting alloys, such as stainless steels, Incoloy alloys, Inconels, Hastelloys, etc., depending on processing and storage conditions. Some are known to be unsuitable for a sea environment because their thermal history results in sensitization leading to intergranular corrosion; others sustain rapid failure because of localized pitting. Only some of the high-temperature-resistant container materials are known to have suitable corrosion resistance, e.g., Inconel 625 and Hastelloys.

An important factor with the encapsulation process is that it can be carried out in a purposely made facility, possibly at the reprocessing site. Stringent inspection techniques can be applied to prove the integrity of the package, and this is not necessarily possible with the high-level waste vitrification stage.

An important public acceptability aspect is that container inspection techniques could be applied or observed by international bodies.

In the early years of a disposal operation, the containers will be thermally hot because of radioactive decay, the heat generated being about 8 kW per ton of irradiated fuel at the end of one year after removal from the reactor, falling to 0.5 kW after 30 years' decay.

Heat removal may present problems at the disposal site; containers may have to be optimized in geometry for heat removal rather than for operational convenience at the reprocessing plant.

If waste containers are disposed of by allowing them to lie on or in the seabed, the solidified waste will at some time be exposed to the surroundings. The most likely breachment mechanism is corrosion. Whether the canister is on or in the sediments at the time of release, a portion of the isotopes will be adsorbed on the sediment.

In assessing the feasibility of any seabed disposal program, then, it is safe to assume that there will be a small failure rate among containers. Although this is expected to be less than 1%, allowance must be made for the exposure of some solidified waste to seawater after only a short cooling period.

Disposal on the Seabed: The simplest method of disposal is to place waste containers directly on the seabed. Emplacement is easy and heat transfer problems are minimized. This method would only be acceptable, however, if the need for a further barrier between the container and the sea were clearly unnecessary.

It is imperative therefore to show that there can be no short-circuit transport of radioactivity from waste containers to man or the ecology. This could be so if the water column and biological processes in deep water act as adequate barriers.

Thus, there must be no significant local currents or local food chains. This information must be well authenticated and the assessment must include an acceptable safety factor so that a temporary unexpected behavior pattern will not alter the basic premise.

Although it may be possible to identify localities where seabed disposal is possible in principle, the sea floor itself may not be suitable or the environment may not be predictable for the time periods of interest.

For example, if the sea floor consisted of very unconsolidated sediments, muds, oozes, etc., or was in an area of frequent turbidity currents, careful, long-term studies of prospective sites would be necessary before feasibility could be shown.

Disposal Under the Seafloor: If an additional barrier between the container and the sea is shown to be necessary and the seafloor proves to be a long-term barrier, then the simplest solution would be to bury the containers in the sediment or underlying rock.

Chemical, physical, and mechanical properties of the sediments and rock, both in the laboratory and in situ, must be completed before feasibility of this option can be fully assessed. If natural barriers are insufficient, it may be possible to design an artificial barrier in the seabed.

Containers buried in sediments or rock will get hotter than those exposed directly to the sea. This may not be important in that sediments or rock are not damaged by high temperature, and indeed in the final outcome a molten magma of sediment/rock with the waste could be acceptable. However, this is probably not the case, so the packages (waste form and canister) will need to be optimized to allow for the poorer heat-transfer properties of the seafloor itself.

The waste could be cooled longer on land in engineered stores—for 10 to 30 years for instance—or the composition of the waste could be altered by diluting the waste with inert material or by removing certain isotopes like cesium-137. The canister could be designed to give the maximum heat-transfer surface.

A more difficult solution would be to improve the local heat-transfer properties of a loose sediment by fluidizing it for a period of years.

The presence of hot sediments would probably affect the rate of corrosion of the can and this would have to be allowed for in design of the final package.

Emplacement: Emplacement on a hard seafloor would present no particular problem, especially since the good heat transfer properties would allow the use of a wide range of container designs. Emplacement on soft mud would be difficult and it would probably be better to accept the fact that initially the container would sink.

If the container is to be placed beneath the seabed, then a number of possible techniques exist:

> If the sediments are hard or if the waste is to be placed in the rock, then holes have to be drilled.
>
> If the sediments are sufficiently soft, then penetrating projectiles might be feasible.
>
> Trenches could be opened in sediments and backfilled.
>
> Soft sediments could be fluidized to allow heavy packages to sink.

It is not intended that the canister be retrievable.

Conclusions: On the Seafloor — High-level liquid wastes can be solidified and encapsulated in containers that would resist corrosion by seawater for some hundreds of years. When the containers fail, the exposed solid would be leached out over a further period of some hundreds of thousands of years.

It is possible to protect the sea from fission products with half-lives of up to 30 years or so, but much longer-lived species will ultimately get into the ocean. The consequences of such release must be assessed.

In the Seafloor — If the seawater is not an adequate barrier, then disposal of packaged wastes into the seafloor may be acceptable. Experimental work is needed to study the physical and chemical properties of the seafloor, the ocean current, corrosion resistant materials, and relevant heat-transfer properties.

Water-Column Research

Further Development of Models: There is need for further development of a reasonably simple model of ocean circulation and mixing processes to allow examination of the relationships between the continuous release of given levels of radioactivity on the ocean bottom and the concentration likely to be found in the ocean generally in space and time, i.e., the first part of the process of estimating the likely doses back to man.

Studies of Variability of Horizontal Currents: Recent studies of horizontal currents in the deep ocean have shown that fluctuating currents are more energetic than mean currents computed over periods from several months to years. There is a suggestion that in the deep ocean the kinetic energy spectrum increases to a maximum at periods of 100 to 200 days; the energy in periods greater than 200 days appears to fall off with time. To obtain stable observations at a single point, it is likely that velocity measurements must be made over a period equal to several times the most energetic period of the spectrum. Such measurements can now be made with moored current meters, but how long measurements are necessary in order to get stationary statistics on the spatial variability of horizontal currents is not known. It is known that large variations in the kinetic energy of the fluctuations can occur over relatively short distances.

Because of the importance of these energetic motions in the horizontal transport of dissolved radionuclides, a better description of them must be obtained. If the measurements are continued for sufficiently long times, it will be possible to learn something about the mean circulation.

A program should be supported to put moored current-meter arrays in the deep ocean. Regions of high priority should be sites in the deep sea that are likely to be considered for waste emplacement or where there is reason to believe that advection may be strong for dynamic reasons, e.g., western boundary currents or intensification of flow because of bottom topography.

There will be significant data input from other scientific programs, such as Polymode in the North Atlantic. Results of these measurements will be incorporated into thinking about deep-sea transport times. In addition, there are the following proposals for long-term current measurement programs:

> Eastern basin of the North Atlantic (under auspices of a subcommittee of SCOR Working Group 34).
>
> Site of U.S. midplate gyre survey (No. 3) in the western Atlantic (U.S. Seabed Disposal Research Program).
>
> Western Pacific (Japan) and regions of restricted (accelerated) bottom-water flow, such as the Samoan Passage in the South Pacific and the Horizon and Clarion Passages south of Hawaii.

Measurement of Small-Scale Horizontal and Vertical Diffusion in the Deep Sea:
At present there are no direct measurements of horizontal or vertical diffusion at depths greater than about 1,000 meters (and only one at this level), and although gross values of the vertical diffusion coefficient (K_z) can be inferred from radon 222 and radium 228 measurements in the deeper layers, a knowledge of the average value and variability of these parameters throughout the water

column, and particularly of K_z, is perhaps the most important missing input data of the circulation and mixing processes model. Hence, work is recommended at a number of selected sites between 1,000 and 5,000 meters in which the acoustically controlled free-swimming submersible developed at the University of Washington would be used to measure diffusion parameters. This technique is already proved, but there also appears to be merit in examining the feasibility of using small radioactive releases, perhaps in the region of about 100 Ci, to measure these parameters over the same depth range.

Useful data might be obtained by suspending scintillation counters at the mean depth of the patch and raising and lowering the counters while the ship drifts slowly across the patch, determining the position of both counters and the ship acoustically relative to bottom-mounted transponders. Once this method is shown to be feasible, it might be advisable to mount a joint experiment to examine the compatibility of the results with those obtained with SPURV, which uses an unmanned submersible and an organic dye. Whatever the method decided upon, a number of measurements will be needed at carefully selected sites to allow examination of the spatial variability of the diffusion parameters. It is expected that these experiments would be done at least over periods of a few weeks to a few months at the long-term current-meter sites and would be supplemented by information from short-term current-meter arrays and float-tracking.

Current-Meter Studies of the Bottom Boundary Layer: Among the problems associated with ocean disposal, one area of ignorance is in the structure of and the dynamics controlling the bottom boundary layer of the ocean.

To be able to predict the fluxes of radionuclides from a source in or on the sediment into the water column, one must be able to measure the vertical eddy flux in the column.

Natural radionuclides, such as radon 222, have been used to estimate the vertical eddy flux in the boundary layer, but their distribution seems to have been affected by the presence of large-scale turbulence and local topographic features. Also, the presence of near-bottom layers well mixed in temperature, salinity, and suspended particles imply that mixing may be more intense than previously expected. The shear stresses acting on the bottom are measurable only by the accurate measurement of velocity profiles. The vertical turbulent flux quantities, for example, $T'W'$ (where T' is the turbulent fluctuation of heat and W' is vertical velocity) must be accurately determined and the variations of these fluxes in time and space described. The frequency and intensity of bursting phenomena are also important as they will reveal whether and how much of the surface sediment can be eroded and put into suspension where it can be transported by the water.

For velocity field measurements needed for determination of turbulent fluxes and the critical shear stress, as well as those needed for study of the structure and dynamics of the benthic boundary layer, certain equipment specifications must be met. These include:

> a low threshold velocity;
> high spatial resolution;
> ability to measure high-frequency turbulent fluctuations; and
> minimum interference with the flow field by the flow-
> measuring device.

The large size of these meters generally prevents deployment closer than a few
meters above the bottom, and the instruments are often subject to mechanical
failure.

The need exists for developing current meters that are capable of making the
necessary measurements in the deep sea. Types of meters that are being developed and have promise for such use include electromagnetic, laser-Doppler back-
scatter and ultrasonic-Doppler current meters, and acoustic vector velocimeters.
Although these devices appear to meet the requirements, none is yet available in
a form that could be deployed as one of a series in a deep-sea near-bottom mooring. More development will be necessary before the required measurements can
be made.

Once available, arrays of these meters should be used to determine the velocity
structure of the deep-sea bottom boundary layer. These measurements will have
to be made in several areas and at many times and as part of a program that includes other measurements, such as near-bottom temperature and salinity profiles
and near-bottom radon 222.

An understanding of the dynamics controlling, and the structure of, the benthic
boundary layer will help in understanding the flux of dissolved and particulate
material from the sediments to the water, as all of this material must pass
through the benthic boundary layer.

Study of the Distribution of Natural and Man-Made Radionuclides: Upward
transport of dissolved substances can be studied by the analysis of measurements
of radium 228 and radon 222 near the sea bottom. Because these species are unstable, their radioactive decay enables us to calculate the time scale of vertical
transfer. Radon 222 (half-life of 3.5 days) is useful up to 50 to 100 meters
from the bottom. The half-life of radium 228 is about 6 years and is useful for
the study of processes with time scales of 3 to 30 years and will influence layers farther off the bottom.

A systematic collection of more radon 222 and radium 228 data, particularly
where there are supporting data regarding the vertical distribution of current velocity and water density, should produce better understanding of mixing processes
taking place within near-bottom layers and in various parts of the ocean. A
measure of regional variations in the intensity of vertical mixing is expected to
become available with this method.

Time scales for large-scale mixing processes in the ocean can be studied by the
analysis of radionuclides, e.g., tritium, helium-3, and carbon-14, in the manner
of the GEOSECS program.

Cycling of Particulate Material: A knowledge of the fluxes of particulate materials through the ocean has significant impact on the understanding of many dynamic processes and adds to the knowledge of oceanic circulation. The composition and distribution of sediments depend upon the particulate flux. The supply
of food derived from the flux of organic matter is a primary control of the structure and diversity of benthic communities. The chemistry of a water mass and
the usefulness of any chemical species in tracing circulation are greatly affected
by the formation, removal and dissolution of particles.

When applied to the problems of radioactive wastes, an understanding of the composition and flux of particulate material will help us determine the paths and cycles in which such wastes will be involved should they enter the water column. For instance, zooplankton feeding and excretion concentrates material and is a major factor in particle transport. Organic or electrochemical bonding of particles will increase the size of particles and therefore their settling rate. Particulates provide adsorption sites for ions in solution and act as scavengers to remove some radionuclides as well as other ions from the water column.

Attempts to determine the age of particles in suspension by means of carbon-14 dating techniques and ratios and decay rates of transuranic nuclides have met with some success and should be continued. However, material used for dating, obtained from large-volume samplers, contains particles which originated at the source of the water mass, as well as particles in vertical flux through the water column. To isolate, analyze, and date the particles which actually contribute to the mass flux, it is necessary to moor properly designed sediment traps at different depths for periods long enough to collect sufficient material for necessary measurements (probably more than 30 days, depending on the sedimentation rate and trap opening).

Measurements of particulate fluxes and concentrations also provide information about vertical eddy fluxes in the bottom 300 to 1,500 meters. The level of the minimum in light-scattering and suspended sedimentary concentration indicates the height above the bottom to which particles are mixed, though horizontal advection must also be considered.

Comparison of the flux of sediment measured by long-term sediment traps at different heights above the bottom indicates the rate of resuspension of bottom sediments. One trap placed at the level of minimum particulate concentration (300 to 1,500 meters above the bottom), as determined by a nephelometer or turbidimeter, measures the particulate input from surface productivity. Sediment traps must simultaneously be moored in the bottom 10 to 100 meters (within the mixed layer if it exists) and minimally at one level between the two traps. The amount of material in the bottom traps in excess of the material in the highest trap is used to calculate the rate of resuspension and vertical mixing of bottom sediment. Further comparison with particulate concentrations as measured by filtration techniques yields residence times for suspended sediments.

Determination of the size distribution of particles caught in large-volume samplers and sediment traps is needed to improve diffusion-advection models. This work should be coordinated with radon 222 and radium 228 vertical distribution studies.

Problems for Which New Measurement Techniques are Needed: The mean rate of vertical advection in deep water can be estimated, from the radiocarbon age and known sources of deep water, at about 3 meters per year. This is only a large-scale long-term steady-state mean. The detailed distribution of vertical velocity in space and time is not known. The prospects for exploring that distribution by geochemical methods or predicting it by means of numerical models seem limited, though probably they are the best approaches available.

Similarly, present estimates of about 1 cm^2 sec^{-1} for vertical eddy diffusion coefficients in deep water are long-term large-scale steady-state averages.

The small-scale mixing processes represented by this eddy diffusion coefficient are being actively studied. They seem to be highly intermittent and variable in space. Nothing is known about how they (and the effective vertical diffusion coefficient) might vary in the long-term in response to climatic change. And although the pattern of horizontal advection and the statistics of mesoscale eddies in the deep water will continue to be explored, it may not prove possible to predict how they might change in response to changes of climate.

Other Problems: Besides the above uncertainties about the numbers for a simple advective-diffusive model, the possibility of extremes or catastrophic events cannot be excluded, though there is only a low likelihood of predicting them or even making up anything like a complete list of possibilities. For example, there is mounting evidence from the deep-sea geologic record regarding major cyclic changes in both the pattern and the effects of abyssal circulation that are controlled by climatic oscillations (periods of about 100,000 to 125,000 years) for an era of intense glacial activity (about 3,000,000 years). The sediment record in some areas implies a much less vigorous circulation that erodes and redistributes sediments during deglaciation.

An even less tractable problem is that of knowing whether under different climatic conditions it is possible for the open North Atlantic to experience deep convective mixing to the bottom in a fashion similar to what is known to occur intermittently in some winters in the northwest Mediterranean.

Biological Problems

Effects of Containers on Ecosystems: Emplacement of a significant amount of packaged solidified waste on or in the seabed may significantly disturb the local environment. Therefore, in-situ studies may be necessary to allow assessment of the environmental effects of such practices, particularly of the chemical effects of container materials on the deep sea ecosystem.

Sediment Trap Experiments: One basic problem in assessing the transfer of radionuclides toward man is the rate of release from the sediment to the water column as a result of biological transport. Release as a result of physical upward transport through convection is not considered here. It is difficult to measure the flux of biological transport out of the sediment, but it can be reasonably assumed that this upward flux is less than or equal to the downward flux of biological material from the overlying layers which provides the basic food supply for benthic organisms. A properly designed sediment trap could be used to catch the rain of material from the surface productive layers. The organic and inorganic fractions of the sediment should be determined.

In-Situ Biocorrosion: The possibility exists that at the pressures and temperatures pertaining, biological processes may cause increased corrosion of the container or waste form. It is felt that the only definitive experiment would be to place typical materials in position in conjunction with a heat and radiation source if these were thought to be relevant. The effects would differ depending on whether emplacement was in or on the sediment, so if a decision on the best position has not been made previously, both environments would be investigated.

Chemical Effects of Wastes on Ecosystems: Without sufficient actual experience on the life of containers in the deep sea, it may be necessary to consider the possibility that a container might break during or immediately after positioning. Depending on the nature of chemical compounds included in the waste, in-situ studies of the resulting chemical effects on the environment will be necessary. If the level of radioactivity is very high and the intensity of radiation and heat generated in the waste is significant, possible synergistic effects of hazardous chemicals, radiation, and heat may have to be properly assessed.

Geologic Considerations Regarding Subseabed Emplacement

Sediment Characteristics: Thickness — The thickness of unlithified sediment needed for an adequate barrier depends largely on the sediment properties, such as mineralogy and permeability. If ideal conditions relative to ion exchange capacity and pore water migration exist, it is envisioned that a few tens of meters may be adequate. However, it will be necessary to determine required burial depths and hence required sediment thickness, on the basis of site-specific conditions. These decisions will require in-situ measurements of both natural pore water flow and of flow induced by thermal and mechanical perturbation of the sediments by the wastes. Consideration must be given to providing adequate barrier thickness beneath, as well as above, emplacement positions. Therefore, it is tentatively suggested that sediment thicknesses ranging from tens to hundreds of meters should be considered, depending on parameters discussed previously.

Lateral Homogeneity — Since any particular disposal site can be expected to receive many waste canisters, it is necessary that sediment properties and conditions be defined over relatively large areas (perhaps 10^4 or 10^5 km^2). Sufficient numbers of long cores, combined with geophysical data and other information, must be obtained to assure that vertical and areal variability is adequately determined. It is likely that sediment properties will vary vertically somewhat at a given specific location, but it is possible, using existing techniques, to select disposal areas in the deep sea regimes with little horizontal variability. Selection of areas with lateral coherence will reduce the number and complexity of specific site investigations, such as coring and drilling, and increase the reliability of performance predictions.

Composition — The high ionic exchange capacity and the low permeability of sediments of clay mineralogy favor clayey deposits as the best media for disposal. The enhanced mobility of transuranics under reducing conditions found in green and gray hemipelagic sediments of high organic content (0.5 to 2%) suggests that the oxidizing condition within red clay deposits of low organic carbon content (less than 0.1%) would be a better environment for waste disposal. Biogenic sediments rich in carbonate and silica are less favored because of their low ionic exchange capacity, higher permeability, and susceptibility to dissolution by ocean water.

Stratigraphy — It is likely that sediment properties vary with depth and that it will be necessary to assure the integrity of barriers so that there will be no possibility of significant short-circuiting. There are some stratigraphic conditions that should be avoided completely and others that must be carefully studied.

> Aquifers: Continuous interbedded layers of highly permeable materials could provide pathways for quick lateral migration of pore water, with potential release through outcroppings at points remote from an actual containment site. Field investigations must be sufficient to assure that the sediment barrier cannot be breached by this mechanism.

Ash Layers: Ash layers resulting from volcanism are often found in deep-sea sediments. The layers are typically less than 1 cm thick but in some cases may be several centimeters thick. The materials are usually chemically altered as they are incorporated into the sediment column, but the effect of ash layers on interstitial water mobility is not fully known. It is quite possible that permeability coefficients in altered ash layers may be considerably less than in the surrounding clay matrix. In addition to studying physical properties of ash layers, it will be necessary to determine their areal extent.

Chert Layers: In some mid-plate/gyre areas, chert (cryptocrystalline silica resulting from in-situ diagenetic silicification of sediments) is found in continuous layers. This material is hard and essentially impervious. A chert layer could act as a sediment barrier by providing an impermeable capping material over waste containers placed beneath the layer. However, the possibility of concentration effects and lateral migration along these deposits should be considered. In addition, the presence of chert layers may affect emplacement procedures.

Continuous Geologic Record — The geologic stability of a waste emplacement site must be predicted for a period several times greater than the toxic life of the wastes. This favors sites at which sediment cores contain a geologic history extending in a continuous, uninterrupted record for at least 5×10^6 years. This criterion suggests a site where tranquil, pelagic sedimentary processes dominate. Regions of catastrophic processes (slumping and turbidity currents) will not contain the necessary record. Contourite deposits (contour-controlled bottom-current deposits) are more complex than pelagic deposits but should not be excluded from consideration because it is probable that a continuous record may be found in some large areas of very thick individual deposits (thousands of meters).

Regional Characteristics: Low Regional Slope of the Seafloor and Enclosed Basins — The angle at which deposits of fine grained sediments become unstable may be very low and the possibility of sediment redistribution is great if local slopes are larger than this angle. Sporadic redistribution of sediments resulting from slumping of sediments on submarine slopes both reduces the predictability of sedimentary processes on the seafloor and raises the possibility of exposure of wastes to the water column. The emplacement site should be in an enclosed basin; however, this basin should not be an abyssal plain, where the primary sedimentary process has been turbidity currents, because the predictability factor of such a regime is quite low. Furthermore, sediment deposits of sufficiently large areas could be deemed suitable if their local geologic setting precluded significant erosion toward a centrally located emplacement site.

Glacial Erratics — During periods of advancing continental glaciers, fields of sea ice and icebergs migrate to lower latitudes more frequently than during warm interglacial periods. Glacial ice carries sediment without regard to particle size. It is therefore possible that ice may deliver large boulders in an area where the delivery of very fine sediments is in equilibrium with the local hydraulic energy levels. The result is an inhomogeneous sediment column, with a random mixture of clays and coarse sediments. Sediments of this type would make the emplacement of waste containers relatively more hazardous and would reduce the predictability of postemplacement effects on the sediment. Therefore, the sediments of the prospective disposal site should be free of ice-rafted debris or other evidence of excessive changes in sedimentary regime resulting from continental glaciers.

Resources and Seabed Utilization — The possibility of the presence of resources on the seafloor or below could induce human interference with the natural environment (e.g., fisheries and mineral exploration and exploitation). To avoid a situation where sediments may be penetrated or removed for the retrieval of valuable minerals or hydrocarbons, exploration for these commodities should be carried out around any prospective site. Attempts should be made to predict economic feasibility of exploitation of some of the rarer metals. There is a possibility that exploitation that is presently noneconomic may become economic in the future.

A possible disturbance of the seafloor by man may occur in the vicinity of shipping lanes or submarine communication cables. In the former case, the seafloor may be interfered with by sinking debris from passing ships or the sinking of ships themselves. The laying of cables and subsequent grappling for repairs may also disturb the sediments.

Geologic Stability — To allow prediction of undisturbed conditions in the next 10^6 years, the disposal site should show evidence that the area has been geologically stable at least during the last 5×10^6 years. Localities of frequent earthquakes, such as are normally found along plate margins, should be avoided, including areas where there is evidence of past volcanic activity. Areas of low-level seismicity, detectable only by ocean bottom seismographs, must be delineated and, if necessary, avoided. Shallow inland seas and continental shelves should be avoided because these areas are influenced by eustatic changes in sea level, which modify local and complex oceanographic and sedimentary regimes.

Emplacement Procedures: (see Figure 4.10) The emplacement process will produce large localized stress-strain conditions, and it will be necessary to determine the dynamic behavior of the sediments under those conditions. The problem centers around the consequent stresses and strains necessary to close the hole above the waste container. This may entail a gradual flow of materials, during which the local disruption is healed. Some of the emplacement concepts and possible effects are discussed below:

Projectile Emplacement — In this concept, the waste container would be in the form of a streamlined projectile which would be controlled in its passage through the water and penetration into the unlithified sediments. The terminal velocity may be of the order of 30 meters per second and penetration exceeding 50 meters may be possible. As the projectile penetrates, the sediment fabric will be ruptured; subsequent sealing of the resulting cavity will be time-dependent. Hole closure may involve a combination of immediate elastic rebound and gradual plastic flow or creep. There are two important assessments to make: (a) the discontinuity in sediment fabric caused by the penetration process; and (b) the sequence of the closure mechanism. If there is no entrapped water in the cavity, and closure is complete, the effects will be minimal. If the closure process entraps a volume of water in the sediment column, it will be necessary to determine its effects on the full range of processes connected with the waste container (i.e., convection, water migration, etc.).

FIGURE 4.10:   SUBSEABED EMPLACEMENT CONCEPTS—SOLIDIFIED HLW

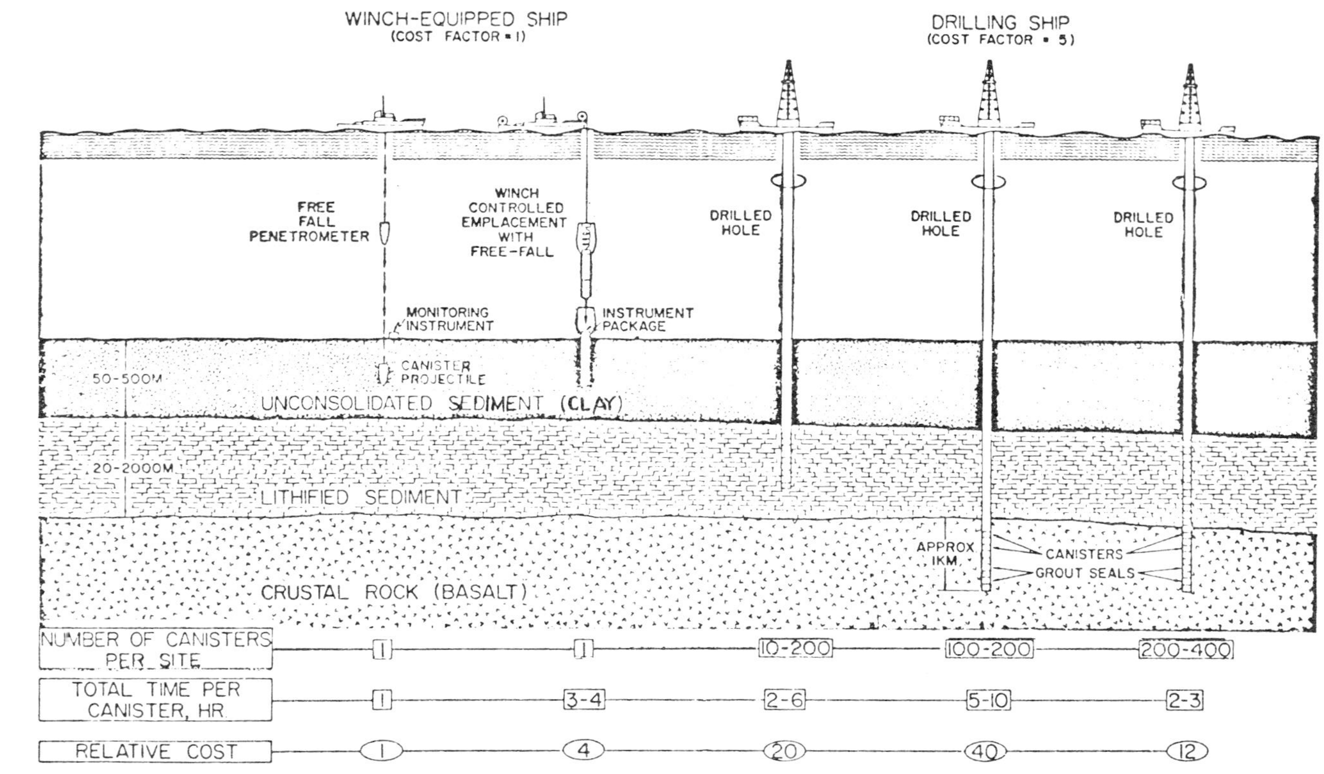

Source: SAND 76-0224

Winch-Controlled Emplacement Device — In this concept, the waste container would be attached to the bottom of a device designed to penetrate into the sediment column the required distance (by free-fall, vibration, or jetting). The waste container would be released before withdrawal of the emplacement device. Sediment behavior during and after withdrawal must be known in order to predict the effects on barrier processes. If necessary or desirable, it would be possible to provide a sealant which can be left to fill the cavity as the emplacement device is withdrawn. In this case the effect of the sealant would also need to be assessed.

Drilled Hole — In this concept a hole would be advanced through rotary drilling techniques. It is likely that many waste containers would be emplaced in each hole with some type of grout seal between canisters. A sealant would be provided between the top container and the sediment surface. The problems and concerns would be similar to those mentioned for the winch-controlled emplacement device with a sealant.

LAND BURIAL OF LOW-LEVEL WASTES

Land burial has been used for disposal of radioactive wastes since the inception of nuclear weapons research in the 1940s. The term burial, as used here, refers to the placement of waste at relatively shallow depths in earth materials, with no intent or provision for ready retrievability at a later date. As such, it is distinguished from geologic disposal and storage techniques by the proximity of the waste to the surface of the earth and the surrounding biosphere. The migration or dispersal of the radioactive and other toxic materials in the waste must be kept to some minimum or acceptable level.

Initially, burial grounds operated by the Atomic Energy Commission handled and disposed of wastes generated within the AEC, as well as waste from private industry. However, during the 1950s much of the waste generated within private industry was delivered to several private companies for sea disposal. Growing pressures against this technique lead to the AEC encouraging the use of burial facilities at Oak Ridge National Laboratory (ORNL), Tennessee, and the National Reactor Test Site (NRTS), Idaho, until private burial grounds could be established. In 1962 a commercial burial ground was opened at Beatty, Nevada, and in 1963 a similar site was opened at Morehead, Kentucky; in May 1963 the AEC discontinued its policy of accepting wastes from private industry. By 1971 six commercial sites were licensed for the handling and disposal of radioactive waste. Figure 4.11 shows the location of major burial grounds.

At present, radioactive wastes in all categories other than high-level waste (produced by the first stage of fuel reprocessing) are buried at both ERDA and commercial burial sites. Only the commercial sites are expected to receive waste from the nuclear fuel cycle, but experience gained at ERDA sites is essential to a thorough understanding of the existing technology for shallow land burial of radioactive waste.

Commercial Burial Practices

The commercial burial sites at Beatty, Nevada; Richland, Washington; Sheffield, Illinois and Morehead, Kentucky are operated by Nuclear Engineering Company.

The West Valley, New York site is operated by Nuclear Fuel Services, and the Barnwell, South Carolina site is operated by Chem-Nuclear Systems. The West Valley site is not presently accepting waste for burial.

FIGURE 4.11: MAJOR GENERATING, STORAGE AND DISPOSAL SITES FOR SOLID LOW-LEVEL RADIOACTIVE WASTE

Source: TID-27341

The waste types presently accepted for burial are given in Table 4.6. In general, the high-activity wastes result from liquid waste treatment processes or from removal of equipment or materials from reactors. Wastes are not accepted for burial if they present a significant nonradioactive hazard such as explosion or fire.

TABLE 4.6: WASTE TYPES RECEIVED AT COMMERCIAL SITES

Low Level	Higher Activity
Paper trash	Ion-exchange resins
Clothing	Filters
Laboratory glassware	Evaporator sludges
Radiopharmaceuticals	Activated materials
Obsolete equipment	Shielding
Building rubble	Solidified liquids

Source: LA-UR-76-1722

Waste packaging is intended to meet Department of Transportation requirements, and in some cases permit above-ground storage of the waste for short time periods.

However, no credit is taken for the containment capability of the package following burial, as the soil material is considered to provide the primary containment barrier.

All sites use open trenches or pits as the primary disposal facility. The trenches are about 7 meters deep, but vary considerably in length and width. Waste is placed in the trench in a sequential fashion, usually starting at one end, and covered with earth at intervals dependent on the waste type and climate. Trenches containing dry wells or specialized casks are used for disposal of some higher-activity wastes. Trenches at the humid eastern sites are commonly sloped to one end, and may contain water collection facilities at the low end for use during or after trench filling. When the entrenched waste accumulates to within a meter or so of the surface, a backfill of uncontaminated earth is placed over the waste and mounded a few feet above the original land surface to encourage precipitation runoff.

The radioactive materials contained in buried waste range from those with very short half-lives, such as radiopharmaceuticals and some fission products, to long-lived materials, such as uranium and plutonium. Present policies restrict the burial of plutonium at concentrations above 10 nCi/g of waste at all but the Richland site, but some such materials remain in place from previous years.

Radionuclides presently buried at commercial sites are divided into three categories. These are Special Nuclear Materials (SNM), which are fissionable materials, such as ^{235}U or ^{239}Pu; Source Materials, such as natural uranium, which can be converted to SNM in a reactor; and By-Product Materials, which include all other radionuclides.

As of January 1974, these wastes composed a volume of over 200,000 m³, containing approximately 2 x 10⁶ Ci of radioactivity (47). Of the total SNM, approximately 80,000 g, or 6%, is ^{239}Pu. The annual volume of wastes received at the various sites is expected to increase dramatically in the coming decade, as the nuclear power industry expands. Various estimates have been made of expected annual volume, ranging up to a maximum of nearly 3 x 10⁶ m³/yr by the year 2000 (48). Something between 100 and 200 acres of land will be required annually to dispose of the projected waste volumes.

The Nuclear Regulatory Commission (NRC) has jurisdiction over the licensing of commercial burial sites, except as it has relinquished that activity to its Agreement States. At present, only the Sheffield site is licensed directly by the NRC, but some activities, such as the handling of specified quantities of special nuclear materials, are regulated by NRC at all sites. Federal regulations require that the title to the land used for burial reside with a state or Federal agency. Thus, commercial burial operators have generally purchased the land and deeded the portions required for burial to the state, which then leases the land to the site operator.

Long-term care of the site following termination of operation rests with the land owner, be it state or Federal. To provide funds for such care, the states collect fees from the burial ground operators. These monies are either placed in a general fund or used to establish a trust fund designated for perpetual site care. Some states require a performance bond, which may decrease in magnitude as the perpetual care fund increases. The funds established through this system are intended to provide for routine monitoring and maintenance following site closure, as well as minor remedial actions that may be required in the event of radionuclide migration. However, the funds are not of sufficient magnitude to provide for major corrective action, including possible waste exhumation, should the need arise in the future.

ERDA Burial Practices

Major ERDA burial grounds are located at Hanford, Washington; at the Idaho National Engineering Laboratory (formerly NRTS), in Los Alamos, New Mexico; at Oak Ridge, Tennessee; and Savannah River, South Carolina (see Figure 4.11). These burial grounds receive wastes generated onsite and by other ERDA facilities. Operations at ERDA burial grounds are similar to those at commercial facilities, with pits or trenches serving as the principal disposal facility. Use is also made of vertical shafts for the disposal of specialized waste materials, such as tritium-contaminated material, waste with high beta/gamma activity, classified material, and animal tissue.

In general, waste packaging at ERDA sites is intended to provide only the containment required during transportation. However, some specialized packaging is used for wastes, such as tritium-contaminated material, with the intent of providing containment after burial.

The wastes handled at ERDA burial grounds are similar to those placed in commercial sites, differing principally in the relative quantities of radionuclides and waste types. The sites are required to report annually to ERDA regarding the total curie content of the buried wastes, as well as the gram quantities of uranium and transuranic radionuclides in the waste. Waste volumes buried at the five major ERDA burial sites total about 1×10^6 m^3 and contain approximately 9×10^6 Ci of radionuclides (49). Unlike the commercial sites, annual inputs to the ERDA sites are expected to remain at about the present level for the forseeable future.

Hydrogeologic Characteristics

The single most important factor affecting the containment capability of a burial ground is the degree to which ground and surface water can contact the waste and subsequently cause migration of radionuclides. As a result, a hydrologic assessment is required as a portion of the licensing procedure for each commercial site. Similar studies are in progress at all existing or planned burial grounds at the major ERDA sites. Climatic and geologic conditions vary dramatically from site to site, with particularly strong contrasts existing between the humid conditions in eastern states and the arid climate of the western states.

Monitoring

The performance of a burial ground, or the degree to which it provides containment of radionuclides, is determined by monitoring programs conducted during and after operation of the burial ground. Monitoring programs at both commercial and ERDA burial grounds are summarized in Table 4.7. The table provides data on the distribution of radionuclides in the soil and rock materials, in the air, water, plants, and animals within or adjacent to the site.

TABLE 4.7: SUMMARY OF MONITORING PROGRAMS AT COMMERCIAL AND ERDA SITES

Medium Sampled	Typical Frequency
Water	Monthly to quarterly
Wells	
Streams and rivers	
Surface runoff	
Soil and rock materials	Quarterly
Near surface	
Borings	
Vegetation	Quarterly
Air	Continuous during operations
Animals, fish, milk, others	Weekly to annually

Source: LA-UR-76-1722

Monitoring programs at ERDA sites are commonly included in a larger environmental monitoring program conducted as part of the overall laboratory operations rather than being specific to the burial grounds. Monitoring programs indicate that, for some sites, changes in environmental levels of radionuclides have resulted from operation of the burial ground. Tritium migration has been separated out as a special case for two reasons. First, tritium is not adsorbed by soil materials, but moves with any migrating water. As such it may serve in a tracer capacity, indicating the extent of water movement through buried waste. Secondly, unlike other radionuclides, tritium does not require a transport medium: it is capable of migrating in the vapor phase, as tritiated water, through unsaturated soil. Thus, migration of tritium does not necessarily indicate that a potential exists for the migration of other radionuclides.

A principal cause of the observed migrations at all sites is the entry of water into a pit or trench through the cover material, rather than by lateral movement of water through the trench. Frequently, the natural materials surrounding the trench are less permeable than the trench fill and cover material. If the climate is sufficiently wet that precipitation can infiltrate the cover material more rapidly than it is conducted away from the trench, water will accumulate in the trench. This bathtub effect may result in an overflow at the ground surface, or in the lateral migration of water and radionuclides, due to the increase in hydrostatic head in the trench. In addition, the saturated conditions in the trench enhance leaching of radionuclides from the burial waste.

Burial sites in the more humid portions of the United States provide sand drains or sumps at the low end of the trench to provide for removal of water. This water is treated, producing a liquid or atmospheric effluent at or below the Maximum Permissible Concentrations for drinking water or breathing air.

Radioactive releases from some burial grounds have resulted in increased levels of radioactivity in the offsite environment. However, the nature of the subsurface system at many of the burial sites is not sufficiently understood to allow a definitive statement of the actual cause of the releases. At the Morehead site, for example, doubt exists as to which of several possible pathways were responsible for the elevated levels observed by the monitoring system.

Role of Shallow Earth Burial in Waste Disposal

Radioactive solid waste, in all categories other than high-level waste and transuranic materials, is currently being disposed of by shallow earth burial. Some radionuclide releases to the environment have resulted from these practices, but none of the releases have endangered public health or safety. Modifications to present practices are possible that could increase the containment capability of burial sites, with the goal of isolating the wastes from man's environment for their hazardous lifetimes.

The near-surface environment is subject to more rapid change than deep geologic formations. Processes that might alter the containment capability of a completed site relate both to the natural hydrogeologic system and to man's activities. The time period for which these influences can be predicted, and thus compensated for through burial site design, is in the range of hundreds to thousands of years. Materials which present significant hazards for longer time periods, such as high-level waste, are not appropriately disposed of through shallow land burial. However, the bulk of radioactive waste generated by the nuclear industry contains radionuclides which are either short-lived or at sufficiently low concentrations as to not present a significant hazard over very long time periods. The great volume of these wastes requires an economical but effective means of disposal. Land burial sites, properly designed and operated, can provide this disposal mode.

Effective use of shallow land burial requires emphasis on stable waste forms, hydrologically-oriented containment engineering, and the natural materials surrounding a burial site. Decisions regarding the relative importance of these mechanisms at a particular site must be based, in part, on economic considerations.

Recommended Principles for Burial of Radioactive Waste

The following recommendations for shallow land burial of low-level radioactively contaminated solid waste were made in 1976 by the Panel on Land Burial of the Committee on Radioactive Waste Management:

Good Housekeeping and Careful Segregation: Segregation and good housekeeping at the source of production must be used to keep the contamination of nonradioactive waste to the lowest practicable level, and to simplify placing waste into suitable categories for particular burial facilities. This will tend to reduce the volume of solid low-level radioactive wastes that are candidates for land burial.

Volume Reduction: Solid radioactive waste should be processed for volume reduction by compaction and incineration, whenever possible, and placed in suitable containers, particularly if retrieval is foreseen. Combustible waste should be incinerated. This process reduces its volume and gives a product that is fireproof and easier to handle. Specific attention should be given to the problems of incineration of organic matter, including the carcasses of experimental animals that have been treated with radioactively labeled substances.

Long-Term Management and Containment: The major goals of a permanent low-level solid waste burial operation are to contain the waste until it decays to an innocuous level, to guarantee that the radioactive waste is under control at all times, and to guarantee that if contaminants migrate from the site, their movements can be predicted so that necessary corrective actions can be taken.

Inaccessibility of Wastes:　In the selection of permanent burial sites and the quantity and type of waste disposed at the sites, it must be realized that it cannot be expected to be able to control access to the site for an indefinite time, and that any hazardous waste remaining at the site after a specified time limit must be as difficult to gain access to as possible.

Monitoring Systems:　A monitoring system must operate at both permanent burial sites and at storage sites so that surface or air contamination will be detected quickly.　Ground and surface water beneath or very near to the burial facilities should be monitored sufficiently often to give the earliest practical warning of failure of any facilities.　"Failure" is defined as significant contamination of the ground or surface water in excess of standards that have been set for a disposal site.　Monitoring should also include adequate biological and ecological monitoring of the biosphere to detect entrance of radionuclides into the local biosphere.

Contingency Plans:　Contingency plans must be made to cover all foreseeable accidents or failures.　They must include plans for corrective action in the event that monitoring shows a hazardous spread of contamination.　If plans for such corrective action cannot be made and put into effect, then the site is not acceptable.

Government Control or Ownership of Sites:　All radioactive waste management sites must be controlled by, or both owned and controlled by, a government agency.　The present practice of controlling public access to such areas must be continued.　The sites should be located in a place remote from human habitation.

Engineered Containment Structures:　Engineered structures for the containment of waste must be designed with the intent to keep water, which can mobilize the contaminants, out of the facility.　They should be located in a place where rupture would not lead to rapid transfer of contaminants into the public domain before they could be recovered.

Early Warnings of Human Activities:　Note must be taken of the operations and activities of instigators or proposers for nearby real estate development, road building, airport construction, and other activities that might change the anticipated probable hazards or accessibility of the low-level waste burial facilities. The appropriate government agencies should be required to report any such considerations to those individuals or commercial venturers and to gather demographic knowledge of the area in which a land burial site is located.　Forecasts of credible usage of the surrounding area by man in the future must be obtained or developed.

Records Management:　Duplicate records of the types, quantities and concentrations of radioactive waste nuclides delivered to a burial site must be made and filed with more than one record bank.　Reports on monitoring results and significant incidents, such as spills or unanticipated release of waste, must be filed with more than one record bank.　These records should show the real (i.e., observed, not calculated) level of contamination of the environment (including the ground area).　These records must be in such a form that they will be useful and available for the effective length of time that the waste burial facility will require human attention.　In addition, such records must be kept not only of the ERDA burial sites but of all sites used for radioactive waste.　Historical records of the use of older burial sites, although often incomplete, should be included in the record.

Cost Analysis: The operation of a waste management system must be governed by a realistic cost-benefit analysis, taking into account the cost in radiation exposure to operators or to the public of any act or failure to act. It must be recognized that the cost in man-rems to the community is the same whether it is absorbed by an operator or by a member of the general public outside the management area.

Transportation Costs and Risks: Although the location of disposal areas in inhospitable regions seems desirable, transportation of waste involves hazards normal to the transportation industry as well as hazards due to radiation. The analysis of any plan that involves transportation must include all hazards and economic data.

Nonexhumation of Radioactive Wastes: Exhumation of waste originally buried without any intent of later retrieval is a potentially very hazardous operation. Such exhumation should not be made unless there is a credible reason to believe that a significant radiation hazard could arise from leaving the waste where it is and that the wastes can be exhumed safely. As a corollary to this recommendation, radioactive waste should not be exhumed and put into temporary engineered storage where the material must await a final decision on permanent disposal.

Federal Definitions and Guidelines: The U.S. government must define the responsibilities and assign the authority for setting guidelines and policies and ensuring the proper coordination between Federal, state, local, and international agencies for the effective management of radioactive waste. Division of responsibilities between the Federal agencies must be done in such a way as to ensure that all significant areas of responsibility are properly covered and none is overlooked. Guidelines and standard procedures may vary from site to site because of local variations in geological conditions and other circumstances. However, a standardized system of definitions, nomenclature, and classification of radioactive wastes is necessary for an effective national and international program of radioactive waste management.

DEEP WELL DISPOSAL OF LOW-LEVEL LIQUID WASTE

An economical method of disposal for large quantities of low-level transuranic contaminated liquid waste has been developed at an ERDA experimental farm located near the northern boundary of the Nevada Test Site (NTS) and operated by the EPA. Since 1972 the farm has been conducting studies to determine the environmental impact of selected transuranic elements, and plutonium-238 was picked as an appropriate study isotope.

In a typical experiment, dairy cows are administered measured amounts of a specific radionuclide either by the ingestion of gelatin capsules or contaminated green feed, or by intravenous injection. During the course of the experiment the involved cows are confined in metabolism stalls which permit total collection of urine and feces, and facilitate the collection of milk and blood samples. Daily collection and analysis of milk, urine and feces is made during all experiments. Additionally, one or more of the dairy cattle may be sacrificed for detailed autopsy and analysis following some particular experiment (50). During an experiment, animal carcasses, fecal matter, urine and milk are collected separately in 55-gallon steel drums and solidified for burial in a Radioactive Material Management Area.

In early planning for the expansion of experiments to include plutonium-238, it was determined that the potential existed for large quantities of low-level transuranic contaminated liquid waste to develop during the course of the project. This waste would primarily be wash and decontamination water resulting from cleanup at the experiment stalls and laboratory area.

Some method of disposal for these liquids had to be found, and the large volumes involved made traditional practices unfeasible for economic reasons. Rather than pumping into a sump or evaporation pond, or attempting a solidification process, the suggestion was made that expended underground nuclear explosion detonation sites be used for disposal of the transuranics. This practice would be in keeping with current ERDA policy of not creating any additional Radioactive Waste Management Areas.

The additional activity resulting from the addition of large volumes of low-level liquid waste to a typical underground cavity would be insignificant compared to the radionuclide inventory present after a nuclear event. The solution seemed ideal from all viewpoints, because:

(a) The hydrogeology of the test site region had been studied at some length by the United States Geological Survey, and had been well defined.

(b) The underground testing program had created scores of possible disposal sites.

(c) The disposal wells had already been drilled as part of the weapons test program and would cost nothing.

(d) The wells had more than enough capacity for any expected volume of waste water from the farm.

(e) The follow-up environmental monitoring would be taken care of routinely by the existing NTS-wide environmental surveillance program conducted by Reynolds Electrical and Engineering Co., Inc. (REECo).

Yucca Flat, the site of over 350 underground nuclear detonations, is an intermontane basin lying at elevations between 3,900 and 5,000 feet above sea level, surrounded by ranges at altitudes of 5,700 feet on the east, 6,900 feet on the north, and 7,600 feet on the west. The region is geologically complex, but studies continuing since the advent of underground testing have developed a considerable body of knowledge in regard to possible radioactive contamination of ground water supplies.

The alluvial fill of the valley varies in thickness from a few hundred feet along the hills to 2,000 feet or more in some central portions. This alluvial layer is underlain by beds of clayey tuff and/or tuffaceous sediments that comprise a tuff aquifer and tuff aquitard which also range up to about 2,000 feet thick. The tuff, in turn, is laid over the lower carbonate aquifer, which is principally dolomite and limestone in the area of the designated disposal wells. Available data indicate that the flow of subsurface water at the level of the water table is toward Yucca Flat from the north, east, and west. The exit from NTS for the subsurface water appears to be to the south-southwest toward the Amargosa Valley, particularly a series of springs in the Ash Meadows area (51). Chemical analysis of spring and well water from the Test Site and from the spring line and in the Amargosa Valley tend to confirm this postulation. Since the water is put to a variety of agricultural uses in the Amargosa Valley, some age determinations become necessary from a health physics viewpoint. A recent report by Issac J. Winograd and William Thordarson, USGS, made these pertinent points (52):

(a) Ground water at the Nevada Test Site drains from the alluvial fill and other strata through the tuff aquifer and Cenozoic tuff aquitard and into the lower carbonate aquifer.

(b) Cenozoic tuff is an aquitard of extremely low transmissivity. Estimates of the average time needed for a water particle to move from the top to the bottom of the tuff aquitard, using an average thickness of 1,000 feet, range from several thousand to as much as 2,000,000 years.

(c) Transport time estimates for water to travel in the lower carbonate aquifer from beneath the Nevada Test Site to the Ash Meadows spring line in the Amargosa Valley range from around 400 years to approximately 30,000 years. The wide bracketing of calculated transit times is due principally to variations of the average fracture porosity of the aquifer as estimated by geologists studying available data.

(d) In summary, then, the time required for water to travel from a detonation point in the Yucca Flats area to the Amargosa Valley is that time required for downward leakage through the tuff aquitard plus that time needed for horizontal transport, and ranges from several thousand to around 2,000,000 years.

(e) There is significant inflow of water from other basins into and through the lower carbonate aquifer underlying the Nevada Test Site. Only a small fraction (on the order of 5%) of the water emerging in the Ash Meadows area is thought to originate in the Yucca Flats basin.

An interesting sidelight is that there are zeolitized zones within the bottom half of the tuff aquitard. Zeolites, of course, have the property of base, or cation, exchange. This same exchange mechanism can be used to soften water by passing hard water through grains of zeolite containing sodium ions. The sodium ions exchange with calcium ions in the water. These zeolite sodium ions (naturally occurring under the Nevada Test Site) also exchange with fission products and fissionable materials, notably plutonium. Studies have indicated that this ion-exchange effect would probably prevent most radionuclides from moving more than a few thousand feet from the point of detonation regardless of flow rates of the ground water (52).

There is also a tendency for atoms of heavy metals to adsorb to the crystal lattice of soil particles in clayey strata. This is probably secondary to the cation-exchange effect in the tuff underlying the Nevada Test Site, but quite possibly could add something to the protection factor.

Public water supplies are thus protected in four major ways: the remoteness of the Nevada Test Site; the extremely low transmissivity of the tuff formations underlying the Yucca Basin coupled with long transit times in the lower carbonate aquifer and the resultant age of water when it does surface; cation exchange in the zeolitized tuff; and dilution of the water percolating down in the Yucca Basin by overwhelming quantities of water flowing through the lower carbonate aquifer from other sources.

Once the approval for down-hole disposal of EPA's waste had been received, implementation quickly followed. A surplus, contaminated 25,000-gallon tank was retrieved from the NTS Radioactive Waste Management Area and two 3,700-gallon fuel tankers were obtained from military surplus. Other than an inspection and repair of all leaks, and the installation of drain valves with quick disconnect couplings and drip-proof caps, this equipment is utilized just as it was received. The 25,000-gallon tank was installed as a temporary holding facility at the farm, and the fuel tankers were marked with appropriate warnings and are used exclusively for the transport of the liquid waste originating at the farm. Matching fittings were installed on two postshot sampling drill holes selected for geologic suitability and proximity to the farm. A centrifugal pump was similarly fitted, to be used to speed the dumping operation and overcome downhole back pressure if encountered.

The actual dumping process is very simple and straightforward. REECo is notified when EPA determines that the storage tank should be emptied (usually at the end of an experiment). Upon notification, one of the disposal trucks is dispatched to the farm.

The tank truck is filled following strict procedures. Personnel wear appropriate anticontamination clothing and visual control of the area and of the fill operation is maintained at all times. The operation is aborted if leaks are observed. Upon completion of the loading operation, hoses and pipes are allowed to drain before disconnects are made and protective caps reinstalled. Portable instrument and smear surveys are then made of the truck and the fill stand. If these reveal negligible contamination, and after the tanker is rechecked for leaks, the teamster is allowed to proceed to the waste dump hole.

To dump down the drill hole, the tank truck is connected via the centrifugal pump to the fittings on the disposal well, and the load is pumped, or allowed to gravity feed through the pump. After unloading the tanker, it and the pump are flushed with clean water before disconnecting. Samples of the waste are drawn during the pump-down to confirm the activity estimates of the disposal for the record, and an additional sample is taken at the end of the flushing process to confirm the relative decontamination of the equipment.

The nearest sampling point for the environmental surveillance program is a 2,500-foot deep industrial water well located approximately 4,000 feet southeast of the disposal hole. This water is sampled monthly and analyzed for pertinent nuclides. As with all other normal water sources on the Test Site, no levels above average background have ever been detected.

At the Nevada Test Site, then, a solution to the problem of disposal of low-level contaminated liquid waste has been found. Approximately 25,000 gallons of low-level plutonium-238 contaminated water have been pumped down-hole thus far.

Factors involved in the decision to dispose of waste water in this fashion were
the availability of deep holes for dumping, the known aspects of the hydrogeology of the region, the low expense of implementation, and an ongoing environmental surveillance program to provide baseline radioanalytical data.

OTHER DISPOSAL CONCEPTS

Ice Sheet Disposal

Concepts for radioactive waste disposal in the major ice sheets of the world
(Greenland and Antarctica) are being evaluated. Potential advantages are great
thicknesses of ice, remoteness from man's activities, and low likelihood for future development. The ice could provide effective direct cooling for the waste
and, at the same time, maintain isolation from man's environment. Like seabed
disposal, ice sheet disposal is not a recommended waste management concept due
to additional technological uncertainties associated with working in the arctic,
and due to international legal considerations.

Three potential ice sheet disposal concepts are being evaluated:

> Meltdown or Free Flow — The waste canister would be placed in
> an individual shallow drilled hole in the ice and allowed to melt
> down through the ice sheet to bedrock.
>
> Anchored Emplacement — The waste canister would be placed in
> an individual shallow drilled hole in the ice but connected to sur-
> face anchors by cables or chains to stop its descent and maintain
> its position (500 to 1,500 feet below the ice surface) for up to
> about 100 years.
>
> Surface Storage/Disposal — The waste canister would be placed
> in a temporary hot cell type storage facility with jack-up piers
> on the ice sheet surface. After about 50 years, the facility
> would be allowed to become covered by accumulating snow and
> would be eventually buried in the ice sheet for final disposal.

Disposal of radioactive waste in ice sheets is considered to have an uncertain potential for isolating waste from man's environment, depending largely on long-
term stability of the ice. Also, international agreements governing this procedure
would undoubtedly be needed.

Extraterrestrial Disposal

Disposal of radioactive waste via rockets is another disposal concept being evaluated. If a stable, nonearth intercept trajectory or orbit can be guaranteed, extraterrestrial disposal offers a method for the complete removal of long-lived
nuclear waste constituents from the earth. The primary unfavorable features are
that the concept deals with only part of the waste due to limitations on weight;
there may be launch safety problems; retrievability and monitoring are difficult;
and there is the possibility of international disagreement.

Since geologic disposal appears adequately safe, and the feasibility and safety of
extraterrestrial disposal remains to be demonstrated, it is not recommended for
a waste management system.

Extraterrestrial disposal of the total waste constituents and of only the actinides were both considered. However, primarily because of the high cost per unit of weight of space transport, space disposal of just the actinides is believed to be most practical. In order to implement this option, partitioning technology will have to be developed so that the waste can be divided into an actinide fraction and an actinide-free fraction. The latter must be pure enough so that the remaining waste can be disposed of under regulations governing waste free of transuranic contamination. The incentives to undertake the effort needed to achieve this are at least uncertain.

Transmutation Concepts

Another possible approach is the use of nuclear processes themselves to change (transmute) the hazardous long-lived radioactive waste constituents into short-lived or nonradioactive isotopes. Transmutation is not a presently feasible approach because of additional technological developments required in separation (i.e., partitioning technology), and in demonstrating fusion reactors.

Transmutation is generally defined as any process whereby a nuclide absorbs or emits radiation and is thereby transformed into another nuclide. Ideally, transmutation could completely eliminate the noxious isotopes. It is theoretically possible to achieve the transmutation. Both fission and fusion reactors show promise for transmutation of actinides. Fusion reactors also may be used to transmute selected fission products because of the potential availability of inexpensive high-energy neutrons.

The transmutation concept of continual recycle of actinides in fission reactors appears to be potentially attractive. The calculations indicate that using existing separation efficiencies, recycling in light water power reactors could achieve an order of magnitude decrease in the short-term actinide toxicity and about a factor of fifty decrease in the long-term toxicity.

From a practical standpoint, however, use of current reprocessing separation efficiency may not provide much improvement because any actinides left in the fraction not transmuted will mean that it will still have to be treated as transuranic contaminated HLW. Thus, the theoretical benefit from transmutation can be achieved only after development of an improved partitioning technology.

REFERENCES

(1) Pacific Northwest Laboratories, *Alternatives for Managing Wastes from Reactors and Post-Fission Operations in the LWR Fuel Cycle,* ERDA-76-43, vol. 3, May 1976.

(2) Duke Power Company, *Oconee-Nuclear Station Units 1, 2, and 3, Final Safety Analysis Report,* Charlotte, North Carolina, May 1969.

(3) U.S. Nuclear Regulatory Commission, *Draft Environmental Statement Related to the Operation of the Barnwell Fuel Recycling and Storage Station,* Division of Materials and Fuel Cycle Facility Licensing, Washington, D.C., May 1975.

(4) U.S. Nuclear Regulatory Commission, *Regulatory Guide 3.24, Guidance on the License Application, Siting, Design, and Plant Protection for an Independent Spent Fuel Storage Installation,* Division of Materials and Fuel Cycle Facility Licensing, Washington, D.C., December 1974.

(5) U.S. Nuclear Regulatory Commission, *Standard Review Plan, Section 9.1.2 Spent Fuel Storage,* Division of Materials and Fuel Cycle Facility Licensing, Washington, D.C., February 1975.

(6) Atlantic Richfield Hanford Company and Kaiser Engineers, *Retrievable Surface Storage Facility Alternative Concepts, Engineering Studies,* ARH-2888 (Rev.), Richland, Washington, July 1974.

(7) D.C. Nelson and D.D. Wodrich, "Retrievable Surface Storage Facility for Commercial High Level Waste," *Nucl. Technol.,* 24, 391, 1974.

(8) W.W. Morgan, "The Management of Spent CANDU Fuel," *Nucl. Technol.,* 24, 409, 1974.

(9) United States Atomic Energy Commission, *Management of Commercial High-Level and Transuranium-Contaminated Wastes,* WASH-1539, September 1974.

(10) C.A. Burchsted and A.B. Fuller, *Design, Construction, and Testing of High-Efficiency Air Filtration Systems for Nuclear Application,* ORNL-NSIC-65, 1970.

(11) United States Atomic Energy Commission, *The Safety of Nuclear Power Reactors (Light Water-Cooled) and Related Facilities,* WASH-1250, July 1973.

(12) K.J. Truss, *Concrete Cannister Demonstration Project Preliminary Analysis Report,* Whiteshell Nuclear Research Establishment, Pinaua, Manitoba, 1975.

(13) P.J. Dyne, *Managing Nuclear Wastes,* AECL-5136, May 1975.

(14) J.A. Morrison, *AECL Experience in Managing Radioactive Wastes from Canadian Nuclear Reactors,* AECL-4707, March 1974.

(15) I.J. Winograd, "Radioactive Waste Storage in the Arid Zone," *Trans. Am. Geophys. Union,* 55, 884, 1974.

(16) W.L. Lennemann, H.E. Parker, and P.J. West, "Management of Radioactive Wastes," *International Atomic Energy Agency Bulletin,* vol. 17, No. 4, August 1975.

(17) United States Atomic Energy Commission, *Generic Environmental Mixed Oxide Fuel (GESMO),* WASH-1327, August 1974.

(18) W.A. Rodger (Nuclear Safety Associates) and J.M. Vallance (Pickard, Lowe and Associates), *The Throwaway Fuel Cycle,* March 1975.

(19) United States Energy Research and Development Administration, *Nuclear Fuel Cycle: A Report by the Fuel Task Force,* ERDA-33, 1975.

(20) M. Resnikoff, "Expensive Enrichment," *Environment,* vol. 17, No. 5, p. 28, July/August 1975.

(21) S.G. Ledford and E.H. Gift, *Economics of PWR Spent Fuel Recovery,* K-OP-185, Rev. 1, September 5, 1975.

(22) Savannah River Laboratory, *Assessment of Nuclear Fuel Reprocessing,* DPST-75-448 (Draft), September 1975.

(23) R.J. Chlorister et al, (Allied General Nuclear Services), *Nuclear Fuel Cycle Closure Alternatives,* April 1976.

(24) A.J. Mooradian, "CANDU Fuel Cycles—Present and Future," presented at the Ninth Annual Conference of the Japan Atomic Industrial Forum (JAIF), Tokyo, March 10-12, 1976.

(25) G.E. Bingham and T.K. Evans, *Final Safety Analysis Report for the Irradiated Fuels Storage Facility,* Allied Chemical Corporation, Idaho Falls, Idaho, December 1975.

(26) S.A. Mayman and W.W. Morgan, *Fuel Cycle Wastes—The Canadian Program,* Atomic Energy of Canada Limited, Whiteshell Nuclear Research Establishment, Pinaua, Manitoba.

(27) P.J. Dyne, *Waste Management in Canadian Nuclear Program,* AECL-5249, August 1975.

(28) L.A. Bertram, "Mechanics Analysis Required for Deep Rock Disposal of Radioactive Wastes," 1975.

(29) R.D. Klett, *Deep Rock Nuclear Waste Disposal Test: Design and Operation,* SAND-74-0042, 1974.

(30) D.W. Larson, D.K. Gartling and H.C. Hardee, *Thermal Response of Buried Radioactive Waste: Conductive Solutions,* SAND-75-0410, 1975.

(31) J.J. Cohen, A.E. Lewis, R.L. Braun, "In Situ Incorporation of Nuclear Waste in Deep Molten Silicate Rock," *Nucl. Technol.,* 13, 76, 1972.

(32) H.C. Claiborne, "Effect of Actinide Removal on the Hazard of High-Level Waste," *Trans. Am. Nucl. Soc. 1974 Ann. Meeting,* Philadelphia, Pennsylvania, 1974, vol. 18, p. 190.

(33) J.A. Adam, *Thermal Considerations in the Disposal of High-Level Radioactive Wastes in a Deep Underground Cavity Created by a Nuclear Explosion,* M.S. thesis, North Carolina State University, Raleigh, North Carolina, 1973.

(34) R.G. Post, J.A. Angelo, *Plowshare—Versatile Tool for By-Product Management,* The University of Arizona, Tempe, Arizona, 1972.

(35) D.H. Stewart, J.R. Sheff, Battelle Memorial Institute, Pacific Northwest Laboratory, Richland, Washington, private communication, 1972.

(36) W.W. Hambleton, "The Unsolved Problem of Nuclear Wastes," *Technology Review,* March/April 1972, p. 15.

(37) C.M. Slansky, J.A. Buckham, "Ultimate Management of Radioactive Liquid Wastes," *Chemical Engineering Progress Symposium Series,* 97, (65), 26, 1969.

(38) *High-Level Radioactive Waste Management Alternatives,* U.S. Atomic Energy Commission, Rept. WASH-1297, 1974.

(39) S.E. Logan, "Deep Self-Burial of Radioactive Wastes by Rock Melting Capsules," presented at the *American Nuclear Society 19th Annual Meeting,* Chicago, Illinois, June 10-14, 1973.

(40) J. Donea, S. Giuliani, L. Kestemont, *Process of Self-Burial of Containers Holding High-Activity Radioactive Waste,* European Atomic Energy Community, Nuclear Research Center, Ispra, Italy, Rept. EUR-4903f, 1972.

(41) A.S. Kubo, *Technology Assessment of High-Level Nuclear Waste Management,* ScD. thesis, Dept. of Nuclear Engineering, Massachusetts Institute of Technology, Cambridge, Massachusetts, 1973.

(42) *Overview of High-Level Radioactive Waste Management Studies,* Battelle Memorial Institute, Pacific Northwest Laboratories, Richland, Washington, Rept. BNWL-1758, 1973.

(43) *Minutes of a Workshop on the Underground Disposal of Radioactive Wastes in Molten Rock,* held at Sandia Laboratories, Albuquerque, New Mexico, October 18-19, 1973, Rept. RS-3150/1702, 1974.

(44) Code of Federal Regulations, Title 10, Part 50, Appendix F.

(45) A.S. Kubo, D.J. Rose, "Disposal of Nuclear Wastes," *Science,* 182, 1205, 1973.

(46) H.C. Hardee, W.N. Sullivan, *Waste Heat Melt Problem,* Sandia Laboratories, Albuquerque, New Mexico, Rept. SLA-73-0575, 1973.

(47) M.F. O'Connel and W.F. Holcomb, "A Summary of Low-Level Radioactive Wastes Buried at Commercial Sites Between 1962-1963, with Projections to the Year 2000," *Radiation Data and Reports,* vol. 15, No. 12, December 1974.

(48) J.O. Blomeke, C.W. Kee and R. Salmon, *Projected Shipments of Special Nuclear Material and Wastes by the Nuclear Power Industry,* Oak Ridge National Laboratory report, ORNL-TM-4631, August 1974.

(49) W.J. Smith, J.L. Warren, M.L. Wheeler and W.J. Whitty, "Alternatives for Managing Waste from Reactors and Post Fission Operations in the LWR Fuel Cycle," Section 24.1, *Burial Grounds,* ERDA-76-43, May 1976.

(50) "EPA Experimental Farm," *Environmental Facts,* Newsletter of the Environmental Protection Agency, Las Vegas, Nevada, April 1976.

(51) I.J. Winograd and W. Thordarson, "Structural Control of Ground Water Movement in Miogeosynclinal Rocks, of South-Central Nevada," *Nevada Test Site,* E.B. Eckel, Ed., USGS Memoir 110, 1968.

(52) I.J. Winograd and W. Thordarson, *Hydrogeologic and Hydrochemical Framework, South-Central Great Basin, Nevada-California, with Special Reference to the Nevada Test Site,* USGS Professional Paper 712-C, 1975.

RECOMMENDATIONS FOR DISPOSAL OF RADIOACTIVE ELEMENTS

The source of the material in this chapter is the NTIS report,
PB-224 588. See p 359 for a complete bibliography.

CARBON-14, COBALT-60, IRIDIUM-192 AND RADIUM-226

Radium-226 is one of the most hazardous materials known. Radium-226 replaces
calcium in the bone structure and is a source of irradiation to the blood-forming
organs. This, along with its long half-life (1,620 years) and high radiation energies, places it in the highest radiotoxicity group. It also has the longest history
of use of any radioactive material, and most of the standards for the effects of
ionizing radiation on man are based on this material. Carbon-14, cobalt-60, and
iridium-192 are moderately dangerous radioactive materials.

Handling, Storage and Transportation

Since these radionuclides are hazardous to man by inhalation, ingestion, or direct
radiation exposure, great care should be exercised in their handling. Handling
should be set up to prevent excessive exposure to personnel. Special procedures
and adequate radiation shielding are required in their handling. Radium-226 is
an alpha emitter. These alpha particles have little penetrating power and can be
stopped by a sheet of writing paper. The beta particles emitted by carbon-14,
cobalt-60, and iridium-192 range in energy from 0.16 to 0.67 Mev and cannot
penetrate more than 0.015 to 0.085 inch of water.

The gamma radiation emitted by cobalt-60, iridium-192, and radium-226 is highly
penetrating and high density shields, such as lead, are required to stop the radiation. For cobalt-60 which emits high energy gamma rays (1.33 Mev), approximately 3.5 inches of lead or 20 inches of concrete are required to reduce its radiation by a factor of 100. To detect and control personnel exposure to their
radiation, all persons working with this material should wear dosimetry devices
which directly indicate the dose. Other commonly used devices are the film
badge and the thermoluminescent dosimeters (TLD).

Specially constructed containers in controlled areas should be used for storing large quantities of these materials. They should be protected by both a primary and a secondary containment barrier. Special monitoring systems and proper warning signs should be located in the general area of the storage facility. Special precautions are required in the storage of radium-226 since approximately one gram of radium produces about 0.001 ml of radon gas per day. Thus, stored radium should be vented to prevent the build-up of radon gas.

Radium-226 is classified as a transport group I radionuclide, cobalt-60 and iridium-192 are classified as transport group III radionuclides, and carbon-14 as a transport group IV radionuclide by the Department of Transportation. The rules and regulations governing their transportation are given in the Code of Federal Regulations (CFR) Title 14—Transportation, Parts 170 to 190 (1). The radium-226 content is limited to 0.001 curies for a Type A package and 20 curies for a Type B package defined in 40CFR173. For cobalt-60 and iridium-192 these limits are increased to 3 and 200 curies. For carbon-14 the limits are 20 and 200 curies. The release rate of these materials is limited to zero under the specified accident conditions for Type A and B packages (1).

Disposal

The disposal of these materials is governed by the AEC Manual Chapter 0524 (2) and 10CFR20 (3). Two sets of standards have been established for the permissible radiation exposure in unrestricted areas. One is for the greatest dose received by an individual and the other for the average dose received by the general population. The standards for the safe release of these materials to the environment in an unrestricted area are contained in 10CFR20 (3) and their release should not result in concentrations in air and water greater than those listed in Table 5.1.

TABLE 5.1: CARBON-14, COBALT-60, IRIDIUM-192, AND RADIUM-226 MAXIMUM PERMISSIBLE CONCENTRATIONS

Radionuclide	Form	Concentration in Air (microcuries/milliliter)	Concentration in Water (microcuries/milliliter)
Carbon-14	Soluble	1×10^{-7}	8×10^{-4}
Carbon-14	Submersion*	1×10^{-6}	--
Cobalt-60	Soluble	1×10^{-8}	5×10^{-5}
Cobalt -60	Insoluble	3×10^{-10}	3×10^{-5}
Iridium-192	Soluble	4×10^{-9}	4×10^{-5}
Iridium-192	Insoluble	9×10^{-10}	4×10^{-5}
Radium-226	Soluble	3×10^{-12}	3×10^{-8}
Radium-226	Insoluble	2×10^{-12}	3×10^{-5}

*Submersion means that values given are for submersion in an infinite cloud of gaseous material.

Source: PB-224 588

These concentrations apply to an individual and their release may be limited if a suitable sample of the population is exposed to one-third concentrations in air or water specified.

Although rarely practiced, the disposal by release in a sanitary sewage system is limited to 0.1 microcurie of radium-226, 10 microcuries of cobalt-60, 100 microcuries of iridium-192, and 1,000 microcuries of carbon-14. The disposal by burial at any one location and time is limited to 100 times the above amounts.

The above regulations for the transportation, release, and disposal of these materials apply only to licensed materials. Radium and other accelerator-produced radionuclides are exempt from AEC control and their regulation and control is a state function. In some states, the possession and use of these materials requires licensing and, in others, no license is required. Thus, the disposal, transportation, and use of these materials in some uses is not subject to the above regulations.

Evaluation of Waste Management Practices

Option No. 1—Land Burial: Land burial of carbon-14, cobalt-60, and iridium-192 wastes, in small concentrations, at approved sites that are acceptable from a geologic standpoint, is an acceptable means of disposal. Their concentration should not be in excess of 10^4 times the maximum permissible concentration for the general population in 10CFR20 (3). All wastes to be disposed of should be in a solid form and encapsulated in a suitable container. The burial site design, geology, and hydrology should be in conformance with the criteria used in selecting and licensing the present commercial burial sites (4). This method of disposal is not considered satisfactory for the disposal of radium-226 because radium-226 has such an extremely long half-life and a high radiotoxicity.

Option No. 2—Near-Surface Solid Storage: The storage of solidified radium-226 wastes in engineered storage facilities offers the best intermediate method for storage of these wastes. The technology for these facilities has been developed and the wastes will be under surveillance and control and can be retrieved, should this be required. The wastes will be stored in stainless steel lined concrete vaults.

Option No. 3—Salt Deposits: This method offers the best potential for the disposal of radium-226 wastes since bedded salt deposits are completely free of circulating ground waters. This method of disposal has been under study by the Oak Ridge National Laboratory since 1957, and, in November of 1970, a committee of the National Academy of Sciences recommended that the use of bedded salt for the disposal of radioactive waste is satisfactory (5). Questions concerning the adequacy of this method have resulted in the need for further development work before it can be accepted as an ultimate method of disposal. The critical problem is the selection of a site that meets the necessary design and geological criteria.

To summarize, small concentrations of carbon-14, cobalt-60, and iridium-192 can be disposed of by land burial at approved sites. Large concentrations of these materials and radium-226 should be disposed of in salt deposits following storage if necessary in near-surface engineered facilities.

Applicability to National Disposal Site

Carbon-14, cobalt-60, iridium-192, and radium-226 are candidates for a National

Disposal Site due to their large commercial usage and high health hazard. The recommended process for the disposal of carbon-14, cobalt-60, and iridium-192 wastes in small concentrations is by land burial. The recommended process for the disposal of radium-226 and large concentrations of the above radionuclides is disposal in salt beds.

CESIUM-134 AND CESIUM-137 (BARIUM-137m)

Cesium-134 and cesium-137 are radioactive isotopes produced by the fission of uranium and plutonium and will exist in combination with other radioisotopes in high-level and low-level radioactive waste streams. Cesium-134 has a fission yield of 8 percent and cesium-137 a fission yield of 6 percent. Cesium-134 is transformed by beta decay into barium-134 which is a stable element. Cesium-137 has a half-life of 30 years and emits only beta particles. Cesium-137 is in radiation equilbrium with its short half-lived daughter (2.6 minutes), barium-137m. Barium-137m, in turn, is transformed by isomeric transition (decay from an excited metastable state to lower state) into barium-137 which is a stable element.

From the viewpoint of waste management, cesium-137, because of its long half-life and high fission yield, is one of the most important fission products produced in nuclear reactors. The only fission product of comparable importance is strontium-90. The importance of cesium-137 is realized by the fact that it accounts for 10 percent of the total fission product activity in nuclear reactor wastes after one year, and for 52 percent of the activity after ten years. Cesium-134 is also of significance due to its high fission yield (8 percent) and high specific power (20 watts/gram).

The projected growth in the production of cesium-134 and cesium-137 will parallel that of the civilian nuclear power program. The expected total accumulated radioactivity of cesium-137 from 1970 to 2020 in millions of curies (6) is:

Year	Radioactivity (megacuries)
1970	5
1980	1,280
1990	6,540
2000	15,600
2020	57,500

Annual production figures for the nuclear weapons program were not available.

Although produced in nuclear power reactors, the primary source of cesium-134 and cesium-137 is in the high-level, aqueous waste streams generated at the spent fuel processing plants. The waste streams obtained from the processing steps are aqueous solutions of inorganic nitrate salts.

The characteristics of cesium-134 and cesium-137 products in the high-level wastes at two different time periods have been determined (Tables 5.2 and 5.3). These radionuclides are also found in the secondary waste streams generated at spent fuel processing plants. The activity in these wastes is quite low (less than a tenth of a curie per gallon).

TABLE 5.2: CESIUM-134 CONTENT IN HIGH-LEVEL FUEL REPROCESSING WASTES

Fission Products	Grams/Tonne	After 1 Year Watts/Tonne	Curies/Tonne	Grams/Tonne	After 10 Years Watts/Tonne	Curies/Tonne
. . . . Light-Water Reactor Fuels*						
[134]Ce	134	1,830	175,200	6	87	8,300
Other fission products	34,966	8,170	2,045,000	35,096	943	308,700
. . . . Fast-Breeder Reactor Fuels*						
[134]Ce	116	220	21,200	1	11	1,000
Other fission products	34,984	12,780	3,408,800	34,899	755	280,000

*Fuel exposed to 33,000 MWD/MTU at 30 MW/MTU.

TABLE 5.3: CESIUM-137 CONTENT IN HIGH-LEVEL FUEL REPROCESSING WASTES

Fission Products	Grams/Tonne	After 1 Year Watts/Tonne	Curies/Tonne	Grams/Tonne	After 10 Years Watts/Tonne	Curies/Tonne
. . . . Light-Water Reactor Fuels*						
[137]Cs-[137m]Ba	1,200	555	203,000	980	451	165,000
Other fission products	32,900	9,445	2,091,000	34,120	579	152,000
. . . . Fast Breeder Reactor Fuels**						
[137]Cs-[137m]Ba	1,230	565	207,000	1,000	453	168,000
Other fission products	33,770	13,235	3,123,000	33,900	303	113,000

*Fuel exposed to 33,000 MWD/MTU at 58 MW/MTU.
**Fuel exposed to 33,000 MWD/MTU at 30 MW/MTU.

Source: PB-224 588

Handling, Storage and Transportation

Cesium-134 and cesium-137 are in the greatest abundance in the mixture of fission products which are separated from the spent fuel during processing. Since both cesium-134 and cesium-137 are hazardous to man by inhalation, ingestion, or direct radiation exposure, care is exercised in their handling. Special procedures and radiation shielding are utilized in their handling.

When separated from the fission product mixture, both isotopes emit low energy beta particles which cannot penetrate more than 0.09 inch of water or 0.007 inch of lead. To stop their highly penetrating gamma rays high-density shields, such as lead, are required. To reduce their gamma radiation by a factor of 10, approximately 1.5 to 2 inches of lead or 8 to 9 inches of concrete is required. To detect and control personnel exposure, all persons working with these materials should wear dosimetry devices which directly indicate the dose. Commonly used devices are the film badge and the thermoluminescent dosimeters (TLD).

Cesium-134 and cesium-137, when separated from the other constituents of the high-level waste stream, are stored in controlled areas in specially constructed containers. They are protected by both a primary and a secondary containment barrier. Special monitoring systems and proper warning signs are located in the general area of the storage facility.

Both cesium-134 and cesium-137 are classified as a transport group III radionuclide by the Department of Transportation, and the rules and regulations governing their transportation are given in the Code of Federal Regulations (CFR) Title 49—Transportation, Parts 170 to 190 (1). Their content is limited to 3 curies for a Type A package and 200 curies for a Type B package defined in 49CFR173. The limits are increased to 20 to 5,000 curies if their physical form meets the requirements of a special form material. Their release rate is limited to zero under the specified accident conditions for Type A and B quantities (1).

The allowable release of radioactivity from packages containing large quantities of these isotopes is limited to gases and contaminated coolant containing total radioactivity exceeding neither 0.1 percent of the total radioactivity of the package nor 10 curies under the hypothetical accident conditions prescribed in 49CFR173.

The bulk of the cesium-134 and cesium-137 arrives at the spent fuel processing plants in the spent fuel assemblies which are contained in Type B packages. The cesium-134 and cesium-137 may be stored as a liquid at the processing plant for up to five years and are then solidified and leave the plant in Type B packages.

Disposal

After solidification at the fuel processing facility, the bulk of the cesium wastes may be held for up to five years before being shipped to a Federal repository. This repository has not yet been defined, but work is in progress, under AEC sponsorship, on engineered surface storage facilities which will be designed to store solidified high-level waste for up to 100 years. In the interim, a satisfactory ultimate disposal scheme, such as disposal in salt deposits, will be evolved. The disposal of cesium-134 and cesium-137 is governed by the AEC Manual Chapter 0524 (2) and 10CFR20 (3). Two sets of standards have been established for the permissible radiation exposure in unrestricted areas. One is for the greatest dose received by an individual and the other for the average dose received by the general population, the latter being about one-third the former. The standards for the safe disposal of cesium-134 and cesium-137 also define their maximum concentrations in air and water (Table 5.4).

The disposal of cesium-134 by release into a sanitary sewage system is limited to 100 microcuries and the disposal of cesium-137 is limited to 100 microcuries. The disposal by burial in the soil at any one location and time is limited to 1,000 microcuries of cesium-134 and 10,000 microcuries of cesium-137 (3).

TABLE 5.4: CESIUM-134 AND CESIUM-137 MAXIMUM PERMISSIBLE CONCENTRATIONS

Isotope	Exposure Group	Form	Concentration in Air (μCi/ml)	Concentration in Water (μCi/ml)
^{134}Cs	Individual	Soluble	1×10^{-9}	9×10^{-6}
^{134}Cs	Individual	Insoluble	4×10^{-10}	4×10^{-5}
^{134}Cs	Population	Soluble	0.33×10^{-9}	3×10^{-6}
^{134}Cs	Population	Insoluble	1.3×10^{-10}	1.3×10^{-5}
^{137}Cs	Individual	Soluble	2×10^{-9}	2×10^{-5}
^{137}Cs	Individual	Insoluble	5×10^{-10}	4×10^{-5}
^{137}Cs	Population	Soluble	0.67×10^{-9}	0.67×10^{-5}
^{137}Cs	Population	Insoluble	1.7×10^{-10}	1.3×10^{-5}

Source: PB-224 588

Evaluation of Recovery Methods

Small quantities of cesium-134 and cesium-137 may be recovered from the high-level waste streams for industrial purposes. Another potential advantage to recovery is that two or more radionuclides with similar characteristics could undergo the same handling, storage, or disposal processes. Although presently rarely recovered, various techniques, in various stages of development, are being investigated.

Option No. 1—Precipitation: Cesium-134 and cesium-137 can be removed from the high-level reactor wastes by precipitating with nickel ferrocyanide. Over 99 percent of the cesium can be removed satisfactorily up to a pH of 10 (7)(8). Cesium can also be removed by precipitating with phosphotungstate (8) or absorbing on alumino-silicate zeolites (9). Cesium can be further purified by ion exchange. Cesium is recovered from the other high-level wastes for commercial purposes or for safety reasons.

Option No. 2—Scavenging-Precipitation Foam Separation: The removal of cesium-134 and cesium-137 from low-level radioactive waste water by scavenging-precipitation foam separation has been studied by Oak Ridge National Laboratory (24). The process consists of two steps: (a) precipitating in a sludge-blanket clarification step; and (b) achieving final decontamination in a foam separation column. The precipitation step includes the use of Grundite clay for the sorption of cesium. The Grundite clay increased the cesium decontamination factor (ratio of initial to final concentration) approximately nine times. The overall decontamination factor for cesium was eight. At present, further development is required to reduce operating costs and increase processing rates.

Option No. 3—Scavenging-Precipitation Ion Exchange: In this process, final decontamination is obtained by ion exchange columns from the scavenging-precipitation process. The process includes a provision for the recycle of the ion exchange waste to the scavenging-precipitation step. All the removed radionuclides are concentrated in the clarifier sludge. With most ion exchange resins sodium is removed with cesium. Cesium can be separated from sodium by phenolic-base cation exchanges at high pH values. For this process, the overall decontamination factor for cesium varied from 100 to 3,000 (6). The use of ion exchange resins with a scavenging-precipitation step is a fairly simple and efficient means of removing cesium-134 and cesium-137 from low-level radioactive aqueous wastes.

Option No. 4—Water Recycle: The water recycle process is used for decontaminating radioactive waste water and recycling the purified water for reuse. This process has been demonstrated at the pilot plant scale by Oak Ridge National Laboratory (10). The steps in the process include: (a) clarification by the addition of coagulants; (b) demineralization by cation-anion exchange; and (c) sorption on granular activated carbon. The majority of the cesium is removed by the cation-anion exchange. The cesium overall decontamination factor was 14,000. The method is an improvement to the treat and discharge methods, but further work is required until full-scale production use is obtained.

Evaluation of Storage/Disposal Options

Option No. 1—Land Burial: Land burial of cesium-134 and cesium-137 in low-level wastes, at approved sites that are acceptable from a geologic and hydrologic standpoint, is an acceptable means of disposal. The cesium-134 and cesium-137 concentrations should not be in excess of 10^4 times their maximum permissible concentrations for the general population in 10CFR20 (3). All cesium wastes to be disposed of should be in a solid form and encapsulated in a suitable container. Liquid wastes should be solidified, preferably using asphalt, in accordance with the prescribed methods.

The burial trenches should be designed not to intercept the ground water table and constructed with a bottom drain and sump for water monitoring. The trenches should be covered with either asphalt or vegetation to limit infiltration of water. The burial site design, geology, and hydrology should be in conformance with the criteria used in selecting and licensing the present commercial burial sites (4). Since land burial sites meeting the criteria are operating on a commercial scale currently, they are considered the most satisfactory method of disposing of dilute concentrations.

Option No. 2—Near-Surface Liquid Storage: Near-surface storage (using carbon steel and stainless steel tanks encased in concrete and buried underground) of high-level, reactor-produced aqueous solutions of cesium and other fission product salts is not considered as a satisfactory means of storage. Aqueous solutions of high-level wastes from reactor fuels have been stored in this manner over the past 25 years. The tanks range in size from 0.33 to 1.3 million gallons and are equipped with devices for measuring temperatures, liquid levels, and leaks. These tanks are considered as an interim storage technique due to a general lack of confidence in their long-term integrity (6). Therefore, the near-surface storage of aqueous solutions of cesium and other fission product salts in steel tanks should only be considered as a near-term storage technique and not as a permanent storage or disposal technique.

Option No. 3—Near-Surface Solid Storage: The storage of high-level solidified cesium-134 and cesium-137 wastes and other waste salts in engineered storage facilities offers the best method for intermediate storage of these concentrated wastes. The advantages of this method are that the wastes will be under surveillance and control and that they can be retrieved, should this be required. The high-level wastes from fuel processing should be solidified and packaged in a suitable container (steel). Of the four high-level solidification processes developed, spray or phosphate glass solidification processes offer better solidified waste characteristics than the pot calcination or fluidized bed calcination processes. The wastes will be stored in stainless steel vaults encased in concrete and will be either air or water cooled.

Option No. 4—Salt Deposits: This method offers the greatest potential for the disposal of high-level wastes. The cesium wastes must be solidified and disposed of in the solid form, encapsulated in a suitable container. The solidified wastes are buried in rooms carved in the salt deposits approximately 1,000 feet below the ground. The salt is a good heat transmitter, provides about the same radioactive shielding as concrete, and can heal its own fractures by plastic flow. The critical problem is the selection of a site that meets the necessary design and geological criteria.

Option No. 5—Bedrock Disposal: The disposal of high-level aqueous solutions of cesium-134 and cesium-137 wastes, along with other radioactive wastes in vaults excavated in crystalline rock over 1,500 feet beneath the ground, has been evaluated by E. I. du Pont de Nemours at their Savannah River Plant near Aiken, South Carolina (11)(12). The wastes would be stored in six tunnels and, once in the tunnels, the wastes will seep into the surrounding rock.

Located above the crystalline rock is the Tuscaloosa formation, a good source of fresh water, which is separated by a layer of clay that would act as a barrier to the leakage of radioactive wastes. An advisory committee appointed by the National Academy of Sciences recommended abandonment of the project. In May 1972, another National Academy of Sciences panel concluded that bedrock storage provides a reasonable prospect for long-term safe storage, but precise information is needed to decide if and where underground storage vaults should be built.

Another method has also been proposed for disposing of liquid wastes by in situ incorporation in molten silicate rock (13). Both of these methods are unproven since sufficient engineering data or exploration has not been completed to verify their suitability. Due to the long half-lives of certain of the elements in the mixed fission products and actinides, it is doubtful that data could ever be accumulated to prove that the geological characteristics of the site over the next few hundred years are acceptable to ensure the absolute safety of such a disposal method.

Option No. 6—Hydraulic Fracturing: The direct disposal of aqueous, low-level radioactive wastes into shale formations has been investigated by Oak Ridge

National Laboratory (14). The method consists of mixing the aqueous wastes with cement and pumping the resulting slurry down a well out into a nearly horizontal fracture in a thick shale formation. Additional work is required to demonstrate that this method of disposal is satisfactory.

To summarize, cesium-134 and cesium-137 can be recovered from the high-level waste streams for separate disposal or reuse. The acceptable method of treatment is solidification, followed by storage in a near-surface engineered storage facility and ultimate disposal in a salt deposit. Cesium should be recovered from the low-level waste streams to minimize the amount directly released to the environment. The recovered cesium should be solidified, preferably using asphalt, and disposed of at approved sites by land burial.

Applicability to National Disposal Sites

Cesium-134 and cesium-137 are candidates for a National Disposal Site due to their public health hazard and projected growth with that of the civilian nuclear power program. From a cost and safety viewpoint, it would be desirable to combine the reprocessing plant and the National Disposal Site. Interim storage in near-surface engineered facilities offers considerable latitude in the site selection. The site selection for ultimate disposal of these wastes is limited to particular geological areas in which salt deposits are present. Temporary, or interim, storage prior to transfer to a National Disposal Site could be accomplished at many site locations.

The recommended treatment for low-level waste streams is recovery followed by solidification with asphalt and disposed by land burial.

HYDROGEN-3

Tritium, an isotope of hydrogen, is produced in nuclear reactors in substantial quantities. It is currently released to the environment in the form of tritiated coolant water from the reactor and tritium wastes resulting from processing of the spent nuclear fuels. Although it is one of the least hazardous radioactive nuclides, the recovery and retention of tritium may be required in the future due to its long half-life for radioactive decay, the rapid rate of expansion in the nuclear power and fuel-reprocessing industries, and its ability to be metabolized in in the form of tritiated water and incorporated into body fluids and tissues.

Tritium is produced in the fission of uranium-235 and plutonium-239 with yields of 0.01 percent and 0.02 percent, respectively. Tritium decays by the emission of a beta particle and an antineutrino to form stable helium-3. The half-life for this process is approximately 12.3 years. The average energy of the beta emitted by tritium is 5.6 kev, which is only about one-hundredth of the energy emitted by most other beta emitters.

Tritium may be produced in nuclear reactors by one or more of the following five methods: fissioning of uranium, neutron capture reactions with boron and lithium added to the reactor coolant, neutron capture reactions with boron in control rods, activation of deuterium in water, and high energy neutron capture reactions with structural materials (15). Fission-product tritium is generally contained by the cladding surrounding each fuel element. While small amounts of tritium were found in the past to leak to the primary coolant with the stainless steel clad elements, the use of zirconium cladding has been found to limit the diffusion of tritium through the cladding to very low quantities. The fission-product tritium generated is contained until the spent fuel is processed when it appears as tritiated water in the fuel processing plant evaporator condensates.

Additional sources of tritium are in the primary coolant loops of pressurized water reactors and high temperature gas cooled reactors. In the pressurized water reactors, boric acid which is dissolved in the primary coolant for reactivity control causes the production of tritium by neutron capture reactions. The production of tritium occurs in high temperature gas cooled reactors by ternary fission and by activation of helium-3 found in trace amounts in the helium coolant.

Although the primary coolant from these reactors is not routinely discharged to the waste disposal system, leakage of coolant from pumps and the water utilized during the refueling operation are discharged to the waste disposal systems. In the reactor power plant waste disposal systems the water undergoes several treatment processes before being released to the environment, but none of these processes is effective in removing tritium. Tritium may compose between 50 and almost 100 percent of the total amount of radioactive material discharged as liquid waste from nuclear reactors. The amount of tritium discharged from reactors in a gaseous form is only about 1 percent of the total tritium discharge.

Handling, Storage and Transportation

Since tritium is moderately hazardous to man by inhalation or ingestion, care must be exercised in its handling. Handling must be set up to prevent excessive exposure to personnel and, in fact, to prevent any unnecessary exposure.

These materials should be stored in controlled areas in specially constructed containers. Tritium, as a gas, as luminous paint, or adsorbed on solid material is classified as a group VII radionuclide by the Department of Transportation. The rules and regulations governing its transportation are given in the Code of Federal Regulations (CFR) Title 49—Transportation, Parts 170 to 190 (1). The transport group VII radionuclide content is limited to 1,000 curies for a Type A package and 50,000 curies for a Type B package defined in 49CFR173. Their release rate is limited to zero under the specified accident conditions for Type A and B quantities (1).

Disposal

The disposal of these materials is governed by the AEC Manual, Chapter 0524 (2) and 10CFR20 (3). Two sets of standards have been established for the permissible radiation exposure in unrestricted areas. One is for the greatest dose received by an individual and the other for the average dose received by the general population. The standards for the safe release of this material to the environment in an unrestricted area are contained in 10CFR20 (3) and its release should not exceed the concentrations in air and water listed in Table 5.5. These release rates apply to an individual and its release may be further limited if a suitable sample of the population is exposed to one-third the concentrations in air or water specified in this report (Table 5.5).

Although rarely practiced, the disposal by release in a sanitary sewage system is limited to 10,000 microcuries (3). The disposal by burial at any one location and time is limited to 100 times the above amount.

TABLE 5.5: TRITIUM ENVIRONMENT RELEASE RATES

Fission Product	Form	Concentration in Air (microcuries/milliliter)	Concentration in Water (microcuries/milliliter)
Hydrogen-3	Soluble	2×10^{-7}	3×10^{-3}
	Insoluble	2×10^{-7}	3×10^{-3}
	Submersion*	4×10^{-5}	---

*Submersion means that the values given are for submersion in a semi-spherical infinite cloud of airborne material.

Source: PB-224 588

Recovery

Tritium is not currently recovered at the reactor site or the spent fuel processing plant site. However, the Oak Ridge National Laboratory is conducting experimental work on recovering tritium during fuel reprocessing for the fast breeder reactor. Voloxidation is the term for the tritium recovery process being investigated in which the fuel, after being chopped in the head-end process, is heated to drive off the tritiated hydrogen. By collecting and condensing the effluent, greater than 99 percent of the tritium is recovered. It is feasible that such a process could be incorporated into the reprocessing of fuel from the present, light water reactors.

Storage/Disposal Options

The current method of tritium disposal is dilution and dispersion, both at the reactor site and the spent fuel processing plant site. Most of this release is in the form of tritiated water in the liquid waste streams and only a very small percent is released to the air. Tritium discharge concentrations are usually much less than 1 percent of the discharge limits.

Because of the inherently small radiological hazard from tritium, the difficulty of removing tritium from the reactor's tritiated water and the small percentage of the allowable concentrations that are currently being released to the environment at the reactor site, tritium removal is probably not practical in the near future. At the spent fuel processing plant, however, a process similar to the voloxidation process should be incorporated in order to recover tritium and dispose of it by one of the following methods.

Option No. 1—Land Burial: Land burial of tritium wastes, in small concentrations, at approved sites that are acceptable from a geologic and hydrologic standpoint, is an acceptable means of disposal. Its concentrations should not be in excess of 10^4 times its maximum permissible concentrations for the general population in 10CFR20 (3). The burial trenches should be designed not to intercept the ground water table and constructed with a bottom drain and sump for water monitoring. The trenches should be covered with either asphalt or vegetation. The burial site design, geology, and hydrology should be in conformance with the criteria used in selecting and licensing the present commercial burial sites (4).

Option No. 2—Near Surface Storage: The storage of high concentrations of these wastes in engineered surface facilities offers the best intermediate method for storage of these wastes. The technology for these facilities has been developed and the wastes will be under surveillance and control and can be retrieved, should this be required. The wastes will be stored in stainless steel lined concrete vaults which will be either air or water cooled.

Option No. 3—Salt Deposits: This method offers the best potential for the disposal of this radionuclide. To summarize, the majority of the tritium generated by the nuclear power reactors is currently released to the environment in the form of liquid wastes. Most of this release occurs at the fuel reprocessing plant, where the tritium could be recovered by voloxidation or a similar process. The tritium should be encapsulated and disposed of in salt deposits following storage in engineered storage facilities.

Applicability to National Disposal Sites

Tritium is a candidate for a National Disposal Site due to its projected growth with that of the civilian nuclear power program. It should be recovered and encapsulated at the fuel reprocessing plant and shipped to the National Disposal Site for disposal in salt beds following storage in engineered storage facilities.

IODINE-129, IODINE-131, KRYPTON-85 AND XENON-133

Iodine and the noble gases, krypton and xenon, are produced during the fission of uranium in nuclear reactors. They represent a potential source of environmental contamination since they are presently released to the environment during reprocessing of the nuclear fuels. Although the release rates at present reprocessing plants are low, the recovery and retention of these gases may be required in the near future due to the rapid rate of expansion of the nuclear power industry and the corresponding expansion in the nuclear fuel reprocessing industry.

In addition to the anticipated growth in this industry, economic incentives exist for reducing fuel decay times prior to reprocessing. Present reprocessing plants handle 150-day decayed light water reactor fuel. With the introduction of the fast-breeder reactors, it is economically desirable to reprocess the fuel after 30 days decay. The shortened decay period will increase the reprocessing plant off-gas handling requirements for the short-lived isotopes such as iodine-131 (8 days) and xenon-133 (5 days). The iodine-131 activity in 30-day-old reactor fuel is 36,000 times higher than it is in 150-day-old reactor fuel.

Of the fission-product halogens, only the isotopes iodine-129 and iodine-131 are physiologically significant after 30 days of postirradiation decay. The iodine-131 contents of reactor fuels are approximately 72,000 and 2 curies per metric ton of reactor fuel after decay times of 30 and 150 days respectively. The iodine-129 content is only about 0.03 curie per metric ton of reactor fuel. Even though the iodine-129 content in reactor wastes is low, it is significant since it has a half-life of 17 million years compared to the iodine-131 half-life of 8 days. Iodine-129 and iodine-131 are also reconcentrated by biological processes in the food chain leading to man. This reconcentration occurs in the grass-cow-milk pathway to the thyroids of small children and man.

Krypton and xenon are both produced in significant quantities in nuclear reactors. These isotopes are chemically inert and, once released, they do not concentrate in body tissues. Of various isotopes of krypton and xenon in reactor fuels, the only two isotopes of significance are krypton-85 and xenon-133. Krypton-85 has a half-life of 10.8 years, a moderate beta radiation energy, a low gamma radiation level, and a fission yield of 1.3 percent. Xenon-133 has a half-life of 5.3 days, a moderate beta and gamma radiation energy, and a fission yield of 6.6 percent.

The projected growth in production of these isotopes will parallel that of the civilian nuclear power program. Present projections indicate that nuclear power will account for 30 percent of the total power production by the year 1980 (16).

Handling, Storage and Transportation

Since these radionuclides are hazardous to man by inhalation, ingestion, or direct radiation exposure, care is exercised in their handling. Handling is set up to prevent excessive exposure to personnel. Special procedures and radiation shielding are utilized in their handling. The beta particles emitted by iodine, xenon, and krypton are of moderate energy and cannot penetrate more than 0.1 inch of water. Their gamma radiation is highly penetrating and high-density shields, such as lead, are required to stop the radiation. To detect and control personnel exposure to their radiation all persons working with this material should wear dosimetry devices which directly indicate the dose. Commonly used devices are the film badge and the thermoluminescent dosimeters (TLD).

These materials are stored in controlled areas in specially constructed containers. They are protected by both a primary and a secondary containment barrier. Special monitoring systems and proper warning signs should be located in the general area of the storage facility.

Iodine-129 and iodine-131 are classified as a transport group III radionuclide and krypton-85 and xenon-133 are classified as a group IV radionuclide in the uncompressed state, at a pressure less than 14.7 psia, and as a group III radionuclide at pressures greater than 14.7 psia by the Department of Transportation. The rules and regulations governing their transportation are given in the Code of Federal Regulations (CFR) Title 49—Transportation, Parts 170 to 190 (1). The transport group III radionuclide content is limited to 3 curies for a Type A package and 200 curies for a Type B package defined in 49CFR173. Their release rate is limited to zero under the specified accident conditions for Type A and B quantities (1).

Disposal

The disposal of these materials is governed by the AEC Manual, Chapter 0524 (2) and 10CFR20 (3). Two sets of standards have been established for the permissible radiation exposure in unrestricted areas. One is for the greatest dose received by an individual and the other for the average dose received by the general population. The standards for the safe release of these materials to the environment in an unrestricted area are contained in 10CFR20 (3) and their release should not cause the concentrations in air and water to exceed the concentrations listed in Table 5.6. These concentrations apply to an individual and their release may be further limited if a suitable sample of the population is exposed to one-third the concentrations in air or water specified in the table.

TABLE 5.6: IODINE, KRYPTON, AND XENON NATIONAL PERMISSIBLE CONCENTRATIONS

Radionuclide	Form	Concentration in Air (microcuries/milliliter)	Concentration in Water (microcuries/milliliter)
Iodine-129	Soluble	2×10^{-11}	6×10^{-8}
Iodine-129	Insoluble	2×10^{-9}	2×10^{-4}
Iodine-131	Soluble	1×10^{-10}	3×10^{-7}
Iodine-131	Insoluble	1×10^{-8}	6×10^{-5}
Krypton-85	Submersion*	3×10^{-7}	---
Xenon-133	Submersion*	3×10^{-7}	---

*Submersion means that the values given are for submersion in a semi-spherical infinite cloud of airborne material.

Source: PB-224 588

Although rarely practiced, the disposal by release in a sanitary sewage system is limited to 1 microcurie for iodine-129, 10 microcuries for iodine-131 and 1,000 microcuries for krypton-85 and xenon-133 (3). The disposal by burial at any one location and time is limited to 100 times the above amounts.

Iodine Recovery

Option No. 1—Caustic Scrubbers: Iodine has been removed from gas streams in many applications by scrubbing with caustic solutions and reacting with silver nitrate impregnated on ceramic packing. Caustic scrubbers are effective in removing approximately 90 percent of the iodine in the off-gas streams and the silver nitrate towers remove about 99 percent of the remaining iodine. In general, caustic scrubbers are not effective in removing organic iodides, such as methyl iodide. Organic iodides can be effectively removed by using an aqueous scrubbing medium of mercuric nitrate in nitric acid if sufficient contact time is allowed for the relatively slow reaction between the absorbed organic iodide and mercury ion.

Option No. 2—Activated Charcoal: Charcoal has been extensively used for the removal of elemental iodine from off-gas streams. Elemental iodine is readily trapped from air streams by activated charcoal, even at relative humidities approaching saturation. Iodine absorption efficiencies of 99.9 percent have been obtained (17). However, organic iodides, such as methyl iodide are not effectively absorbed. To remove organic iodides, the activated charcoal beds should be preceded by a catalytic oxidation step to completely oxidize all organic vapors in the gas stream. The addition of the oxidation step prior to the charcoal bed is effective for removing organic iodides, but the bed life may be limited to a few months or a residence time of a few seconds may be required which is not practical in a large gas flow system. The charcoal beds are quite effective for the removal of trace quantities of elemental iodine and their use should probably be limited to the final iodine removal step.

Option No. 3—Impregnated Charcoal: Special impregnated charcoals have been developed for removing methyl iodide. The impregnated charcoals presently under consideration are: (a) iodized with one or more iodine-containing substances, to provide I-131, I-127 exchange capability; or (b) triethylenediamine-impregnated to effect removal by the utilization of the alkyl halide-organic amine reaction. Methyl iodine retention efficiencies greater than 99 percent have been obtained with both these impregnated charcoals (18).

The removal efficiencies for the iodine-impregnated charcoal refer only to I-131. The performance of these impregnated charcoal beds is drastically reduced by poisoning due to impurities in the air stream (19). Iodine-impregnated charcoal and triethylinediamine-impregnated charcoal are satisfactory absorbents for organic iodides, such as methyl iodide, but an upstream filter system may be required to remove impurities in the air.

Option No. 4—Silver Zeolite: Inorganic and organic forms of iodine are both effectively removed from the off-gas stream by silver zeolite. The silver zeolite has ideal properties of nonflammability, capability of iodine absorption up to 500°C, retention of absorbed iodine up to 1000°C, and rapid reaction times. The zeolite is a synthetic zeolite absorber (Linde Molecular Sieve 13X) in which the normal sodium zeolite matrix has been converted to silver zeolite. The silver zeolite retention efficiency of elemental iodine and methyl iodide is greater than 99.8 percent at temperatures varying from room temperature to 500°C (20).

The efficiency of the silver zeolite is adversely affected by long-term exposure to organic contaminants in the air stream. The zeolite life expectancy can be extended by preceding the zeolite with an oxidation step. This is economically desirable since the basic cost of silver zeolite is 3 to 10 times the cost of charcoal. Silver zeolite is the most efficient absorber of all airborne iodine species and is particularly applicable to the large gas flows encountered in nuclear fuel reprocessing plants.

Krypton and Xenon Recovery

Option No. 1—Activated Charcoal: Krypton and xenon can be retained by adsorption beds of activated charcoal. In this process, the krypton and xenon are cooled by means of cold trays and are then passed through a constant temperature (25°C) charcoal column. These absorption beds are used to provide holdup time which is long compared to the half-life of the isotopes. This process is effective for the 5.3 day xenon-133 but not for 10.8 year krypton-85. For xenon-133 the gas could be held up in a charcoal bed for a period of 18 days to reduce its activity 10 times or a period of 54 days for a reduction of 1,000 times.

Option No. 2—Cryogenic Distillation: A cryogenic distillation process for the recovery of krypton and xenon from off-gas streams generated during the processing of spent fuels has been applied at the Idaho Chemical Processing Plant (21).

In this process, the off-gas stream is first treated prior to entering the cryogenic system by passing the gas through a catalytic conversion system which removes nitrous oxide and hydrogen. In an off-gas stream with a high hydrogen content, a special catalytic unit and condensor would also be required to reduce the hydrogen concentration in the off-gas to levels below the explosive limit. The off-gas is then fed to the cryogenic system which is cooled by liquid nitrogen. This system consists primarily of two regenerators, a distillation unit, and a batch still. The gas stream is cooled to $-260°F$ in the two regenerators prior to entering the distillation column.

In the distillation column, krypton and xenon are condensed and absorbed by liquid nitrogen which flows through the plates in the column. The krypton and xenon are dissolved in the liquid nitrogen and are transferred to a batch still where they are separated and purified in a fractionation step. This process has been applied in actual plant operation at the Idaho Chemical Processing Plant and is capable of processing up to 20 scfm of off-gas. The process has the potential for recovering 99 percent of the gases with 4 to 10 percent of nitrogen and oxygen impurities in the final product. Actual overall krypton-xenon recovery efficiencies have varied from 30 to 60 percent. The low recovery efficiencies were due to difficulties in operation of the gas cleanup step and to leaks within the product bottling system.

Option No. 3—Fluorocarbon Solvents: Krypton and xenon can be effectively removed from contaminated air streams by selective absorption in fluorocarbon solvents, such as refrigerant-12 (dichlorodifluoromethane). This process has been tested on a pilot plant scale at processing rates up to 20 scfm of gas by Union Carbide at their Oak Ridge, Tennessee facility (22). In this process, the removal of krypton and xenon is accomplished by contacting the gas stream with a stream of fluorocarbon solvent (refrigerant-11 or -12) in a packed absorber column at low temperatures ($-177°F$ to $-21°F$) and high pressure (169 to 437 psia).

Besides krypton and xenon substantial amounts of oxygen, nitrogen, and argon are dissolved in the solvent. Since krypton and xenon are substantially more soluble in the solvent than these other gases, the absorber column is followed by a fractionator system where most of the dissolved nitrogen, argon, and oxygen is driven off. The solvent is then driven to a stripper system where the product gas concentrated in krypton and xenon is evolved and the solvent is recycled back to the absorber.

Based on the pilot plant test data, a plant operating with an absorber at $-30°F$ and 425 psia should give a krypton and xenon removal efficiency of over 99.7 percent with refrigerant-12 as the solvent (22). For application to fuel reprocessing plants with feed rates in excess of 100 scfm, some scale-up of the pilot plant facility is required. This method offers the best potential for the effective removal of krypton and xenon from off-gas streams. It has been successfully operated on a pilot-plant scale but has not yet been applied in actual plant operations.

Storage/Disposal Options

Option No. 1—Land Burial: Land burial of iodine, krypton, and xenon wastes, in small concentrations, at approved sites that are acceptable from a geologic and hydrologic standpoint, is an acceptable means of disposal. Their concentrations should not be in excess of 10^4 times their maximum permissible concentrations for the general population in 10CFR20 (3). The burial trenches should be designed not to intercept the ground water table and constructed with a bottom drain and sump for water monitoring. The trenches should be covered with either asphalt or vegetation. The burial site design, geology, and hydrology should be in conformance with the criteria used in selecting and licensing the present commercial burial sites (4). Since land burial has successfully been practiced on a commercial scale, it should be considered as the most satisfactory method of disposing of low concentrations of these wastes.

Option No. 2—Near-Surface Storage: The storage of high concentrations of these wastes in engineered storage facilities offers the best immediate method for storage of these wastes. The wastes will be stored in stainless steel lined concrete vaults which will be either air or water cooled.

Option No. 3—Salt Deposits: This method offers the best potential for the disposal of this radionuclide. To summarize, krypton, xenon, and iodine are released in the off-gas streams during processing of the spent fuels. The iodine can be removed from the off-gas stream by silver zeolite. Krypton and xenon can be effectively removed by cryogenic distillation or by selective absorption in fluorcarbon solvents.

Following recovery, the wastes can be held to allow the short-lived isotopes such as iodine-131 and xenon-133 to decay. For iodine-131 and xenon-133, their activity is reduced by a factor of 10,000 in 106 days and 71 days, respectively. The long-lived isotopes krypton-85 (10.8 years) and iodine-129 (17 million years) should be encapsulated and disposed of in salt deposits following storage in near-surface engineered facilities. The wastes to be disposed of can be encapsulated in high-pressure cylinders or solidified. A method for solidifying the gases xenon and krypton includes dispersion in glasses or resins and entrapment in molecular sieves or small pressurized steel bulbs which are in turn encased in epoxy resin.

Applicability to National Disposal Sites

Iodine-129 and krypton-85 are candidates for a National Disposal Site due to their long half-lives and projected growth with that of the civilian nuclear power program. Iodine-131 and xenon-133 are not candidates for a National Disposal Site because of their short half-lives.

The recommended treatment for the recovery of iodine-129 and iodine-131 from reprocessing plant off-gas streams is by the use of silver zeolite. Krypton-85 and xenon-133 can be removed by cryogenic distillation or absorption in fluorocarbon solvents. Iodine-131 and xenon-133 can be stored in engineered storage facilities to allow their radioactive decay and then disposed of in low concentrations in land burial facilities. Iodine-129 and xenon-133 should be stored in engineered storage facilites followed by disposal in salt beds.

PLUTONIUM-238, PLUTONIUM-239, PLUTONIUM-240, PLUTONIUM-241, AMERICIUM-241, AMERICIUM-243, CURIUM-242 AND CURIUM-244

Plutonium, americium, and curium are artificially produced radionuclides that do not exist in nature except for very small amounts of plutonium. These elements are characterized by their high radiotoxicity, long half-life, and ability to fission.

TABLE 5.7: ACTIVITY OF PLUTONIUM, AMERICIUM, AND CURIUM PRODUCTS PRESENT IN WASTES GENERATED BY THE PROCESSING OF SPENT REACTOR FUELS*

Light Water Reactor Fuels
(Fuel Exposed to 33,000 MWD/MTU at 30 MW/MTU)

Actinide Products		After 1 year		Grams/Tonne	After 100 years	Curies/Tonne
	Grams/Tonne	Watts/Tonne	Curies/Tonne		Watts/Tonne	
Pu^{238}, Pu^{239}, Pu^{240}, Pu^{241}	48	2.7	641	71	2	64
Am^{241}, Am^{243}	144	6.2	189	140	6	179
Cm^{242}, Cm^{244}	32	306	8,430	1	2	57
Other Actinides	5,436	0.1	30	5,438	0.1	25

Fast-Breeder Reactor Fuels
(Fuel Exposed to 33,000 MWD/MTU at 30 MW/MTU)

Actinide Products		After 1 year		Grams/Tonne	After 100 years	Curies/Tonne
	Grams/Tonne	Watts/Tonne	Curies/Tonne		Watts/Tonne	
Pu^{238}, Pu^{239}, Pu^{240}, Pu^{241}	428	11	3,203	412	8	264
Am^{241}, Am^{243}	742	54	1,620	794	48	1,469
Cm^{242}, Cm^{244}	20	625	17,000	1	3	72
Other Actinides	4,550	2	277	4,637	1	165

* Assumes that 0.5 percent of the plutonium in the spent fuel is lost to waste.

Source: PB-224 588

Plutonium-239 is the most important among these elements because of its use in nuclear weapons and the place it holds as the key material in the development of fast-breeder reactors for the civilian nuclear power program. Plutonium-239 is readily fissionable with neutrons, and one pound of this material is equivalent to about ten billion watt hours of heat energy.

Americium, curium, and plutonium (except for Pu-241) decay by the emission of high energy, 5 to 6 Mev, alpha particles. Plutonium-241 decays by beta emission to form americium-241. All these elements eventually decay to lead, and the time required is on the order of millions of years. The three long-lived isotopes in the americium-243 decay chain are: Pu-239, half-life 2.4×10^4 years; U-235, half-life 7.1×10^8 years; and Pu-231, half-life 3.3×10^4 years.

The projected growth in production of these isotopes will parallel that of the civilian nuclear power program. The production of plutonium-239 will increase with the planned introduction of the fast-breeder reactors in the 1980s.

The majority of the transuranium elements are produced in nuclear reactors and are contained within the spent fuel elements. The spent fuel elements are processed for the recovery of the usable fission materials (uranium and plutonium) which are refabricated into new fuel elements. Solvent extraction using nitric acid is the means currently used for the first-stage removal of the fissionable materials. Approximately 99 percent of the plutonium is removed during the processing. The other radionuclides along with the unrecovered plutonium are contained in the aqueous effluent from the reprocessing step. The composition of fuel elements prior to processing has been tabulated (6), and the activity of plutonium, americium, and curium products present in the waste stream following the processing step are shown in Table 5.7.

Handling, Storage and Transportation

These radionuclides are all alpha emitters except for plutonium-241. The alpha particles have little penetrating power and can be stopped by a sheet of writing paper. They are extremely hazardous if inhaled. Semiremote handling techniques are required to protect personnel from radiation injury and adequate radiation shielding is required to avoid excessive radiation exposure. These materials are safely handled in glove boxes. The beta particles emitted by plutonium-241 are of low energy (0.021 Mev) and cannot penetrate more than 0.002 inch of water.

These radionuclides are stored in controlled areas in specially constructed containers. Extreme precautions are taken to avoid the release of any material. Plutonium is stored at low temperatures and in dry air to avoid corrosion. The storage container is often protected by both a primary and a secondary containment barrier. Precautions are also taken to prevent the formation of a critical mass. Plutonium in a liquid solution is more likely to become critical than solid plutonium. The container design and arrangement must preclude any possibility of obtaining nuclear criticality. Special monitoring systems and proper warning signs are located in the general area of the storage facility.

Plutonium, americium, and curium are classified as a transport group I radionuclide by the Department of Transportation, and the rules and regulations governing their transportation are given in the Code of Federal Regulations (CFR) Title 49—Transportation, Parts 170 to 190 (1). Their content is limited to 0.001 curie for a Type A package and 20 curies for a Type B package defined in 49CFR173. The limits are increased to 20 and 500 curies if their physical form meets the requirements of a special form material. Their release rate is limited to zero under the specified accident conditions for Type A and B quantities (1). Plutonium-239 is designated as a fissile material by the Department of Transportation and must meet the controls required to provide nuclear criticality safety.

Disposal

The disposal of these materials is governed by the AEC Manual Chapter 0524 (2) and 10CFR20 (3). Two sets of standards have been established for the permissible radiation exposure in unrestricted areas. One is for the greatest dose received by an individual and the other for the average dose received by the general population. The standards for the safe release of these materials to the environment in an unrestricted area are contained in 10CFR20 (3) and their concentrations in air and water should not exceed the values listed in Table 5.8. These concentrations apply to an individual and their release may be limited if a suitable sample of the population is exposed to one-third the concentrations in air or water specified.

TABLE 5.8: PLUTONIUM, AMERICIUM, AND CURIUM MAXIMUM PERMISSIBLE CONCENTRATIONS*

Radionuclide	Form	Concentration in Air (microcuries/milliliter)	Concentration in Water (microcuries/milliliter)
Plutonium-238	Soluble	7×10^{-14}	5×10^{-6}
Plutonium-238	Insoluble	1×10^{-12}	3×10^{-5}
Plutonium-239	Soluble	6×10^{-14}	5×10^{-6}
Plutonium-239	Insoluble	1×10^{-12}	3×10^{-5}
Plutonium-240	Soluble	6×10^{-14}	5×10^{-6}
Plutonium-240	Insoluble	1×10^{-12}	3×10^{-5}
Plutonium-241	Soluble	3×10^{-12}	2×10^{-4}
Plutonium-241	Insoluble	1×10^{-9}	1×10^{-3}
Americium-241	Soluble	2×10^{-13}	4×10^{-6}
Americium-241	Insoluble	4×10^{-12}	2×10^{-5}
Americium-243	Soluble	2×10^{-13}	4×10^{-6}
Americium-243	Insoluble	4×10^{-12}	3×10^{-5}
Curium-242	Soluble	4×10^{-12}	2×10^{-5}
Curium-242	Insoluble	6×10^{-12}	3×10^{-5}
Curium-244	Soluble	3×10^{-13}	7×10^{-6}
Curium-244	Insoluble	3×10^{-12}	3×10^{-5}

*Maximum permissible concentrations apply to an individual in an unrestricted area.

Source: PB-224 588

Although not practiced, the disposal by release in a sanitary sewage system is limited to 0.1 microcurie (3). The disposal by burial at any one location and time is limited to 10 microcuries (3).

Recovery Option

Solvent Extraction: Solvent extraction processes have successfully been used for well over a decade as the standard method for the recovery of up to 99.8 percent of the plutonium and uranium in spent reactor fuel elements. The basis for separation in this process is the differences in solubility of these materials in the organic and aqueous phase. Nearly all major fuel reprocessing facilities use the Purex solvent extraction process. These facilities include those operated under contract to the Atomic Energy Commission (AEC) and Nuclear Fuel Services facility at West Valley, New York, which is privately owned and operated.

In the Purex process, the spent fuel elements are dissolved in nitric acid and an organic compound, tributyl phosphate (TBP) in an inert hydrocarbon is added. When this organic mixture is added, the TBP extracts both the uranium and the plutonium into the organic phase while the fission products and other wastes remain in the aqueous phase. Plutonium and uranium are separated by reducing the plutonium to the trivalent form, usually with ferrous sulfamate, and contacting with nitric acid. In the Aquafluor process used at General Electric's Morris, Illinois plant, the uranium and plutonium separation will be accomplished by ion exchange. The ion exchange separation step will limit the plant's capacity since they must be designed to prevent the accumulation of a critical mass.

Storage/Disposal Options

Option No. 1—Land Burial: Disposal of these high-level radionuclides, in small concentrations, by direct burial in unlined trenches is not considered a satisfactory means of disposal. These types of wastes are presently disposed of in this manner at approved sites. This method of disposal does not provide for any type of secondary containment barrier to prevent their release by leaching with the local ground water. Since these materials have an extremely long half-life and a high radiotoxicity, this method of disposal is not satisfactory.

Option No. 2—Near-Surface Liquid Storage: Near-surface storage of these radionuclides in stainless steel tanks encased in concrete and buried underground is not considered as a satisfactory means of disposal. Aqueous solutions of spent fuel reprocessing wastes have been stored in this manner over the past 25 years. These tanks are considered as an interim storage technique due to a general lack of confidence in their long-term integrity (6). Therefore, the near-surface storage of these wastes in steel tanks should only be considered as an interim storage technique.

Option No. 3—Near-Surface Solid Storage: The storage of solidified wastes in engineered surface facilities offers the best immediate method for storage of these wastes. The necessary technology for these facilities has been developed. The wastes will be under constant surveillance and control and can be retrieved, should this be required. The wastes should be solidified and encapsulated in a suitable container (steel). Of the four high-level solidification processes developed for reactor-produced wastes, spray or phosphate glass solidification offer better solidified waste characteristics than the pot calcination or fluidized bed calcination processes. The wastes will be stored in stainless-steel-lined concrete vaults which will be either air or water cooled.

Option No. 4—Salt Deposits: This method offers the best potential for the disposal of solidified, high-level wastes since bedded salt deposits are completely free of circulating ground waters. The wastes must be solidified and disposed of in the solid form encapsulated in a suitable container. The solidified wastes are buried in rooms carved in the salt deposits approximately 1,000 feet below the ground. The critical problem is the selection of a site that meets the necessary design and geological criteria for the mixture of fission products and actinide wastes.

Option No. 5—Bedrock Disposal: The disposal of high-level liquid wastes along with other spent fuel processing wastes in vaults excavated in crystalline rock over 1,500 feet beneath the ground has been evaluated by E. I. Du Pont de Nemours at their Savannah River Plant near Aiken, South Carolina (11)(12). Further work is necessary.

To summarize, plutonium, americium, and curium are hazardous, long-lived isotopes for which extreme precautions are required in their disposal. Plutonium is separated from the spent fuel processing wastes by solvent extraction for reuse. Recovery factors as high as 99.8 percent can be obtained. The americium, curium, and remaining plutonium are contained with the final reprocessing wastes. The acceptable methods of treatment are spray or phosphate glass solidification and disposal in either salt deposits or near-surface engineered facilities. Plutonium, americium, and curium wastes from research laboratories and other sources should also be solidified, encapsulated, and disposed of in the same manner.

Applicability to National Disposal Site

Plutonium, americium, and curium are candidates for a National Disposal Site due to their long half-lives, health hazard, and projected growth with that of the civilian nuclear power program. From a cost and safety viewpoint, it would be desirable to combine the reprocessing plant and the National Disposal Site. Interim storage in near-surface, engineered storage facilities offers considerable latitude in the site selection. The site selection for ultimate disposal of these wastes is limited to particular geological areas in which salt deposits are present.

RUTHENIUM-106 (RHODIUM-106), CERIUM-144 (PRASEODYMIUM-144) AND PROMETHIUM-147

Ruthenium-106, cerium-144, and promethium-147 are radioactive isotopes that are produced in nuclear reactors by the fission of uranium. Ruthenium-106 has a half-life of 368 days and emits only beta particles. It exists in radiation equilibrium with its short half-lived daughter, rhodium-106. Rhodium-106 has a half-life of 30 seconds and emits both gamma rays and high energy (3.55 Mcv) beta particles. Cerium-144 is both a beta and gamma emitter and has a half-life of 284 days. It decays by beta emission to form praseodymium-144 which has a half-life of 17.3 minutes. Promethium-147 has a half-life of 2.62 years and decays by the emission of a 0.224 beta particle to form stable samarium-147. These three isotopes, Ru-106, Ce-144, and Pm-147 have moderately long half-lives (284 days to 2.6 years) and account for a majority of the total fission product

TABLE 5.9: RUTHENIUM-106, CERIUM-144, AND PROMETHIUM-147 CONTENT IN HIGH-LEVEL FUEL REPROCESSING WASTE

Light-Water Reactor Fuels
(Fuel Exposed to 33,000 MWD/MTU at 30 MW/MTU)

| Fission Products | Grams/Tonne | After 1 year | | | After 10 years | |
		Watts/Tonne	Curies/Tonne	Grams/Tonne	Watts/Tonne	Curies/Tonne
$Ru^{106} - Rh^{106}$	81	2,656	546,000	0.16	5	1,100
$Ce^{144} - Pr^{144}$	143	3,800	912,000	0.05	1.5	300
Pm^{147}	92	44	85,000	8	4	8,000
Other Fission Products	34,784	3,500	679,000	35,002	1,019	307,600

Fast-Breeder Reactor Fuels
(Fuel Exposed to 33,000 MWD/MTU at 58 MW/MTU)

| Fission Products | Grams/Tonne | After 1 year | | | After 10 years | |
		Watts/Tonne	Curies/Tonne	Grams/Tonne	Watts/Tonne	Curies/Tonne
$Ru^{106} - Rh^{106}$	204	6,660	1,370,000	0.41	14	2,760
$Ce^{144} - Pr^{144}$	177	4,700	1,128,000	0.06	2	370
Pm^{147}	300	150	279,000	28	13	26,000
Other Fission Products	34,219	2,290	653,000	34,872	737	251,870

Source: PB-224 588

activity and heat content in the spent fuel processing. After one year of radioactive decay, they are responsible for 65 percent of the total heat content and for 70 percent of the total activity in the high-level waste. Since these isotopes are principally produced in nuclear reactors, their projected growth will parallel that of the civilian nuclear power program.

Promethium-147 is used in radioisotope power generators and as a beta source in the radioactive gauge field. It is also used in the preparation of self-illuminating materials and devices for signs and signals. At the present time, promethium-147 is separated from the reactor-produced waste streams and distributed by the AEC through the Oak Ridge National Laboratory, Isotope Sales Department.

In the future when its production is warranted by market demands, it would be available in large quantities from commercial firms operating chemical reprocessing plants. The commercial use for ruthenium-106 and cerium-144 is very limited.

Ruthenium-106, cerium-144, and promethium-147 are only a few of the many fission products produced in nuclear reactors and are contained within the spent fuel elements. The spent fuel elements are removed from the reactor and processed for the recovery of usable fissionable materials. These radionuclides and the other high-level wastes are contained in the aqueous effluent from the spent-fuel processing step.

The range of chemical compositions of these waste streams has been tabulated (23). These waste streams are primarily aqueous solutions of inorganic nitrate salts and any differences in their composition occurs mainly in the amounts and types of salts added during the processing step. The amount and activity of Ru-106, Ce-144, and Pm-147 present in the high-level waste streams after 1 year and 10 years are shown in Table 5.9.

These radionuclides can also be found in the secondary waste streams generated at the spent fuel processing plant. The volume of these waste streams can be quite large but the radionuclide activity in these streams is quite low. Ruthenium-106, which usually oxidizes and volatilizes during the fuel reprocessing step, is also found in the off-gas streams.

Handling, Storage and Transportation

Since all three radionuclides are hazardous to man by inhalation, ingestion, or direct radiation exposure, care is exercised in their handling. Special procedures and radiation shielding are utilized in their handling. The beta particles emitted by Ru-106, Ce-144, and Pm-147 are of low energy and cannot penetrate more than 0.02 inch of water. The daughter products of Ru-106 and Ce-144 emit high energy beta particles which can penetrate up to 0.8 inch of water. The principal shielding problem associated with their daughter's beta radiation is the bremsstrahlung effect which results from the presence of highly penetrating rays resulting from the deflection of the beta particles by the nuclei in the shielding medium.

Since the bremsstrahlung effect increases with atomic number, shielding materials of high atomic weight are not satisfactory. Ru-106, Ce-144, and their daughters also emit gamma radiation. The gamma radiation is highly penetrating and high-density shields, such as lead, are required to stop the radiation. To detect and control personnel exposure to their radiation, all persons working with this material should wear dosimetry devices which directly indicate the dose. Commonly used devices are the film badge and the thermoluminescent dosimeters (TLD). These materials are stored in controlled reservations in specially constructed containers which are protected by both a primary and a secondary containment barrier. Special monitoring systems and proper warning signs are located in the general area of the storage facility.

Ruthenium-106 and cerium-144 are classified as a transport group III radionuclide and promethium-147 as a transport group IV radionuclide by the Department of Transportation. The rules and regulations governing their transportation are given in the Code of Federal Regulations (CFR) Title 49–Transportation, Parts 170 to 190 (1).

The Ru-106 and Ce-144 content is limited to 3 curies for a Type A package and 200 curies for a Type B package defined in 49CFR173. The Pm-147 content is limited to 20 curies for a Type A package and 200 curies for a Type B package. These limits for each radionuclide are increased to 20 and 5,000 curies if their physical form meets the requirements of a special form material. Their release rate is limited to zero under the specified accident conditions for Type A and B quantities (1). The allowable release of radioactivity from packages containing large quantities of these materials is limited to gases and contaminated coolant containing total radioactivity exceeding neither 0.1 percent of the total radioactivity of the package nor 10 curies under the hypothetical accident conditions prescribed in 49CFR173.

Disposal

The disposal of these materials is governed by the AEC Manual Chapter 0524 (2) and 10CFR20 (3). Two sets of standards have been established for the permissible radiation exposure in unrestricted areas. One is for the greatest dose received by an individual and the other for the average dose received by the general population, the latter being about one-third the former. These standards also define their maximum concentrations in air and water for exposure to either an individual or to the general population. For an individual, the safe release of these materials to the environment in an unrestricted area should not exceed the concentrations in microcuries per milliliter (μc/ml) listed below in Table 5.10.

TABLE 5.10: RUTHENIUM, CERIUM AND PROMETHIUM MAXIMUM PERMISSIBLE CONCENTRATIONS

Isotope	Form	Concentration in Air	Concentration in Water
Ru-106	Soluble	3×10^{-9} μc/ml	1×10^{-5} μc/ml
Ru-106	Insoluble	2×10^{-10} μc/ml	1×10^{-5} μc/ml
Ce-144	Soluble	3×10^{-10} μc/ml	1×10^{-5} μc/ml
Ce-144	Insoluble	2×10^{-10} μc/ml	1×10^{-5} μc/ml
Pm-147	Soluble	2×10^{-9} μc/ml	2×10^{-4} μc/ml
Pm-147	Insoluble	3×10^{-9} μc/ml	2×10^{-4} μc/ml

Source: PB-224 588

The concentrations for the safe release of these materials to the general population are one-third the above values. The disposal by release into a sanitary sewage system is limited to 10 microcuries for Ru-106 and Ce-144 and to 100 microcuries for Pm-147. The disposal by burial in the soil at any one location and time is limited to 1,000 microcuries for Ru-106 and Ce-144 and to 10,000 microcuries for Pm-147 (3).

Recovery Options

Option No. 1—Volatilization-Absorption: Ruthenium-106 removal from solutions can be accomplished by oxidizing the ruthenium to its tetroxide state and volatilizing it by simultaneous heating and gas sparging. In the tetroxide state, ruthenium boils at 100°C. Ruthenium can be removed from uranium nitrate solutions by sparging with air containing ozone. Ruthenium decontamination factors (ratio of initial to final concentration) of about 100 were obtained during an eight-hour sparge with 1 percent ozone in air (8). Ruthenium can also be oxidized to the volatile form by hot concentrated nitric acid. The volatized ruthenium is then recovered by absorption. Two such absorbents are pyrolusite and activated carbon. Decontamination factors with pyrolusite alone are less than 100 (8). Activated carbon used subsequent to pyrolusite absorption increases the decontamination factor 10 times.

Option No. 2—Scavenging-Precipitation Foam Separation: The removal of radionuclides from low-level radioactive waste water by scavenging-precipitation form separation has been studied by Oak Ridge National Laboratory (24). The process consists of two steps: (a) precipitating in a sludge-blanket clarification step, and (b) achieving final decontamination in a foam-separation column. Processing rates of 300 gallons per hour (gph) for scavenging-precipitation and 120 gph for foam separation have been achieved. Cerium-144 decontamination factors of greater than 20 have been obtained (24). Ruthenium-106 decontamination factors of only four were obtained. The low decontamination factor is expected since ruthenium is one of the most difficult radionuclides to remove by scavenging.

Option No. 3—Scavenging-Precipitation Ion Exchange: In this process, final decontamination is obtained by ion exchange columns from the scavenging-precipitation process. The process includes a provision for the recycle of the ion exchange waste to the scavenging-precipitation step. All the removed radionuclides are concentrated in the clarifier sludge. For cerium-144 and promethium-147 the overall decontamination factors varied from 20 to 700, and for ruthenium-106 the overall decontamination factor varied from 1.5 to 8 (6). For cerium-144 and promethium-147, the use of ion exchange resins with a scavenging-precipitation step is a fairly simple and efficient means of removing these radionuclides from low-level, radioactive aqueous wastes. For ruthenium-106 wastes, this method is not satisfactory since ruthenium in the liquid form is not very susceptible to ion exchange.

Option No. 4—Water Recycle: The water recycle process is used for decontaminating radioactive waste water and recycling the purified water for reuse. This process has been demonstrated at the pilot plant scale by Oak Ridge National Laboratory (10). The steps in the process include: (a) clarification by controlled addition of coagulants, (b) demineralization by cation-anion exchange, and (c) sorption on granular activated carbon. For optimum removal of the ruthenium-106 during the coagulation-clarification step, the pH of the waste must range between 7 and 8 for proper alum floc formation. High decontamination factors for ruthenium-106 can only be obtained during the final step by sorption on activated carbon. The overall decontamination factor for ruthenium-106 was 1,230 and for cerium-144, it was 800. The method is an improvement to the treat-and-discharge methods, but further work is required until full-scale production use is obtained.

Storage/Disposal Options

Option No. 1—Land Burial: Land burial of Ru-106, Ce-144 and Pm-147 wastes, in small concentrations, at approved sites that are acceptable from a geologic and hydrologic standpoint is an acceptable means of disposal. Their concentrations should not be in excess of 10^4 times the maximum permissible concentration for the general population in 10CFR20 (3). All wastes to be disposed of should be in a solid form and encapsulated in a suitable container. Liquid wastes should be solidified, preferably using asphalt.

The burial trenches should be designed not to intercept the ground water table and constructed with a bottom drain and sump for water monitoring. The trenches should be covered with either asphalt or vegetation to limit infiltration of water. The burial site design, geology, and hydrology should be in conformance with the criteria used in selecting and licensing the present commercial burial sites (4). Since land burial is successfully practiced on a commercial scale and since these isotopes have moderate half-lives, it should be considered as the most satisfactory method of disposing of all dilute concentrations of these wastes.

Option No. 2—Near-Surface Liquid Storage: Near-surface storage of aqueous wastes in stainless steel tanks encased in concrete and buried underground is not considered as a satisfactory means of disposal. Aqueous wastes have been stored in this manner over the past 25 years. The tanks range in size from 0.33 to 1.3 million gallons and are generally equipped with devices for measuring temperatures, liquid levels, and leaks. At the present time, these tanks are considered as an interim storage technique due to a general lack of confidence in their long-term integrity (6). Since present regulations require the solidification of all reactor wastes within 5 years following reprocessing (3), the near-surface storage of these wastes in steel tanks should only be considered as a near-term storage technique and not as a permanent storage or disposal technique.

Option No. 3—Near-Surface Solid Storage: The storage of high-level solidified Ru-106, Ce-144, and Pm-147 wastes in engineered surface facilities offers the best immediate method for storage of these wastes. The aqueous wastes should be solidified and packaged in a suitable container (steel). Of the four high-level solidification processes developed, spray and phosphate glass solidification offer the best solidified waste characteristics.

The wastes will be stored in stainless steel-lined concrete vaults which will be either air or water cooled. The periodic replacement of waste containers probably will not be required since their activity is reduced by a factor of 1,000 in 10 years for Ru-106 wastes, in 21 years for Ce-144 wastes, and in 26 years for Pm-147 wastes. The replacement of the waste containers might be required if other long-lived isotopes are also present.

Option No. 4—Salt Deposits: This method offers the best potential for the disposal of wastes from fuel reprocessing. The reactor-produced wastes must be solidified and disposed of in the solid form encapsulated in a suitable container. The preferred solidification process is spray solidification. The solidified wastes are then buried in rooms carved in the salt deposits approximately 1,000 feet below the ground. Salt deposits should be considered for the disposal of these wastes at approved sites that are acceptable from a design and geological standpoint.

Option No. 5—Bedrock Disposal: The disposal of liquid wastes containing ^{106}Ru, ^{144}Ce and ^{147}Pm along with other spent fuel processing wastes in vaults excavated in crystalline rock over 1,500 feet beneath the ground has been evaluated by E. I. du Pont de Nemours at their Savannah River Plant near Aiken, South Carolina (11)(12).

Option No. 6—Hydraulic Fracturing: The direct disposal of aqueous low-level radioactive wastes into shale formations has been investigated by Oak Ridge National Laboratory (14), but additional work is required to demonstrate that this method of disposal is satisfactory.

To summarize, ruthenium-106, cerium-144, and promethium-147 are radioactive isotopes that account for a majority of the initial activity present in spent fuel processing wastes. The acceptable method of treatment is solidification and interim storage in near-surface engineered facilities followed by permanent disposal in salt deposits. Cerium-144 and promethium-147 can be recovered from low-level aqueous waste streams by scavenging-precipitation ion exchange. Ruthenium-106 can be recovered by volatization and absorption on pyrolusite and activated carbon. Generally, during the reprocessing of uranium and plutonium, it is desirable to prevent the volatization of ruthenium. This can be accomplished during the leaching operation by adding a reducing agent such as sodium nitrite or nitrogen dioxide to suppress the ruthenium volatization.

Applicability to National Disposal Sites

Ruthenium-106, cerium-144, and promethium-147 are candidates for a National Disposal Site. From a cost and safety viewpoint, it would be desirable to combine the reprocessing plant and the National Disposal Site. The recommended treatment for the high-level wastes is solidification by either the spray or phosphate glass solidification processes, interim storage in near-surface engineered facilities, and ultimate disposal in salt deposits.

STRONTIUM-90 (YTTRIUM-90)

Strontium-90 is a radioactive isotope principally produced by the fission of uranium and plutonium. It has a half-life of 28 years and emits a 0.546 Mev beta particle. Strontium-90 produces a short half-lived daughter, yttrium-90. It has a half-life of 64 hours and emits a 2.27 Mev beta particle. Yttrium-90, in turn, produces zirconium-90 which is a stable element.

Strontium-90 has attracted great interest as a public health hazard since it is the most biologically significant of the radioactive fission products produced in either nuclear weapon tests or nuclear reactors. Its biological significance is derived from several factors: (a) it has a large fission yield (5.9%), (b) it has a long effective half-life, and (c) it tends to deposit and concentrate in the bone tissue (due to the fact that strontium is chemically similar to calcium). In the wastes generated at nuclear power plants, strontium-90 and its daughter, yttrium-90 are responsible for approximately 7 percent and 38 percent of the total fission product activity in the wastes after 1 year and 10 years, respectively. The projected growth in the production of strontium-90 will parallel that of the civilian nuclear power program.

Strontium-90 is a long-lived high-energy beta emitter. It is used in military and space applications as an energy source for small auxiliary power units. It is also used in the radioactive gauge field. Beta gauges activated by strontium-90 are extensively used in industry to measure and control the thickness or density of thin materials (paper, textiles, etc.) or coatings.

In the field of nuclear medicine, strontium-90 is used as a source of beta radiation for the treatment of skin diseases and eye disorders and also as a medical blood irradiator. The strontium-90 used in the above applications is obtained from the waste streams in the Richland, Washington chemical separations plant and is distributed through the Oak Ridge National Laboratory, Isotopes Sales Department. In the future, when its production is warranted by market demands, it could be available in large quantities from commercial firms operating chemical processing plants.

Although produced in nuclear power reactors, the primary source of strontium-90 is in the high-level aqueous waste streams generated at the spent fuel processing plants. The range of chemical compositions of the various types of waste streams obtained from the processing step have been tabulated (23) and, in all cases, are primarily aqueous solutions of inorganic nitrate salts. Strontium-90 is normally found in the waste stream in the form of a nitrate or oxide. The characteristics of strontium-90 products in the high-level wastes at two different time periods are shown in Table 5.11.

Strontium-90 is also found in the low-level waste streams generated at spent fuel processing facilities. The strontium-90 activity in these waste streams is from a few hundredths to several tenths of a curie per gallon of waste. Strontium-90 is also found in the wastes resulting from research laboratories and medical and industrial applications. Generally, the strontium-90 waste from these sources is found in the solid form.

Handling, Storage and Transportation

Since strontium-90 is hazardous to man by inhalation, ingestion, or direct radiation exposure, care is exercised in its handling. Special procedures and radiation shielding are required in the handling of strontium-90. The beta particles emitted by strontium-90 generally cannot penetrate more than one centimeter of metal or glass. The principal problem associated with the beta radiation is the bremsstrahlung effect which results from the presence of highly penetrating rays resulting from the deflection of beta particles by the nuclei in the shielding medium.

Since the bremsstrahlung effect increases with atomic number, shielding materials of high atomic weight are not satisfactory. Lead and high-density concrete are commonly used shielding materials for strontium-90 radiation. To detect and control personnel exposure to strontium-90 radiation, all persons working with this material should wear dosimetry devices which directly indicate the dose. Commonly used devices are the film badge and the thermoluminescent dosimeters.

TABLE 5.11: STRONTIUM-90 CONTENT IN HIGH-LEVEL FUEL REPROCESSING WASTES

Light-Water Reactor Fuels
(Fuel Exposed to 33,000 MWD/MTU at 30 MW/MTU)

Fission Products	After 1 year				After 10 years	
	Grams/Tonne	Watts/Tonne	Curies/Tonne	Grams/Tonne	Watts/Tonne	Curies/Tonne
$Sr^{90} - Y^{90}$	534	530	151,000	427	425	120,900
Other Fission Products	34,566	9,470	2,071,000	34,673	605	196,100

Fast-Breeder Reactor Fuels
(Fuel Exposed to 33,000 MWD/MTU at 58 MW/MTU)

Fission Products	After 1 year				After 10 years	
	Grams/Tonne	Watts/Tonne	Curies/Tonne	Grams/Tonne	Watts/Tonne	Curies/Tonne
$Sr^{90} - Y^{90}$	300	298	85,000	241	239	68,000
Other Fission Products	34,600	13,502	3,345,000	34,658	529	213,000

Source: PB-224 588

Strontium-90 is stored in controlled reservations in specially constructed containers. The strontium-90 is protected by both a primary and a secondary containment barrier. Special monitoring systems and proper warning signs are located in the general area of the storage facility.

Strontium-90 is classified as a transport group II radionuclide by the Department of Transportation, and the rules and regulations governing their transportation are given in the Code of Federal Regulations (CFR) Title 49–Transportation, Parts 170 to 190 (1). The strontium-90 content is limited to 0.05 curie for a Type A package and 20 curies for a Type B package defined in 49CFR173. The limits are increased to 20 and 500 curies if the physical form of the strontium-90 meets the requirements of a special form material.

The strontium-90 release rate is limited to zero under the specified accident conditions for Type A and B quantities (1). The allowable release of radioactivity from packages containing large quantities of strontium-90 is limited to gases and contaminated coolant containing total radioactivity exceeding neither 0.1 percent of the total radioactivity of the package nor 0.5 curie under the hypothetical accident conditions prescribed in 49CFR173.

Disposal

The disposal of strontium-90 is governed by the AEC Manual, Chapter 0524 (2) and 10CFR20 (3). Two sets of standards have been established for the permissible radiation exposure in unrestricted areas. One is for the greatest dose received by an individual and the other for the average dose received by the general population. The radiation protection standards for these two groups are given below. The standards for strontium-90 also define its maximum concentrations in air and water in microcuries per milliliter (μc/ml) listed below in Table 5.12.

TABLE 5.12: STRONTIUM-90 MAXIMUM PERMISSIBLE CONCENTRATIONS

Exposure Group	Strontium-90 Form	Concentration in Air	Concentration in Water
Individual	Soluble	3×10^{-11} μc/ml	3×10^{-7} μc/ml
Individual	Insoluble	2×10^{-10} μc/ml	4×10^{-5} μc/ml
Population	Soluble	1×10^{-11} μc/ml	1×10^{-7} μc/ml
Population	Insoluble	0.67×10^{-10} μc/ml	1.3×10^{-5} μc/ml

Source: PB-224 588

Although rarely utilized, the disposal of strontium-90 by release into a sanitary sewage system is limited to 1 microcurie or a concentration of 1×10^{-5} microcurie per milliliter of water in the soluble form or 1×10^{-3} microcurie per milliliter of water in the insoluble form. The disposal of strontium-90 by burial in the soil at any one location and time is limited to 100 microcuries of strontium-90 (3).

Recovery Options

Option No. 1—Precipitation: Strontium-90 can be removed from high-level processing wastes following the spent fuel processing step by precipitating as a carbonate or fluoride (25) or absorbing on aluminosilicate zeolites (9). The standard procedure for the recovery of strontium-90 is to precipitate as a carbonate and then add titanium dioxide and heat to drive off the CO_2, forming strontium titanite. The strontium-90 can be recovered from the other high-level wastes for commerical purposes or to provide isolation of the strontium-90 for heat transfer and safety reasons.

Option No. 2—Scavenging-Precipitation Foam Separation: The removal of strontium-90 from radioactive waste water by scavenging-precipitation foam separation has been studied by Oak Ridge National Laboratory (24). Strontium-90 is precipitated by the addition of calcium carbonate or calcium phosphate. Strontium-90 decontamination factors (ratio of initial to final concentration) of 1,050 have been obtained (24). Further development is required to reduce operating costs and increase processing rates.

Option No. 3—Scavenging-Precipitation Ion Exchange: In the phenolic ion exchange column, calcium and magnesium are removed with the strontium. The overall decontamination factors for strontium-90 varied from 1,200 to 12,000 (6). The use of ion exchange resins with a scavenging-precipitation step is a fairly simple and efficient means of removing strontium-90 from low-level radioactive aqueous wastes.

Option No. 4—Water Recycle: The majority of the strontium-90 is removed by the cation-anion exchange. The strontium-90 overall decontamination factor was 5,700. The method is an improvement to the treat-and-discharge methods, but further work is required until full-scale production use is obtained.

Storage/Disposal Options

Option No. 1—Land Burial: Land burial of strontium-90 wastes, in small concentrations, at approved sites that are acceptable from a geologic and hydrologic standpoint, is an acceptable means of disposal. Strontium-90 concentration should not be in excess of 10^4 times the maximum permissible concentration for the general population in 10CFR20 (3). All strontium-90 wastes to be disposed of should be in a solid form and packaged in a suitable container. Liquid strontium-90 wastes should be solidified, preferably using asphalt.

The burial trenches should be designed not to intercept the ground water table and constructed with a bottom drain and sump for water monitoring. The trenches should be covered with either asphalt or vegetation to limit infiltration of water. The burial site design, geology, and hydrology should be in conformance with the criteria used in selecting and licensing the present commercial burial sites (4). Since land burial has successfully been practiced on a commercial scale, it should be considered as the most satisfactory method of disposing of all dilute concentrations of strontium-90 wastes.

Option No. 2—Near-Surface Liquid Storage: Near-surface storage of reactor-produced aqueous solutions of strontium-90 and other fission product salts, which are stored in carbon steel or stainless steel tanks encased in concrete and buried underground, is not considered as a satisfactory means of disposal. Since present regulations require the solidification of all reactor wastes within 5 years following reprocessing (3), the near-surface storage of aqueous solutions of strontium-90 and other fission product salts in steel tanks should only be considered as a near-term storage technique and not as a permanent storage or disposal technique.

Option No. 3—Near-Surface Solid Storage: The storage of solidified, high-level wastes containing strontium-90 and other waste salts in engineered storage facilities offers the best intermediate method for storage of these wastes. Of the four high-level solidification processes developed, spray or phosphate glass solidification offers the best solidified waste characteristics. The wastes will be stored in stainless-steel-lined concrete vaults which will be either air or water cooled.

Option No. 4—Salt Deposits: This method offers the best potential for the disposal of wastes from fuel reprocessing. The wastes containing strontium-90 must be solidified and disposed of in the solid form and packaged in a suitable container. The solidified wastes are then buried in rooms carved in the salt deposits approximately 1,000 feet below the ground. The critical problem is the selection of a site that meets the necessary design and geological criteria for the mixture of fission product and actinide wastes.

Option No. 5—Bedrock Disposal: The disposal of low-level aqueous solutions of strontium-90 wastes along with other spent fuel reprocessing wastes in vaults excavated in crystalline rock over 1,500 feet beneath the ground has been evaluated by E. I. du Pont de Nemours, at their Savannah River Plant near Aiken, South Carolina (11)(12).

Another method has also been proposed for disposing of liquid wastes by in situ incorporation in molten silicate rock (13). Due to the long half-lives of certain of the elements of the mixed fission products and actinides and high biological hazard, it is doubtful that data could ever be accumulated to prove that the geological characteristics of the site over the next hundred years are acceptable to ensure the absolute safety of such a disposal method.

Option No. 6—Hydraulic Fracturing: Since strontium-90 is leached from cement and no additional encapsulation is provided and since it has a long half-life, this method of disposal is not recommended.

To summarize, strontium-90 can be recovered from the high-level waste streams from the spent fuel processing facilities for separate disposal or reuse. The separation of strontium-90 from these waste streams will probably be required with the introduction of the fast-breeder reactor due to its high heat content. The acceptable method of treatment is spray or phosphate glass solidification, followed by storage in near-surface engineered facilities. Finally, these wastes should be recovered from the low-level waste streams to minimize the amount directly released to the environment. An adequate method of recovery is scavenging-precipitation ion exchange. The recovered strontium-90 can then be solidified, preferably using asphalt, and disposed of at approved sites by land burial.

Applicability to National Disposal Sites

Strontium-90 is a candidate for a National Disposal Site due to its health hazard and its projected growth with that of the civilian nuclear power program. From a cost and safety viewpoint, it would be desirable to combine the reprocessing plant and the National Disposal Site. Interim storage in near-surface engineered facilities offers considerable latitude in the storage site selection. The site selection for ultimate disposal of these wastes is limited to particular geological areas in which salt deposits are present.

The recommended treatment for strontium-90 in low-level waste streams is recovery by scavenging-precipitation ion exchange followed by solidification with asphalt and disposal by land burial. For the high-level strontium-90 wastes, the recommended processes are recovery by either spray or phosphate glass solidification, interim storage of the solidified waste in near-surface engineered storage facilities, and disposal in salt beds.

ZIRCONIUM-95 AND NIOBIUM-95

Zirconium-95 is a radioactive isotope principally produced in nuclear reactors by the fission of uranium and plutonium. Zirconium-95 has a half-life of 65 days and emits both beta particles and gamma rays. Zirconium-95 produces, by beta decay, niobium-95. Niobium-95 has a half-life of 35 days, emits a 0.160 Mev beta particle and a 0.765 Mev gamma ray. Niobium-95 is transformed by beta decay into molybdenum-95 which is a stable element.

Zirconium-95 and niobium-95 are both short-lived isotopes and account for 25 percent of the total activity in spent fuel processing wastes after 90 days. Their activity decreases to less than 1 percent of the total after 1 year. They also present considerable difficulty in the separation of uranium and plutonium from these elements since they both form radioactive colloids in solution. They also tend to be absorbed on surfaces such as container walls. The projected growth in zirconium-95 and niobium-95 will parallel that of the civilian nuclear power program.

The primary sources of zirconium-95 and niobium-95 wastes are the spent fuel processing plants. These isotopes and the other resultant high-level radioactive wastes are contained in the aqueous effluent from the spent fuel processing step. The range of chemical compositions of the various types of waste streams obtained from the spent fuel processing step have been tabulated (23).

The activity of zirconium-95 and niobium-95 in the spent reactor wastes at two different times is given in Tables 5.13 and 5.14. Due to their short half-lives, their activity after 10 years is reduced to zero. They are also found in very small quantities in the low-level radioactive waste streams generated at spent fuel processing facilities.

The physical and chemical properties of zirconium-95 and niobium-95 are as follows: Zirconium-95 is a member of the Group IV series of metals, all of which have a characteristic +4 oxidation state. Zirconium-95 is not readily soluble in nitric acid but dissolves at substantial rates in both hydrofluoric and moderately concentrated sulfuric acid.

Niobium-95 exists in the +5 oxidation state in a majority of its compounds. Niobium-95 is a passive metal and does not dissolve in most acids when pure, but is attacked by these acids if impurities are present. Both isotopes tend to form radioactive colloids in solution which do not act as true solutes but tend to be absorbed on surfaces.

TABLE 5.13: ZIRCONIUM-95 CONTENT IN HIGH-LEVEL FUEL REPROCESSING WASTES

Light-Water Reactor Fuels
(Fuel Exposed to 33,000 MWD/MTU at 30 MW/MTU)

Fission Products	Grams/Tonne	After 1 year Watts/Tonne	Curies/Tonne	Grams/Tonne	After 10 years Watts/Tonne	Curies/Tonne
Zr^{95}	1.3	146	28,000	0	0	0
Other Fission Products	35,099	9,854	2,194,000	35,100	1030	317,000

Fast-Breeder Reactor Fuels
(Fuel Exposed to 33,000 MWD/MTU at 58 MW/MTU)

Fission Products	Grams/Tonne	After 1 year Watts/Tonne	Curies/Tonne	Grams/Tonne	After 10 years Watts/Tonne	Curies/Tonne
Zr^{95}	630	310	59,000	0	0	0
Other Fission Products	34,897	13,490	3,371,000	34,900	776	281,000

Source PB-224 588

TABLE 5.14: NIOBIUM-95 CONTENT IN HIGH-LEVEL FUEL REPROCESSING WASTES

Light-Water Reactor Fuels
(Fuel Exposed to 33,000 MWD/MTU at 30 MW/MTU)

Fission Products	Grams/Tonne	After 1 year		Grams/Tonne	After 10 years	
		Watts/Tonne	Curies/Tonne		Watts/Tonne	Curies/Tonne
Nb^{95}	2	285	59,000	0	0	0
Other Fission Products	35,098	9,715	2,163,000	35,100	1030	317,000

Fast-Breeder Reactor Fuels
(Fuel Exposed to 33,000 MWD/MTU at 58 MW/MTU)

Fission Products	Grams/Tonne	After 1 year		Grams/Tonne	After 10 years	
		Watts/Tonne	Curies/Tonne		Watts/Tonne	Curies/Tonne
Nb^{95}	3	600	125,000	0	0	0
Other Fission Products	34,897	13,200	3,305,000	34,900	776	281,000

Source: PB-224 588

Handling, Storage and Transportation

Since zirconium-95 and niobium-95 are hazardous to man by inhalation, ingestion, or direct radiation exposure, care is exercised in their handling. Special procedures and radiation shielding are utilized in their handling. The beta particles emitted by them generally cannot penetrate more than 0.1 inch of water or 0.04 inch of glass. Their gamma rays are highly penetrating. Lead and concrete are commonly-used shielding materials. To detect and control personnel exposure, all persons working with these materials should wear dosimetry devices which directly indicate the dose. Commonly-used devices are the film badge and the thermoluminescent dosimeters.

Zirconium-95 and niobium-95 are stored in controlled areas in specially constructed containers. They are protected by both a primary and a secondary containment barrier. Special monitoring systems and proper warning signs are located in the general area of the storage facility.

Zirconium-95 is classified as a transport group III and niobium-95 as a transport group IV radionuclide by the Department of Transportation, and the rules and regulations governing their transportation are given in the Code of Federal Regulations (CFR) Title 49—Transportation, Parts 170 to 190 (1). The zirconium-95 content is limited to 3 curies for a Type A package and 200 curies for a Type B package defined in 49CFR173.

For niobium-95, the limits are 20 curies for a Type A package and 200 curies for a Type B package. The limits are increased to 20 and 5,000 curies if their physical form meets the requirements of a special form material. Their release rate is limited to zero under the specified accident conditions for Type A and B quantities (1). The allowable release of radioactivity from packages containing large quantities of these isotopes is limited to gases and contaminated coolant containing total radioactivity exceeding neither 0.1 percent of the total radioactivity of the package nor 10 curies under the hypothetical accident conditions prescribed in 49CFR173.

Disposal

The disposal of zirconium-95 and niobium-95 is governed by the AEC Manual Chapter 0524 (2) and 10CFR20 (3). Two sets of standards have been established for the permissible radiation exposure in unrestricted areas. One is for the greatest dose received by an individual and the other for the average dose received by the general population, the latter being one-third the former.

TABLE 5.15: ZIRCONIUM-95 AND NIOBIUM-95 MAXIMUM PERMISSIBLE CONCENTRATIONS

Isotope	Exposure Group	Form	Concentration in Air microcuries/milliliter	Concentration in Water microcuries/milliliter
Zirconium-95	Individual	Soluble	4×10^{-9}	6×10^{-5}
Zirconium-95	Individual	Insoluble	1×10^{-9}	6×10^{-5}

(continued)

FIGURE 5.15: (continued)

Isotope	Exposure Group	Form	Concentration in Air microcuries/milliliter	Concentration in Water microcuries/milliliter
Zirconium-95	Population	Soluble	1.3×10^{-9}	2×10^{-5}
Zirconium-95	Population	Insoluble	0.33×10^{-9}	2×10^{-5}
Niobium-95	Individual	Soluble	2×10^{-8}	1×10^{-4}
Niobium-95	Individual	Insoluble	3×10^{-9}	1×10^{-4}
Niobium-95	Population	Soluble	0.67×10^{-8}	0.33×10^{-4}
Niobium-95	Population	Insoluble	1×10^{-9}	0.33×10^{-4}

Source: PB-244 588

The standards for the safe disposal of zirconium-95 and niobium-95 also define their maximum concentrations in air and water. Their concentrations in an unrestricted area should not exceed the concentrations in microcuries per milliliter included in Table 5.15.

Although rarely practiced, the disposal of zirconium-95 and niobium-95 into a sanitary sewage system is limited to 100 microcuries. Their disposal by burial in the soil at any one location and time is limited to 10,000 microcuries (3).

Recovery Options

Option No. 1—Scavenging-Precipitation Foam Separation: The removal of zirconium-95 and niobium-95 from low-level radioactive waste water by scavenging-precipitation foam separation has been studied by Oak Ridge National Laboratory (24). Decontamination factors (ratio of initial to final concentration) greater than 50 have been obtained for both isotopes (24). At the present time, further development is required to reduce operating costs and increase processing rates.

Option No. 2—Scavenging-Precipitation Ion Exchange: The overall decontamination factors for these isotopes varied from 10 to 150 (6). The use of ion exchange resins with a scavenging-precipitation step is a fairly simple and efficient means of removing these isotopes from low-level radioactive aqueous wastes.

Option No. 3—Water Recycle: The process has been demonstrated at the pilot plant scale by Oak Ridge National Laboratory (10). Overall decontamination factors of 350 were obtained. This is the minimum value since the concentrations of zirconium-95 and niobium-95 were reduced to the analytical limits of detection (background). This method is an improvement to the treat-and-discharge methods, but further work is required until full-scale production use is obtained.

Storage/Disposal Options

Option No. 1—Land Burial: Land burial of low-level zirconium-95 and niobium-95 wastes, in small concentrations, at approved sites meeting the geologic and hydrologic criteria, is an acceptable means of disposal. Their concentrations should not be in excess of 10^4 times their maximum permissible concentrations for the general population in 10CFR20 (3). All wastes to be disposed of should be in a solid form and encased in a suitable container. Liquid wastes should be solidified, preferably using asphalt.

The burial trenches should be designed not to intercept the ground water table and constructed with a bottom drain and sump for water monitoring. The trenches should be covered with either asphalt or vegetation. The burial site design, geology, and hydrology should be in conformance with the criteria used in selecting and licensing the present commercial burial sites (4). Since land burial has successfully been practiced on a commercial scale and both zirconium-95 and niobium-95 have short half-lives, it should be considered as the most satisfactory method of disposing of all dilute concentrations of these wastes.

Option No. 2—Near-Surface Liquid Storage: Near-surface storage of reactor-produced aqueous solutions containing zirconium-95 and niobium-95 high-level waste in carbon steel and stainless steel tanks encased in concrete and buried underground is a satisfactory means of short-term storage. Since present regulations require the solidification of all reactor wastes within 5 years following reprocessing (3), the near-surface storage of aqueous solutions of zirconium-95 and niobium-95 and other fission product salts in steel tanks should only be considered as a near-term storage technique and not as a permanent storage or disposal technique.

Option No. 3—Near-Surface Solid Storage: The storage of high-level solidified zirconium-95 and niobium-95 wastes and other waste salts in engineered storage facilities offers the best intermediate method for storage of these wastes. Of the four high-level solidification processes developed, spray or phosphate glass solidification processes offer better solidified waste characteristics than the pot calcination or the fluidized bed calcination processes. The wastes will be stored in stainless-steel-lined concrete vaults which will be air or water cooled. The periodic replacement of the waste containers probably will not be required since the activity of zirconium-95 is reduced by a factor of 100 in 132 days and niobium-95 activity is reduced by a factor of 100 in 232 days. The replacement of the containers may be required if other long-lived isotopes are present.

Option No. 4—Salt Deposits: This method offers the best potential for the disposal of wastes from fuel reprocessing since bedded salt deposits are completely free of circulating ground waters. The wastes must be solidified and packaged in a suitable container. The solidified wastes are buried in rooms carved in the salt deposits, approximately 1,000 feet below the ground.

Option No. 5—Bedrock Disposal: The disposal of high-level aqueous solutions containing zirconium-95 and niobium-95, along with other wastes from spent fuel reprocessing wastes in vaults excavated in crystalline rock over 1,500 feet beneath the ground, has been evaluated by E. I. du Pont de Nemours at their Savannah River Plant near Aiken, South Carolina (11)(12). Another method has also been proposed for disposing of liquid wastes by in situ incorporation in molten silicate rock (13).

Option No. 6—Hydraulic Fracturing: Since zirconium-95 and niobium-95 are leached from cement and no additional encapsulation is provided, this method of disposal requires additional work to determine its suitability.

To summarize, both zirconium-95 and niobium-95 are short-lived isotopes with a high initial activity but whose activity decreases by a factor of 1,000 within 2 years. These isotopes will probably be contained with the long-lived fuel reprocessing wastes since their recovery is not required due to their short half-life and limited commercial use. The acceptable method of treatment is solidification followed by storage in near-surface engineered facilities and disposal in salt deposits.

Applicability to National Disposal Sites

Zirconium-95 and niobium-95 are both candidates for a National Disposal Site due to their projected growth with that of the civilian nuclear power program. From a cost and safety viewpoint, it would be desirable to combine the reprocessing plant and the National Disposal Site. Interim storage in near-surface engineered facilities offers considerable latitude in the site selection. The site selection for ultimate disposal of these wastes is limited to particular geological areas in which salt deposits are present.

The recommended treatment for high-level aqueous wastes containing zirconium-95 and niobium-95 is solidification using either spray or phosphate glass solidification processes. If required for economic reasons, recovery prior to solidification could be effected by scavenging-precipitation ion exchange. The solidified wastes should then be held in interim storage in engineered storage facilities, and finally disposed of in salt deposits.

REFERENCES

(1) *Code of Federal Regulations,* "Title 49, Transportation," Parts 1 to 199. (Revised as of January 1, 1971). Washington, U.S. Government Printing Office, 1971.

(2) *AEC Manual,* Chapter 0524, "Standards for Radiation Protection," May 12, 1964.

(3) *Code of Federal Regulations,* "Title 10—Atomic Energy." (Revised as of January 1, 1971), Washington, U.S. Government Printing Office, 1971.

(4) R.J. Morton, *Land Burial of Solid Radioactive Wastes: Study of Commercial Operations and Facilities,* Environmental and Sanitary Engineering Branch, Atomic Energy Commission, WASH-1143, 1968.

(5) *Report by the Committee on Radioactive Waste Management, Disposal of Solid Radioactive Wastes in Bedded Salt Deposits,* National Academy of Sciences–National Research Council, Washington, 1970.

(6) Staff of the Oak Ridge National Laboratory, *Siting of Fuel Reprocessing Plants and Waste Management Facilities,* ORNL-4451, 1970.

(7) H. Etherington, ed., *Nuclear Engineering Handbook,* New York, McGraw-Hill Book Company, Inc., 1960.

(8) C.J. Touhill, B.W. Mercer, and A.J. Shuckrow, *Treatment of Waste Solidification Condensates,* Battelle Northwest, BNWL-723, 1968.

(9) W.H. Regan, ed., *Proceedings: The Solidification and Long-Term Storage of Highly Radioactive Wastes, Richland, Washington, February 14–18, 1966,* Atomic Energy Commission, CONF-660208.

(10) W.C. Yee, F. DeLora, and W.E. Shuckley, *Low-Level Radioactive Waste Treatment: The Water Recycle Process.* Oak Ridge National Laboratory, ORNL-4472, 1970.

(11) R.M. Girdler, *Storage of Liquid Radioactive Wastes at the Savannah River Plant,* Du Pont de Nemours and Company, Aiken, South Carolina, DPSPU-69-30-9, July 1969.

(12) J.B. Cooke, et al, *Report on the Proposal of E.I. Du Pont de Nemours and Company for the Permanent Storage of Radioactive Separation Process Wastes in Bedrock on the Savannah River,* Du Pont de Nemours (E.I.) and Company, Aiken, South Carolina, DPST-69-444, May 1970.

(13) J.J. Cohen, A.E. Lewis, and R.L. Braun, "In Situ Incorporation of Nuclear Waste in Deep Molten Silicate Rock," *Nuclear Technology,* 13:76-87, April 1972.

(14) R.C. Struxness, et al, *Engineering Development of Hydraulic Fracturing as a Method for Permanent Disposal of Radioactive Wastes,* Oak Ridge National Laboratory, ORNL-4259, 1968.

(15) *Selected Materials on Environmental Effects of Producing Electric Power,* Report of the Joint Committee on Atomic Energy, Congress of the United States, 91st Congress, 1st session, August 1969, Washington, U.S. Government Printing Office.

(16) Atomic Energy Commission, *Forecast of Growth of Nuclear Power,* January 1971, Washington, U.S. Government Printing Office, WASH-1139, 1971.

(17) R.C. Milham and L.R. Jones, *Iodine Retention Studies,* Du Pont de Nemours, (E.I.) and Company, Savannah River Laboratory, DP-1234, 1970.

(18) R.E. Adams, R.D. Ackley, and W.E. Browning, *Removal of Radioactive Methyliodine from Steam-Air Systems,* Oak Ridge National Laboratory, ORNL-4040, 1967.

(19) R.D. Ackley, Z. Combs, and R.E. Adams, *Aging, Weathering, and Poisoning of Impregnated Charcoals Used for Trapping Radioiodine,* Oak Ridge National Laboratory, ORNL-TM-2860, 1970.

(20) S.G. Forbes and R.F. Berger, *Fission Product Sampling and Decontamination Development Program,* Phillips Petroleum Company, Idaho Operations Office, IDO-172581, 1969.

(21) C.L. Bendixsen and G.F. Offutt, *Rare Gas Recovery Facility at the Idaho Chemical Processing Plant,* Idaho Nuclear Corp., Idaho Falls, Idaho, IN-1221, 1969.

(22) M.J. Stephenson, J.R. Merriman, and D.I. Duthorn, *Experimental Investigation of the Removal of Krypton and Xenon from Contaminated Gas Streams by Selective Absorption in Fluorocarbon Solvents,* Union Carbide Corp., Oak Ridge Gaseous Diffusion Plant, K-1780, 1970.

(23) K.J. Schneider, *Status of Technology in the United States for Solidification of Highly Radioactive Liquid Wastes,* Battelle Northwest, BNWL-820, 1968.

(24) L.J. King, A. Shimozato, and J.M. Holmes, *Pilot Plant Studies of the Decontamination of Low-Level Process Waste by a Scavenging-Precipitation Foam Separation Process,* Oak Ridge National Laboratory, ORNL-3803, 1968.

(25) J.R. LaRiviere, "Packaging and Storing Radioactive Wastes," *Chemical Engineering Progress,* 66:42-44, February 1970.

RADIOACTIVE WASTE MANAGEMENT AT U.S. GOVERNMENT FACILITIES

The sources of the material in this chapter were the following reports: ERDA-1538, ARH-SA-269, BNWL-1913, SRO-TWM-76-1 and DP-1366. See p 359 for a complete bibliography.

This chapter gives an overall view of integrated radioactive waste management at two major U.S. government facilities—Hanford Reservation in Washington and the Savannah River Plant in South Carolina.

WASTE MANAGEMENT OPERATIONS AT HANFORD RESERVATION

The semiarid Hanford site occupies about 570 square miles or 365,000 acres of the southeastern part of the State of Washington. In early 1943, the United States Army Corps of Engineers selected the Hanford site as the location for reactor and chemical separation facilities for the production and purification of plutonium for possible use in nuclear weapons (Manhattan Project). A total of eight graphite-moderated reactors using the Columbia River water for once-through cooling, and a new type of dual purpose reactor (N Reactor) using a recirculating water coolant and producing both plutonium and steam for electricity, were eventually built along the Columbia River. Today, only the N Reactor remains in operation.

The Hanford site operating areas are identified by area numbers. Each area identified as a controlled or limited access area is totally enclosed in a high antipersonnel fence. The six 100 Areas, bordering directly on the Columbia River in the northernmost portion of the Hanford Reservation, are where the plutonium production reactors were built. At one time in the early 1960's, nine production reactors were operating. Currently, only the dual-purpose N Reactor is operating to provide plutonium for military purposes and ERDA research and development purposes and to provide the by-product steam used for electrical power production in the Washington Public Power Supply System (WPPSS) Plant.

In the middle of the site, on a plateau about 7 miles from the Columbia River,

FIGURE 6.1: MAP OF HANFORD RESERVATION

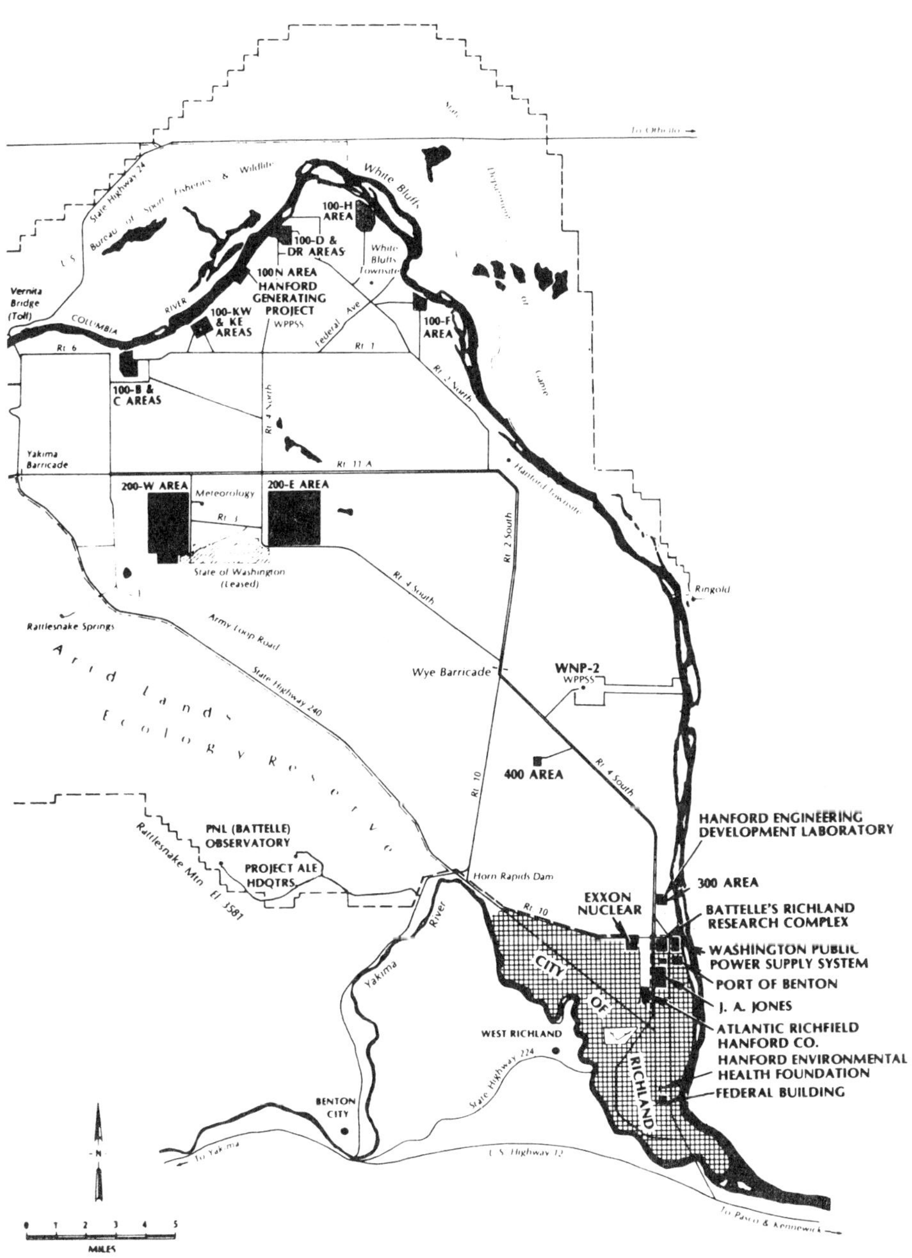

Source: ERDA-1538

are the two 200 Areas where the fuel and waste processing and waste storage activities are located. The location of the 200 Areas was chosen because it provides the most isolation from the site boundaries and is the most removed from both surface and subsurface water. Under this area, the groundwater table is some 150 to 300 feet below the surface.

The 300 Area is nearest to Richland. The research and development laboratories and the fuel fabrication facility for the N Reactor are located in this area.

The 400 Area is about 7 miles north of the 300 Area and is the site of the Fast Flux Test Reactor now under construction.

The area of the Hanford Reservation not included as one of the 100, 200, 300, or 400 Areas is designated as the 600 Area. Figure 6.1 is a map of the Hanford Reservation showing the locations of these various operational areas and other features.

100 Areas

The shutdown of eight production reactors and the recent (1973) elimination of solid waste burials in the 100 Areas significantly decreased waste management activity in the 100 Areas. Most of the 100 Areas now generate essentially no waste. Waste management facilities of the 100 Areas are arbitrarily divided into the active and inactive areas.

N Reactor differs significantly from all earlier Hanford reactors. At N Reactor, the primary coolant circulates through the reactor and absorbs heat from the fuel elements but is not returned directly to the river. Instead, heat is transferred through heat exchangers to a secondary coolant loop which ordinarily produces the steam to power the electrical generators of the WPPSS. By not releasing cooling water from the reactor core directly to the river, the amount of radioactivity released to the river is greatly reduced.

Radioactive Liquid Waste Facilities: 1301-N Crib and Trench — In the 100-N liquid waste system, the major source of radioactive liquid waste released to the ground is the continuous bleed of primary coolant of the N Reactor. The cooling water is circulated through the reactor process tubes and over the fuel element surfaces repeatedly. Induced activity is minimized by purifying this water before injection into the primary coolant loop. However, the water accumulates radioactivity in the reactor coolant loop as follows:

> Induced activity in dissolved piping corrosion materials (iron, chromium, manganese, zirconium and cobalt)
> Fission product releases from fuel elements with failed cladding
> Fission products from the fissioning of uranium residue resulting from fuel cladding failures.

Radioactivity in the coolant loop is reduced by continuous bleed-off of part of the coolant and replacement of it with treated river water. The average bleed rate of 1,000 gpm allows the entire primary coolant volume of 150,000 gallons to be replaced every 2.5 hours. Ammonium hydroxide is added to primary coolant water during operation to raise the pH to the alkaline side of neutral. The bleed-off coolant water is then released to the 1301-N crib which measures about 125 by 290 feet.

The total flow to the crib during normal reactor operation is 2,500,000 gpd from:

    36-inch contaminated drain with about 2,000,000 gpd during reactor operations
    12-inch line of about 500,000 gpd during reactor operations
    3-inch drain from waste disposal valve pit, 350 gpd
    6-inch auxiliary decontamination line from 109-N, intermittent.

The overflow from the crib is directed to the adjacent 1,600-ft long by 50-ft wide trench. Some flow from the crib and trench forms a surface seepage at the river bank referred to as the N Area Riverbank Springs. About one-half of the discharged water reaches the river in 4 to 10 days. The remainder follows longer flow paths and requires weeks to months to reach the river.

Most radioactive species discharged to the crib are retained in the soil by chemical reactions, precipitation, and ion exchange. Since the pH is on the alkaline side, the hydroxides and basic salts of zirconium, cerium, and other rare earth and transition elements precipitate and are retained in the soil by filtration. Due largely to the type of clay minerals present in the region, ion exchange of the cations retains the long-lived radionuclides (^{60}Co, ^{90}Sr, ^{137}Cs) in the soil.

102-Inch Discharge Line to the River — A 102-inch diameter discharge line terminates in the Columbia River about 500 feet out from water's edge during low volume river flows of about 50,000 cfs. The waste consists primarily of steam condenser cooling water discharged via the 102-inch line to the Columbia River channel. (Flow rates down the line with the reactor operating are about 225,000 gpm and with reactor shutdown are about 100,000 gpm.)

Decontamination Waste (1310-N Facility) — Approximately 600,000 gallons of liquid waste are shipped to 200 Areas each year. The bulk of the waste is generated once a year when the N Reactor primary coolant piping is internally decontaminated to remove radioactive corrosion products and deposited materials. Decontamination is a once-through process in which 260,000 gallons of 8 weight percent phosphoric acid is flushed through the primary coolant loop and routed to the 1310-N tank. The primary coolant system is then rinsed with water and NaOH is added to the 1310-N tank to neutralize the phosphoric acid, increasing the 1310-N tank's waste volume to approximately 600,000 gallons. The resulting intermediate level liquid waste is shipped to the 200 West Area in 20,000-gallon railway tank cars for evaporation to a solid and for in-tank storage.

Radioactive Solid Waste Facilities: Solid waste from 100-N Area was buried in trenches in other 100 Areas in the past. The waste contained dried salts and oxides of the activation and fission products found in solution in the primary coolant water. A few metallic components, taken from the active reactor zone, containing zirconium, iron, and steel activated to ^{95}ZrNb, ^{60}Co, and ^{54}Mn were also buried in these facilities. Waste containing transuranium elements and/or high concentration (10 mCi/g) of long-lived radionuclides (>5 year half-life) was shipped to the 200 Areas for burial. Since December 31, 1973, all solid waste has been shipped to the 200 Areas for disposal with the following exceptions.

Contaminated Equipment Storage—An area was set aside for temporarily storing contaminated equipment removed from radiation zones prior to reuse or disposal.

Dummy Storage Silos—Silos are in active use for storage of irradiated process spacers, called dummies because they resemble fuel elements but contain no uranium and are used as spacers. No contamination exists on the ground surface. Radiation level at the top of pit 1 is 25 mR/hr and at pit 2 is 35 mR/hr. The dose rate at the radiation zone boundary is <1 mR/hr. Dummies are periodically removed.

100-K Area Silos—Two silos in the 100-K Area (each 8 ft in diameter and 24 ft deep) are used for disposal of metallic scrap from N Reactor's fuel storage basin. This waste consists of irradiated fuel element support hardware, bent spacers and other metallic refuse found in the storage basin. Ruptured fuel element fragments are not included in this waste.

Radioactive Gaseous Waste Facilities: The tritium and radioiodines formed in the reactor primary water are bled out of the primary loop into the ventilation exhaust air. Air that leaks in and is purged out of the reactor core graphite gas atmosphere contains ^{40}Ar which is activated to ^{41}Ar. This outleakage is also carried away by the building ventilation exhaust air. The ventilation air and gases exhausted from the building pass through high efficiency particulate air filters (HEPA) and charcoal absorbers before being discharged to the atmosphere from the 200-ft high ventilation stack at 150,000 cfm.

Some of the 100 Area gaseous radioactive waste originates in the 105/109 N facility as activation salts that are dried on the outside of the reactor piping and surrounding surfaces. Some of the radioactive salts are entrained by the air currents and are removed by high efficiency filters prior to discharge through the ventilation stack.

Inactive Areas: Radioactive waste management facilities in the five 100 Areas in which the reactors were retired are largely the terminated cribs, trenches, burial areas and the contaminated facilities such as the reactors, spent fuel basins and effluent holdup basins. The status of these facilities is constantly monitored and the wastes are maintained in their terminated state pending permanent disposition. The 183-H basin is currently used for receiving the neutralized chemical waste from the uranium. Very little work with radionuclides is currently in progress and only a small quantity of low-level radioactive waste is still being generated. Any radioactive solid waste generated is now transferred to the 200 Area for disposal.

No facilities or equipment for waste management of radioactive gases are needed or are now in service except at the 144-F animal farm facility performing plutonium inhalation studies. High efficiency particulate air filters are installed at 144-F at the experimental enclosures and on the stack, which has a normal flow of 6,500 cfm to the atmosphere. The stack exhaust is continuously sampled.

200 Areas

Due to the high rates of heat generation in the high-level boiling Current Acid Waste (CAW), produced from the Purex Plant processing of irradiated reactor fuel, solidification is not feasible without special processing because excessive temperatures would be encountered in the solids formed. With half-lives of about 30 years, ^{137}Cs and ^{90}Sr contribute much of the heat through their radioactive decay. Initially, decay of other fission products in the waste also generates sufficient heat to make the CAW boil even in the absence of ^{137}Cs and ^{90}Sr. However, due to the relatively short half-lives of the other radionuclides in the waste, the rate of heat generation drops to levels acceptable for solidification within about 5 years. Consequently, CAW is processed (in B Plant) to remove a majority of the cesium and strontium from the liquid waste; the cesium and strontium are encapsulated as a solid for long periods in water cooled storage. The remaining liquid waste is neutralized and stored in double-wall tanks until sufficient radioactive decay allows solidification to salt cake.

Dilute nonboiling liquid waste is stored in tanks and/or evaporated to solid cakes in an effort to minimize liquid waste storage inventories and convert the waste to immobile forms as soon as practicable.

The following subsections treat these processes in more detail, as well as the disposal of the less radioactive waste generated in the fuel and waste processing operations. Table 6.1 indicates the sources and products of the waste output from the 200 Areas operations. As indicated in the table, waste is also received from other areas.

TABLE 6.1: RADIOACTIVE WASTE SOURCES

	Type of Waste					
	Liquid				Solid	Gaseous
	High-Level		Inter-			
		Non-	mediate	Low		
Source	Boiling	Boiling	Level	Level		
200 EAST AREA						
Purex Plant	X	X	X	X	X	X
B Plant	X	X	X	X	X	X
244-AR Vault	-	-	X	X	X	X
241-A, AX, AY Tank Farms	-	-	X	X	X	X
Critical Mass Laboratory	-	-	-	X	X	X
Semiworks	-	-	-	-	-	X
CR Vault	-	-	-	-	-	X
ITS-1 and 2 Concentrators	-	-	X	X	X	X
241-BX, BY and C Tank Farms	-	-	-	-	-	X
200 WEST AREA						
T Plant	-	X	X	X	X	X
242-T Concentrator	-	-	X	X	X	X
Laundry	-	-	-	X	X	X
231-Z	-	-	-	X	X	X
U Plant	-	-	-	X	-	X
Z Plant	-	X	X	X	X	X
UO$_3$ Plant	-	-	X	X	X	X
222-S Laboratory	-	X	X	X	X	X
241-SX Tank Farm	-	-	-	X	-	X
Redox Plant	-	-	-	-	-	X
241-TX Tank Farm	-	-	-	-	-	X
242S Evaporator	-	-	-	X	X	X
FROM OTHER CONTRACTORS						
Pacific Northwest Laboratory	-	X	X	X	X	X
United Nuclear Industries, Inc.	-	X	X	X	X	X
Westinghouse Hanford Company	-	-	X	X	X	X
J. A. Jones Construction Company	-	-	-	-	X	-

Source: ERDA-1538

B Plant Waste Fractionization and Encapsulation: B Plant removes ^{90}Sr and
^{137}Cs from high-level radioactive waste resulting from the processing of Hanford
reactor fuels. The isolated strontium and cesium fractions are further purified
and are doubly encapsulated and stored in monitored water basins which provide both radiation shielding and cooling for the capsules.

Both currently-generated and stored self-boiling wastes are processed by B Plant.
A flowchart for these operations is shown in Figure 6.2. CAW from the Purex
Plant is routed underground to B Plant via the 244-AR Vault. The solids are
removed from the current acid waste and treated for removal of strontium,
while the supernatant solution is processed for cesium removal by precipitation
with phosphotungstic acid. The recovered cesium is further processed by ion
exchange. Strontium is recovered by processing the acid waste through the
solvent extraction system. Stored waste is processed similarly, with the exception that precipitation has already separated solutions into sludges and supernatant liquid in the storage tanks. After processing of the supernatant liquids
for cesium removal, the alkaline sludges are sluiced from the tanks, acidified,
and transferred by underground piping to B Plant for strontium removal. After
clarification, the strontium is recovered by solvent extraction in four extraction
columns. The overall strontium recovery rate to date is approximately 85%;
the current recovery rate is over 90%. The overall recovery rate of cesium is
about 88%; currently, cesium recovery rate averages about 93%.

Following recovery in B Plant, cesium and strontium solutions are purified and
stored as liquids in stainless steel tanks equipped for cooling coils. The cesium
and strontium are encapsulated in the Waste Encapsulation and Storage Facility
(WESF). The cesium solution is transferred to WESF and converted to cesium
chloride solution by the addition of concentrated hydrochloric acid. This solution is evaporated to yield solid cesium chloride which is melted. The molten
material is vacuum transferred to 316 stainless steel or Hastelloy C-276 cylinders
(capsules) where it solidifies as cesium chloride salt. After endcaps are welded
on the capsules, they are leak checked with helium, decontaminated and enclosed in outer capsules. Endcaps are welded on the outer capsules and the
welds tested ultrasonically. Completed capsules are transferred to the underwater storage basin.

The strontium solution is transferred to WESF where the strontium is precipitated as strontium fluoride by the addition of sodium fluoride. The strontium
fluoride is separated by filtration, sintered and the dry powder packed into
Hastelloy cylinders. Encapsulation proceeds as described for cesium. The filtrate
from the precipitation step is returned to B Plant for rework or disposal.

Results of the compatibility of cesium chloride and strontium fluoride with
their respective capsule materials have been studied (1). A conservative extra-
polation of corrosion test results shows that strontium and cesium inner capsules
will maintain their integrity for at least 600 years when stored under water.
In this period, radionuclide activity will have decayed to less than 200 millicuries
per capsule. When an ultimate disposal method for Hanford radioactive waste
is available, the strontium and cesium capsules can be sent to that facility or
perhaps to a facility provided for commercial waste disposal. The lifetime of
the capsule storage basins in the WESF is expected to substantially exceed the
time required to develop an ultimate disposal method. There is no existing plan
to convert the capsule storage basins to an "ultimate disposal site."

FIGURE 6.2: WASTE FRACTIONIZATION FLOWSHEET

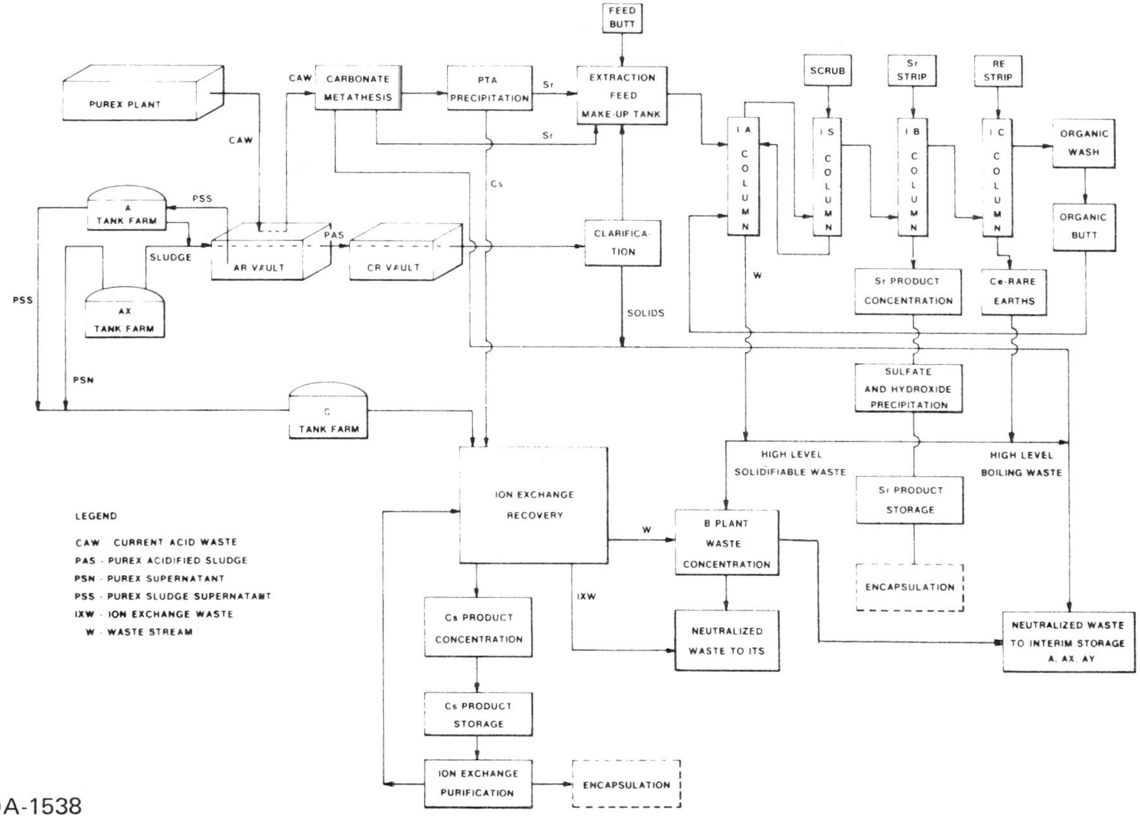

Source: ERDA-1538

All liquid effluent streams will be continuously monitored and routinely sampled. Storage basin water will be circulated within a given storage cell. If radioactivity is detected, the water will be transferred to B Plant for rework or disposal and the source of the contaminations determined and appropriate repairs made. Cooling water streams will empty into the B Plant cooling water header. If radioactivity is detected, the cooling water will be diverted and transferred to B Plant for disposition. Gaseous effluents will be passed through HEPA filters to remove radioactive particulates, monitored, and exhausted to the atmosphere via a stack. Solid radioactive waste will be packed in barrels and buried.

Tank Farms Operations: High-level radioactive waste, produced in the chemical processing of irradiated Hanford reactor fuels, is stored in large underground tanks. Self-boiling waste produced by Purex Plant is stored in the 200 East boiling waste storage tank complex. Boiling waste produced at the Redox Plant is stored in part of the SX tank farm. High-level waste is neutralized with sodium hydroxide or sodium carbonate, and then stored as an alkaline slurry. The neutralization precipitates a major portion of the fission products, resulting in a solids or sludge layer covered by supernatant liquid.

The accurate measurement of liquid level is difficult in tanks with active bottoms systems containing large amounts of liquid surface, build-up of salt cake at the vertical walls, and bottoms salt cake. From information obtained through operation of the three Hanford evaporative systems, a possible worst case type transfer between a "clean" tank and one with a large amount of salt cake is estimated to differ by a factor of 1.5 between the inches transferred and the inches received. Conversely, for a transfer between clean tanks without salt formation on the walls, one standard deviation uncertainty might be expected to be within 2,000 gallons. Therefore, the magnitude of a detectable loss is very dependent upon the current knowledge of salt and its disposition within the tanks.

Other problems are related to surface crusts that present an irregular and often mobile profile of the tank waste-vapor space interface. For example, crustal movements have caused liquid level variations as great as six inches upon occasion. Also, a dry surface crust can render the conductivity probe inoperative. Under abnormal (unplanned) conditions, foaming can interfere with surface measurement until foaming subsides. Special facilities, such as well encased weight factor and conductivity probe measurement systems, are being studied as possible corrections for these latter problems.

High-level waste stored in the A and AX tank farms is presently being processed by B Plant to remove the long-lived fission products. This yields nonboiling waste which can be solidified for in-tank storage as solid salt cake. For B Plant processing, the supernatant liquid is pumped off a given tank, exposing the sludge. The sludge is then broken into a slurry by slicing with water or supernate, using specially designed high pressure nozzles. The slurry is routed to B Plant for cesium removal. The solids are then dissolved by acidification, and that solution is sent to B Plant for strontium removal.

Supernatant liquid from tanks in the SX tank farm has been processed by B Plant. Sludge temperatures in certain SX tanks are being controlled by air cooling, with offgases being partially condensed and returned to the tanks or filtered and discharged to the atmosphere. Sluicing of the sludges is not planned due to the age of the tanks and the possibility of liquid leakage.

No boiling waste is being generated at present since the Purex Plant is not in operation. The only planned new source of boiling waste will be Purex CAW resulting from future processing of irradiated N Reactor fuel. This waste will be routed directly to B Plant for removal of long-lived fission products. However, due to decay heat from the remaining short-lived fission products, the processed waste will self-boil. Storage as a liquid will be required until radioactive decay reduces the concentration of the shorter half-life radionuclides. Such processed waste will be suitable for solidification about 5 years after its generation.

The first tanks were for nonboiling waste. Figure 6.3 is a schematic drawing of a typical nonboiling tank design. These tanks are vented to the atmosphere through air-cooled reflux condensers. Instrumentation was provided to measure (1) the sludge and supernate temperature and (2) the liquid and sludge levels in the tanks. A grid of dry wells in each tank farm is used to monitor the soil for radioactive materials, thus serving as a secondary leak detection system. Well sites and well design were dictated by the specific site conditions and the suggested flow path of liquids both above the water table and within the groundwater.

FIGURE 6.3:　TYPICAL STORAGE TANK FOR NONBOILING WASTE

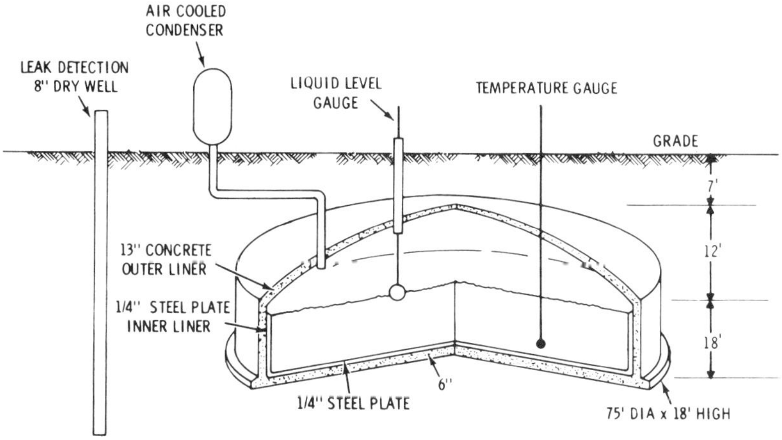

Source:　ERDA-1538

Newer tanks were built to contain waste with greater heat generating characteristics. These tanks are of the same general construction as those for nonboiling waste. However, additional features are provided to permit self-concentration. The vapors are routed through headers to a de-entrainer and water-cooled condensers, which are vented to the atmosphere after passing through a wire mesh de-entrainer, high efficiency filters, and exhaust blowers. As shown in the schematic of Figure 6.4, condensates can be sent to an underground disposal site or can be returned to the waste tank to avoid overconcentration. These tanks are provided with airlift circulators for the agitation of the tank contents to prevent localized temperature buildup in the supernate and the resultant bumping caused by sudden steam release.

FIGURE 6.4: TYPICAL HIGH-LEVEL BOILING WASTE STORAGE FACILITY

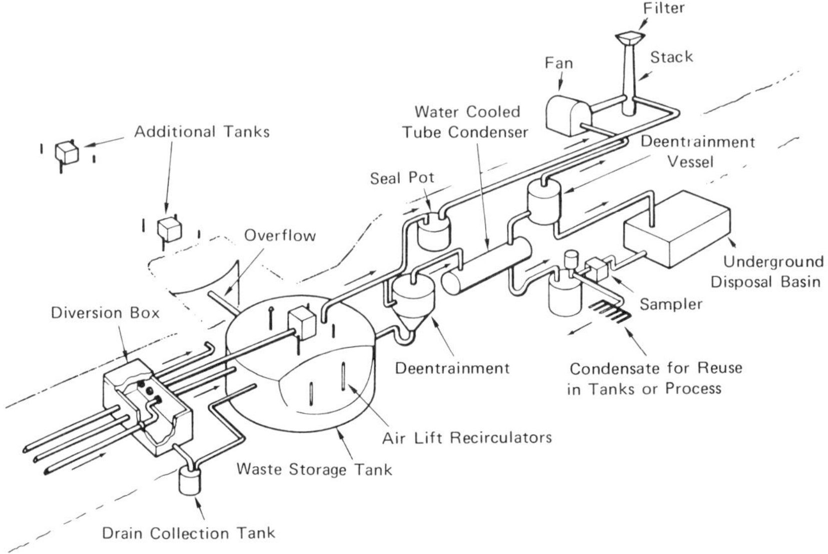

Source: ERDA-1538

Waste Solidification and Stabilization: To ensure continued waste containment
for long periods, nonboiling high-level liquid radioactive waste is being converted
to salt cake. This is accomplished by evaporation and crystallization, and solids
accumulation in existing storage tanks. Organic liquid waste is also stored in
tanks pending acquisition of equipment to process it. The waste is primarily
degraded solvent which can no longer be used in solvent extraction processes.

The primary function of the In-Tank Solidification (ITS) systems is to concentrate the nonboiling waste to produce a partially mobile salt cake. Ideally, the
feed stream would be concentrated as far as possible so that the inventory of
concentrated bottoms is minimized and the maximum amount of solids is formed
and precipitated. Each of the ITS systems includes a heat exchanger for water
evaporation and a series of bottoms tanks (bottoms loop). After concentration
in the evaporator, the bottoms stream is routed through the bottoms loop. Heat
losses to the ground and the tank ventilation systems progressively lower the
bottoms temperature so that eventually the stream is saturated and solids are
formed and deposited in the bottoms tanks. Some crystal growth occurs in
tanks upon cooling but most occurs in the crystallizer. The saturated bottoms
supernate stream is blended with fresh feed and recycled back to the evaporator.

When a tank is filled with salt cake, an estimated 50% of the salt cake volume
is occupied by interstitial liquids. In addition supernatant liquid may be above
the salt cake. The supernatant liquid and as much of the interstitial liquid as
possible will be pumped from the tanks via salt wells reaching to the bottom
of the tanks. To remove all of the interstitial liquid may not be practical because of capillary holdup in the salt cake and limits to removing a residual heel
from the bottom of the salt well. The interstitial liquid may remain associated

with the pumped-out salt cake. As the salt cake then warms up from the energy released because of radioactive decay, a tendency for concentration by evaporation will occur reducing further the volume of residual liquid. The current development programs involve determination of the limits of removal of interstitial liquid by pumping from the salt wells. This effort will test the predictions.

Upon failure of the carbon steel liners of tanks containing salt cake, the residual liquids will be subjected to potential leakage to the soils underlying these tanks. Estimation of the volume of such leakage is difficult since much of the liquid will continue to be held in the salt cake by capillary action. The small volume which may leak would be held in the soils underlying the tanks and held there by capillary action, much as has been observed in the leakage which has previously occurred. Vertical monitoring wells and horizontal laterals under tanks would detect such leakage.

An underground storage tank which is filled with salt cake from one of ITS systems, and which has the maximum amount of interstitial or residual liquor removed, is considered stabilized for interim storage. Until the waste is stabilized the integrity of the tank and process lines are of primary importance and under continual surveillance.

Liquid level measurements are taken in active tanks (having recent transfers to or from) twice per shift and in inactive tanks once per shift. Material balances are made every 2 hours during transfers, and a final material balance is made at the completion of the transfer and drain back from line holdup. In-tank temperatures in some tanks are monitored every shift.

A network of dry wells is in place or planned for every tank except those in the AY and AZ farms, which have internal leak detection systems as a part of their construction. Similar systems are associated with the AX farm tanks, but single-wall construction has predicated the additional protection provided by dry wells. The AX leak detection systems consist of a pattern of drain channels in the underlying concrete base slabs which route leakage to 60-ft deep wells situated adjacent to each tank. Each well is equipped with weight factor/specific gravity and radiation monitoring instrumentation and pumpout facilities. The AY and AZ farm tanks, being of double-wall construction, have the unique feature of leak detection and pumpout facilities within the annular space between the two steel walls. In addition, all A farm tanks and nine SX farm tanks have systems of three underlying laterals about 10 feet beneath the concrete bases. These dry wells and laterals are monitored for radiation increases on a frequency determined by the status of the tank, e.g., whether the tank is active, inactive, or in restricted use.

Subsequent to the stabilization of solid salt cake in a tank for interim storage, the plan is to core-drill, analyze, and characterize the salt cake. Then the tank will be isolated by cutting and blanking all process piping to and from the tank, blanking all risers and equipping the tank with a filtered ventilation system. Subsequent surveillance of a stabilized, isolated tank will consist of periodically obtaining in-tank photographs, sampling and analyzing the vapor space, monitoring the in-tank temperatures, and monitoring the dry wells and laterals.

Liquid Waste Streams and Transfer Lines: The 200 Area activities of fuels processing, waste partitioning and plutonium processing require the transfer of large quantities of high-level radioactive waste within and between the areas.

Stainless steel underground waste transfer lines in concrete encasements are in place between 200 East and 200 West Areas for transfer of radioactive liquid waste between these areas. These lines are used to transfer neutralized high-level wastes between 200 East and 200 West Areas for processing, underground storage, evaporation and solidification.

Cribs and Ponds: Certain liquid effluents are released to the ground via underground structures called cribs. These effluent streams are primarily process and steam condensates which have a potential for releasing radioactivity upon process upset or equipment failure. The amount of radioactivity normally in these streams is very small; the concentration is generally less than 0.05 μCi/ml (except tritium). While some of these streams have been classified as intermediate-level liquid waste (5 x 10^{-5} to 10 μCi/ml), other streams have radioactivity concentrations well below ERDA's ERDAM-0524 Table II guides for uncontrolled areas and have diversion capability if process upsets occur. All of these streams are within controlled areas.

A crib is constructed by digging a ditch about 20 feet deep and up to 1,400 feet long, backfilling with rock and covering with an impermeable membrane and soil. A pipe running the length of the crib is designed to distribute the liquid uniformly along the crib length. The released liquid percolates through the soil. The soil column between the bottom of the crib and the groundwater is 150 to 300 feet thick and contains up to 50% silts and sand, having some clay content. Since percolation rates through these materials are slow, the liquid spreads laterally and involves much more soil than that directly beneath the crib. Formerly, when aqueous waste containing organic or complexing agents was discharged to the ground, the volume was limited to what was believed the soil could retain above the groundwater by capillarity.

A typical crib disposal site is shown in Figure 6.5. While the coarse gravels have little capacity to sorb or filter radioactive materials, the clays have good ion exchange properties and make good filter beds. The soil columns in the 200 Areas exhibit cation exchange capabilities of 2.5 to 10 milliequivalents per 100 grams of soil. The ion exchange capacity of the clays varies widely with the type of ion being sorbed. For example, tritium and nitrate ions are only slightly sorbed, if at all. Ruthenium is held relatively well, but a fraction of the ruthenium is of such ionic form that little sorption takes place. The tritium, nitrate, and ruthenium then flow to the groundwater at essentially the same rate as the downward percolating water from the crib. These materials may at times enter the groundwater directly below a crib at concentrations above the appropriate limits for drinking water but are rapidly diluted below such limits. Cesium and strontium are tightly held by the soil, most being held within a few tens of feet below a typical crib. Plutonium is held very tightly by the soils, with essentially all of it held within 10 feet of the point of release.

An extensive network of wells is provided for sampling groundwaters. Groundwaters associated with waste disposal sites are routinely sampled and analyzed. About half of the wells are sampled quarterly and about half are sampled semi-annually. Some sampling is done by lowering a container on a cable down the well to the water table; some is by pumping from the well with samples collected from the pumped water.

Since sampling by the pumping method is preferred, wells are being modified and equipment is being installed to provide most sampling by this technique.

FIGURE 6.5: SCHEMATIC OF TYPICAL CRIB FOR INTERMEDIATE-LEVEL WASTE

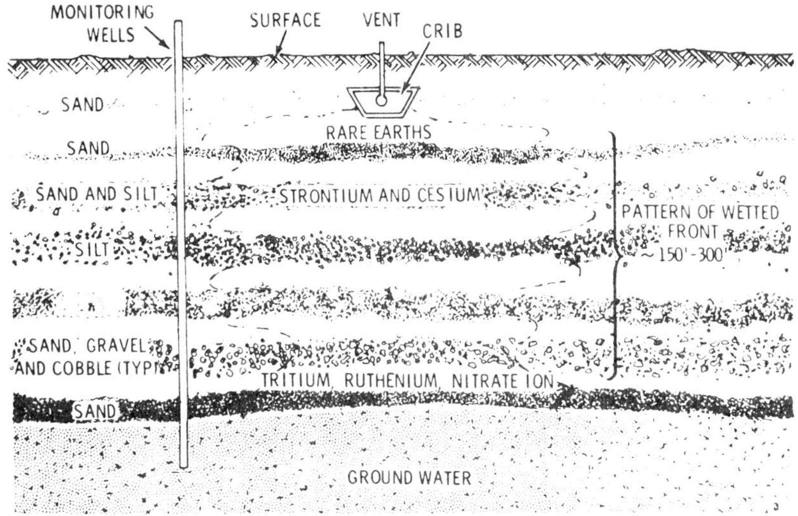

Source: ERDA-1538

Samples are analyzed under an audited quality control program. When long-lived radionuclides, such as ^{90}Sr, ^{60}Co and ^{137}Cs, are detected in the groundwaters at concentrations approaching 1/10 of the concentration guide (*Atomic Energy Commission Manual*) for drinking water, the crib site is deactivated, and the process effluents are routed to a new crib. Only a few cribs were used to the point of reaching this limit. In those cases, the radionuclides are still observed to be bound to the soils near the point of entry to the groundwater system.

A total of 177 cribs have been provided for disposal of liquid waste since startup of the 200 Areas facilities. Of these, 144 have been deactivated, 8 have not been used, 10 are in standby, and 15 are currently being used for disposal of liquid waste from normal processing operations. The shutdown of most Hanford reactors and the resultant decrease in reactor fuels processing, coupled with additional waste treatment of several intermediate-level liquid waste streams, resulted in a major reduction in the quantity of radionuclides discharged to the ground via cribs during the past several years.

Approximately 80% of the process condensate overheads from concentration of Purex first cycle high-level waste is routinely recycled to the process. Total recycle of this stream is planned. Ammonia scrubber waste, previously routed to a crib, will be collected and routed to a waste evaporator for decontamination. Funds were requested to provide recycle of another Purex process condensate stream (from concentration of back-cycle waste) and to provide improved monitoring, diversion, and collection of Purex steam condensate for cleanup in the event this stream should become contaminated due to failure of process vessel heating coils. Process condensate overheads from the B Plant high-level waste concentrator will be batch collected, sampled, and analyzed prior to discharge, when the currently funded project is completed. If above discharge limits, this waste stream will be recycled. Improved monitoring, diversion, and collection of B Plant steam condensate for cleanup will also be provided by this project.

Process condensate overheads from the self-boiling 241-A, AX, and AY waste storage tanks are treated via an ion exchange column to remove ^{137}Cs and ^{90}Sr. Process condensate overheads from the ITS units 1 and 2 are also treated via an ion-exchange to remove ^{137}Cs and ^{90}Sr. Process condensate overheads from the ITS units 1 and 2 are also treated via an ion-exchange to remove ^{137}Cs prior to discharge.

Process condensates from the 242-T Waste Evaporator are batch collected and monitored; they can be recycled, if necessary. Liquid waste from the operation of Z Plant, containing essentially all of the plutonium-bearing liquid waste from that operation, is routed to underground waste storage tanks via the 242-T Waste Evaporator. Currently no appreciable quantity of plutonium is discharged directly to the ground. The plutonium discharged to ground in the 200 Areas with the Purex Plant processing about 900 tons of uranium per year is currently estimated to be 13 g/yr. The actual plutonium discharged to ground during the first 6 months of CY-1974 was less than 5 grams (based on limits of analytical detection). Process condensates from the new 242-S Waste Evaporator-Crystallizer are treated via ion-exchange to remove ^{137}Cs prior to discharge. All waste from this facility meets surface discharge limits. When currently funded plant modifications are in place, the total quantity of radioactive materials discharged to the ground via cribs should be reduced to less than 100 curies and 13 grams of calculated plutonium annually. Conceptual designs of facilities are proceeding to reduce total annual discharges to less than 60 curies (except tritium) and alternatively to meet drinking water standards for all discharges.

Buried Storage: More than 5 million cubic feet of contaminated solid waste of widely diverse nature and contaminated to varying degrees with radioactive materials have been buried on the 200 Area plateau since 1944. This waste was buried in 19 sites using approximately 155 acres of land. Prior to 1968, essentially all of the waste buried at these sites was generated by the 200 Areas operating facilities. Since 1968, waste generated by 300 Area operations was also buried in the 200 Areas sites. Beginning in December 1973, waste from reactor operations in the 100 Areas was sent to the 200 Areas for burial. Small volumes of solid waste, generated by offsite AEC operations, were and will continue to be buried there.

Solid radioactive waste buried in the ground is considered to be in long-term storage. An exception is that waste buried after April 30, 1970, which contains or is suspected of containing transuranium nuclides; this is considered to be in interim storage (20 years). Large items of solid waste, such as failed equipment from locations where the presence of transuranium nuclides can be safely ruled out, are packaged in cardboard cartons, wood boxes, or steel or fiber drums and buried in the so-called "dry waste" trenches.

The various waste containers used provide containment of nontransuranic radioactive contamination and minimize radiation exposure to personnel during temporary storage, handling, shipment, and burial operations. Once buried, no reliability is placed on the containers for confinement or retrievability of these materials. Although burial of the waste containers by backfilling the trench is normally done at the close of the day's receipt of solid waste, waste trenches are immediately backfilled whenever the dose rate at the edge of the trench reaches 100 mR/hr. Solid waste is normally covered with 10 to 20 feet of earth to prevent uptake of radionuclides by plant life or disturbance by burrowing animals. An exception, waste contained in concrete boxes or small drums, may have a minimum dirt cover of 4 feet provided that radiation levels at grade are less than 1 mR/hr. Periodic routine surveillance of filled burial trenches is provided to assure that the burial grounds are maintained to meet existing standards.

Solid waste, containing or suspected of containing transuranium nuclides buried after April 30, 1970, is packaged and buried in compliance with ERDAM-0511. This directive states that "such wastes shall be segregated from other solid wastes and shall be packaged and buried so that they can be readily retrievable, as contamination-free packages, within an interim period of 20 years."

Formerly, the waste was packaged in iron drums and iron or concrete boxes and buried in special trenches. Subsequent evaluation of iron drums directly buried in Hanford soils indicated that failures could occur in less than 20 years and retrieval, as contamination-free packages, might not be possible. Two alternates to direct burial were implemented on a test basis, either of which will protect the containers from direct contact with the soil and will permit ease of retrieval. A prototype concrete V-trench was built and filled with transuranic-bearing waste drums, as shown in Figure 6.6. A metal cover and several feet of earth cover isolate the drums from the environment. This alternate provides protection of the drums from the soil and allows sampling of the storage trench atmosphere for radioactive materials and combustible gases, either of which would indicate drum failure.

A simpler alternate, pad storage, is also being tested. The transuranic-bearing solid waste is segregated as combustible and noncombustible at the point of origin and placed in labeled drums. The segregated waste drums are placed on the storage pad in a stack four drums high; each layer of drums is separated by plywood treated with fire retardant. When drums are stacked to a volume of 24 x 12 feet, the stack will be covered with plywood and plastic-reinforced nylon sheeting prior to covering with 4 feet of earth. Capability for sampling the storage atmosphere is also provided. (A schematic is shown in Figure 6.7.)

FIGURE 6.6: CONCRETE LINED V-TRENCH FOR TRU WASTE

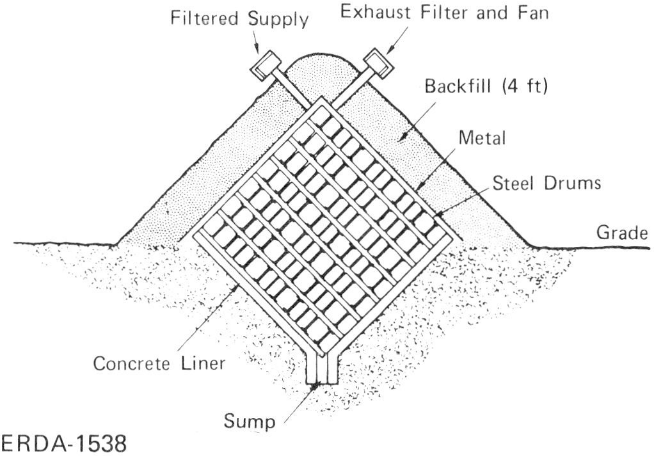

Source: ERDA-1538

FIGURE 6.7: THE TRANSURANIC SLAB

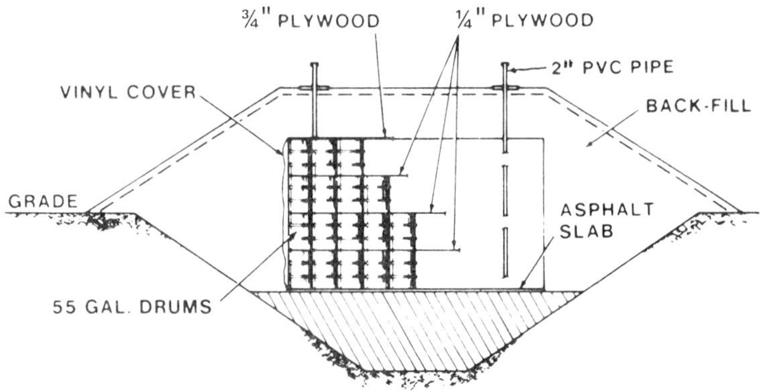

Source: ERDA-1538

Small items of transuranic-bearing solid waste, containing fission products capable of generating high dose rate gamma emissions, are packaged and shipped in shielded casks. This waste is discharged from the casks to caissons and the caissons can be retrieved.

Some of the Purex processing equipment is so large and becomes so contaminated in service, that its transport to a burial ground would require abnormally high exposure of operating personnel. Such equipment is placed on flatcars and pushed into the solid waste storage tunnels appended to the Purex Plant.

Currently, two methods for reducing the volume of solid waste are being practiced. Some waste from the laboratories and from the plutonium finishing operation is compacted prior to storage to provide a volume reduction of approximately four to one.

Ventilation Equipment: Gaseous effluents from the 200 Areas facilities are limited to airstreams containing relatively low concentrations of radionuclides, either in gaseous or entrained particulate form. Multiple filtration is relied upon to remove particulate matter, while a "silver reactor" and/or wet scrubbers are used to remove radionuclides such as radioiodine and oxides of nitrogen. Figure 6.8 shows a typical ventilation flow pattern for a chemical processing plant. Air is drawn through a washer and filter and supplied to a processing area. The air flows, sequentially, from less contaminated to more contaminated zones. After passing through the most contaminated zone, the air flows through an exhaust duct to HEPA filters and then through a stack to the atmosphere.

Filtration of exhaust gases is not provided near the location where radioactivity enters the air at each hood, cell, or glove box if central exhaust filtration is provided. Filtration is not provided where moisture or corrosive fumes must be removed from the exhaust to protect the integrity and efficiency of the filters, but the filtration is provided as near as practical to the source. The design limitations which affect the location of such filtration include the space available, ease of changing filters and protection of filters from fire through distance. The

FIGURE 6.8: SCHEMATIC OF VENTILATION OF TYPICAL SEPARATIONS PLANT

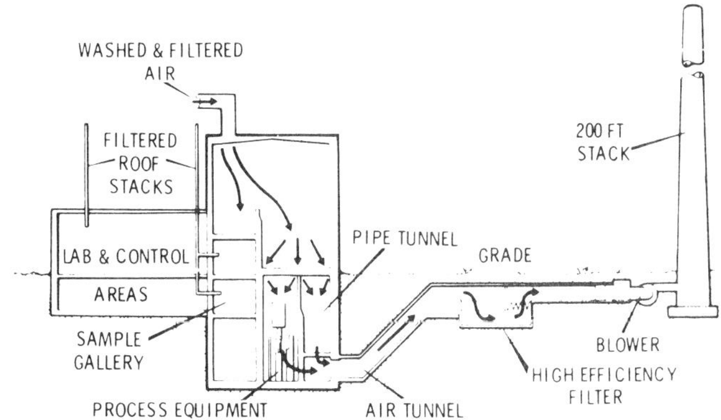

Source: ERDA-1538

important fact is that all exhaust gases from hoods, glove boxes, and cells are filtered before release to the environment.

In the large canyon-type processing building, the cell exhaust gases are collected in structurally integrated tunnels because of the large quantities of air necessary for directional-flow contamination control. It is more practical to combine all of these gases and provide a separate shielded structural filter facility. Where especially toxic or radioactive off-gases from processes are encountered, special off-gas systems with gas treatment and filtering equipment within the cells are provided. These are usually mixed with the main exhaust system before the final filters.

Process chemicals and radioactive contaminants are entrained by air that comes in close contact with process materials. These gases are segregated into separate vent systems according to process requirements and compatibility of the process gases. The gases are processed for removal of noxious materials in a variety of ways: iodine absorber (Purex); ammonia scrubber (Purex, B Plant); NO_x absorber (Purex, UO_3 Plant) and HF scrubber (Z Plant). Once the noxious materials are reduced to within acceptable levels, the process vent gases are exhausted to the ventilation system for additional filtration and return to the atmosphere.

Summary of Low-Level Waste Management in 200 Areas

Solid Waste: Solid waste is essentially dry material. If moisture is present, it is sorbed by the material. It is confined to marked burial sites, permanently recorded, where the dry Hanford soil holds the radionuclides in confinement. Six and one-quarter million cubic feet of solid waste have been buried here since startup in 1944, in 30 sites covering 168 acres, and containing about 1.92 million curies of radionuclides through 1975. Evidence at Hanford exists to show that radionuclide migration under solid waste disposal sites does not occur. Planned studies will further characterize in detail the soils under the burial sites in the 200 Areas.

Burial practices vary with the type of waste being handled, as follows:

> Failed process equipment such as tanks, columns, agitators and pumps, normally contaminated to high radiation intensities, is packaged in wood, metal and concrete boxes, or in plastic. Spread of contamination to the environment is unlikely during transportation to the burial ground. The boxes are covered with 10 to 20 feet of earth. Radiation at ground level over these burial sites is maintained at less than one millirem per hour. The potential for the contamination uptake by plants or animals is lessened by frequent surveillance and corrective measures.
>
> Very large failed equipment systems from the Purex Plant are stored in an underground railroad tunnel.
>
> Construction debris, cleanup materials, laboratory supplies, and small tools contaminated to low radiation intensities are placed in cardboard, wood or metal boxes. At least four feet of earth cover is used to ensure that radiation at grade level is less than one millirem per hour.
>
> Transuranic contaminated trash and scrap materials have been segregated since 1970 as combustible or noncombustible and packaged in steel drums for storage in separate modules. Larger items of transuranic waste are packaged in steel or in fiber glass-impregnated plywood boxes. In both cases burial under four feet of earth is conducted anticipating retrieval within 20 years as contamination-free packages. The total plutonium accumulated from 1944 through 1975 is about 400 kilograms in 28 solid waste burial sites.

Gaseous Waste: Emissions to the atmosphere from the production and waste management facilities are limited to air streams with which process gases may be combined. These contain relatively low concentrations of radionuclides, either in gaseous or entrained particulate form. The composition of the gas streams varies from facility to facility. Most of the nitrogen oxides developing from plutonium separation and uranium recovery processes are scrubbed out with water in the acid recovery process; iodine-131 is reduced to innocuous levels by decay in aged fuel elements. Particulate material is removed by scrubbing or filtration through sand, fiber glass, and other high efficiency media. A continuous flow from 64 principal process building stacks, totaling 1.33×10^{13} liters, was monitored in 1975.

Other air exhausts (e.g., personnel-occupied areas) are not filtered, but are sampled on a routine basis. Starting in fiscal year 1975 and continuing in fiscal year 1976, a program has been underway to more precisely analyze and measure the particle sizes and radioactivity in the stack gases compared to the amount of radioactivity measured by the installed systems. Deficiencies have been identified in monitoring systems and these studies change in the configurations to improve measurement precision. Even so, Manual Chapter 0524 Table II concentration guides are being met.

Liquid Waste: Most of the liquid effluent is cooling water used in condensers and cooling coils. These streams, 1.47×10^{10} liters in 1975 containing less than 3.21 beta curies, are routed to open ponds where the water percolates to the water table. Radiation monitoring instruments are installed on all such streams of high contamination risk to divert the streams to retention facilities if excessive radionuclide concentrations develop. These volumes may then be reprocessed. Retired ponds, and the ditches leading to them, have been covered with several feet of soil and revegetated with shallow-rotted grasses. Liquid levels in currently-used ponds are maintained to minimize access by plant and animal life to contaminated sediments.

Other potential effluent streams arise from the condensation of process vapors, steam, and evaporator overheads. These streams are treated before release to reduce the radionuclides to "as low as technically and economically practical" levels. These effluents are then directed to underground cribs where the sorptive capacity of the soil retains the minute quantity of long-lived radionuclides remaining.

Except for tritium, releases to cribs are within Chapter 0524 Table II concentration guides for all but three condensate streams. These are process and steam condensate from the high-level waste tank farm. In the latter case operating problems with ion-exchange facilities have not been resolved and continue under investigation. When the Purex Plant resumes operation in 1978, recently installed recycle and evaporative capacity is expected to bring process condensates and ammonia scrubber wastes into adherence with Table II concentration guides.

The close surveillance which these radionuclides in the soil require is provided by an extensive groundwater sampling and analysis program involving 106 wells in the 200 Areas, as well as site monitoring by Battelle-Northwest. During 1975 a total of 760 water samples from these wells were analyzed. When long-lived radionuclides such as strontium-90, cobalt-60 or cesium-137 are detected in the groundwater at concentrations approaching one-tenth of Chapter 0524 Table II concentration guides, the crib site is deactivated. Migration to the water table is thus slowed and ultimately ceases. Breakthrough of these radionuclides is not expected in future operations. Tritium does percolate to the water table at essentially the same rate as water discharged to the crib. It may enter the groundwater directly below the crib at greater than the concentration guides in Table II but is rapidly diluted by the groundwater to less than such guides.

The above handling of wastes is supported by policy statements and guided by internal procedures to strengthen the overall control of the program. These documents specify holdup of work or operations that could exceed concentration guides, identification and forceful correction of deviations from procedures, management review of corrective action, and follow-up through inspections and audits to ensure corrective action is implemented. Practices designed to reduce the potential for occasional spreads of radioactivity to the environment have been instituted.

In 1974 functions related to monitoring for the surveillance over radioactivity were moved to an audit group separate from production to provide greater emphasis and to assure adherence to requirements for minimizing release of radioactive materials to the environment. Approximately 100 people are engaged full time in this effort. The scope of this work includes surveillance over all 200 Area radioactive waste disposal sites, as well as locations where inadvertent spills have occurred. In 1975 this surveillance included over 250 separate locations and 75 miles of road systems at frequencies varying from monthly to annually. Over 2,200 sample analyses were made, primarily air and water, but including animal tissue and vegetation. In addition to this effort, nearly 9,000 samples of all types of wastes are taken annually by production personnel for inventory purposes. This is in addition to the environmental monitoring program by Battelle-Northwest.

This separate audit function is believed to be an essential step in the assessment of management practices. It amounts to a double-check and support to those who have the primary assignment for data-taking and compilation and helps to prevent omissions or unsatisfactory conditions.

300 Area

Radioactive Waste Management: The 300 Area liquid waste handling systems
provide for retention of some liquids containing above trace amounts of radioactivity for transport to and subsequent treatment in the 200 Area of the 100-H
Area. Other liquids containing only trace amounts of radioactivity are discharged
to ponds. The 183-H solar evaporation basin was first used in 1973 on an intermittent basis and became fully operational in 1975. It should accommodate that
portion of the fuel processing effluent (except for water rinses and scrubbers)
containing uranium, copper, fluorides, nitrates, chromium(VI), and sulfates. The
plutonium and fission product waste is generated in laboratories and test facilities conducting research and development work on separation processes, waste
solidification and reactor fuels.

Solid and liquid radioactive wastes are now sent to the 200 Areas for disposal,
except for (1) some low-level uranium-contaminated liquid waste sent to the
183-H basin or to 300 Area ponds and (2) slightly contaminated animal waste
sent to 300 Area ponds. Radioactive airborne effluent wastes in the 300 Area
are filtered and monitored prior to release through stacks, except for a uranium
fume exhaust from 333 Building that passes through a water scrubber unit before
being sampled and discharged to the atmosphere.

Radioactive Waste Liquids: All radioactive liquid waste generated in the 300
Area is retained and shipped to 200 Areas for disposal, with two minor exceptions
noted above. Some effluents contaminated with uranium are released to the
process water ponds. Discharges to ponds during 1972 were 2.8 X 10^6 gpd,
consisting of about 11,000 gpd from the life sciences facility and the remainder
from the fuel fabrication area, the contaminated liquid waste collection system,
the retention waste system and the normally uncontaminated waste system.

Temporary storage of radioactive liquids prior to shipment to the 200 Areas is
provided by the 340, 340-A, and 340-B Waste Retention and Neutralization
Facility. The three buildings, constituting a total area of 3,800 ft^2, provide
facilities also for chemical neutralization and transfer of the radioactive waste
into tankers. The main building houses a large rectangular concrete pit 21 feet
deep containing two stainless steel, 15,000 gallon neutralization tanks, valves,
and transfer pumps. The pit area, which is covered by removable concrete blocks,
adjoins the operation gallery and sampling room. The annex building, 340-A,
contains six 8,000 gallon stainless steel waste storage tanks. Three 5,000 gallon
tanker trucks can be accommodated in the load-out addition. Two 20,000 gallon
tank cars can be accommodated in 340-B, a railroad tank car load-out facility. (2)

Normally Uncontaminated (Retention) Liquid Waste: A second liquid waste
retention system, the 307 Retention Basins, provides a collection point for
normally uncontaminated liquid waste from laboratories which may, inadvertently,
contain low-level radioactive waste material. The four basins are of reinforced
concrete construction with the tops just above grade. Capacity of each is about
50,000 gallons. Basins 1 and 2 are provided with a neoprene lining to simplify
decontamination if required. Basins 3 and 4 have bare concrete surfaces and
are used only in case of emergency. A continuous in-line β/γ liquid effluent
monitor samples the waste stream just ahead of the influent into the 307 Basins.
This monitor signals an alarm in the nearby 325 Building equipment room and
automatically shuts off the basin's discharge pump to a pond when a radionuclide concentration is detected.

Basins 1 and 2 are provided with a series of baffles which form a serpentine flow path. The time for free flow through the basin by the serpentine route is sufficiently long to allow an adequate counting interval for the sample in the monitor. The discharge pump can be shut down automatically by this equipment prior to any release of radioactivity to the ground. Normally the contents of the basins are discharged to the ponds. A continuous sampler removes a representative sample from the influent lines to the basins. Samples are taken to the laboratory weekly for alpha and beta/gamma analysis for backup to the monitor.

Two buildings on this system are provided with wastewater diverter systems (similar to the wastewater system described above). In these buildings, cooling water from equipment containing large quantities of radioactive materials is monitored by the wastewater diverter system monitor prior to release to the retention waste system. If a rupture in a cooling coil should allow contamination of the cooling water, the wastewater diverter will signal an alarm and divert the contaminated water to a 20,000 gallon underground catch tank for verification and appropriate disposal without taxing the 307 Basin waste management complex. Underground piping connects the normally uncontaminated process waste from the buildings in a second separate sewer system discharging into the first of the four 307 Facility retention basins.

High-Level Radioactive Liquid Waste: High-level radioactive liquid process waste is retained, using shielded stainless steel casks, for transfer to the 200 Areas. Cask capacities range from half liter to 500 gallons. Normally these are transported directly from the laboratory generating the waste via the highway to the 200 Areas for disposal. Special vehicles are required, as the large casks weigh up to 17 tons.

Radioactive Solid Waste: Active facilities for long-term management of solid radioactive waste do not exist in the 300 Area as no solid radioactive waste is now buried or incinerated in this area. All radioactive solid waste is transferred via truck and highway from the laboratories or facility where radiation work is performed to the 200 Areas for disposal. Each transfer of solid waste to the 200 Areas is performed according to written procedures and is permanently recorded.

All radioactive solid waste is divided into transuranic and nontransuranic solid waste as it accumulates. Since transuranic waste contains some radioactivity from elements whose atomic number is greater than 92, this waste is handled separately and in special disposal containers. Since the nontransuranic waste does not contain radioactivity from transuranic elements, packaging varies from low-level radioactive waste sealed in plastic-lined cardboard containers to high-level radioactive waste requiring large concrete casks for containers.

Retired Solid Waste Storage Facilities: Approximately 7.5×10^5 ft^3 of solid waste containing approximately 10 Ci of uranium contamination have been placed into the 300 Area burial sites since fuel fabrication began at Hanford. Thorium contaminated waste is less than 0.5 Ci and is segregated from uranium (i.e., a different trench). Most of the 300 Area contaminated waste is in a matrix of bonding materials on fabrication components, scrap material and equipment as surface contamination.

Of the approximately 7.5×10^5 ft^3 of solid waste, approximately 90,000 ft^3 are surface contaminated metallic, concrete, and asphalt materials. The remaining approximately 6.6 x 10^5 ft^3 occupy approximately 1 x 10^5 ft^3 of underground space. The volume reduction occurs as a result of the weight of the solid waste and the soil overburden.

Radioactive Gaseous Waste: Each laboratory or facility in which radioactive materials are handled or processed is equipped with its individual exhaust treatment system. Wherever practical, airborne radioactive materials are removed from exhaust gases near the hood, glove box, or cell in which they are generated. For glove boxes and manipulator hot cells, all recent new designs have provided for filters directly at the air exhaust outlet from the compartment. Some were installed years ago on the duct work between the glove box and the first filter where the contamination has not been enough of a problem to warrant a modification.

The exhaust treatment systems in every facility, except 333 Building, utilize HEPA filters which are tested to meet at least a 99.97% efficiency for particles 0.3 μm and greater in diameter. The filters are tested following installation and on a scheduled frequency thereafter. Filters are scheduled for an annual routine dioctyl phthalate (DOP) test but are checked promptly if any indication of filter malfunction occurs. Over many years, two filter failures have been experienced and resulted in local contamination with no activity released offsite. Occasionally, filters are replaced because of plugging, which reduces air flow below acceptable levels. Where required, a charcoal absorber is used as a collector for radioactive iodine. All gaseous streams are sampled daily and are also constantly monitored if the air stream has a potential for becoming highly contaminated. Plutonium contaminated exhausts are double or triple filtered with sampling provided between the filters and sampling or monitoring after the final filter.

An alarming device is part of the monitoring system and would give immediate notification of release due to loss of filter integrity. Failure of any major filter system in active status would be detected within a short time (less than an hour). All expended filters from the facilities or laboratories in which radioactive materials are handled or processed are treated as radioactive solid waste.

The wet uranium fume exhaust from the abrasive cut-off saw in the 333 Building passes through a water scrubber unit before being sampled and discharged to the atmosphere. The water from the scrubber unit is sent to the 300 Area process ponds.

Battelle, Northwest Laboratory has undertaken to evaluate and define the status of waste materials disposed to the 300 Area and has evaluated geophysical techniques for the detection and mapping of waste materials. Various survey techniques have accurately located buried materials in and adjacent to disposal trenches. Metal detector, thermal infrared, magnetometer, acoustic transmission, acoustic refraction, and acoustic reflection techniques have been field-tested and evaluated. Ground-penetrating radar systems have been identified as a promising technique.

Numerous drilling procedures, equipment, and techniques have been evaluated for use in drilling of contaminated burial ground facilities. A small-diameter drill system is being used to collect sediment samples from various locations adjacent to the 300 North Burial Ground.

Two instrumented wells have been installed at the 300 North Burial Ground.

These wells will be used for neutron moisture and gamma density probe access. Sediment samples are currently being characterized as to various physical parameters concerning fluid transport. Computer codes have been written for analysis of physical parameters delineated by various laboratory techniques.

WASTE MANAGEMENT PLAN AT THE SAVANNAH RIVER PLANT

The Savannah River Plant (SRP) was established in 1950 by the U.S. Atomic Energy Commission to produce nuclear materials for national defense (Figure 6.9). Nuclear materials are produced at this site by transmutation of elements in large nuclear reactors that are moderated and cooled by heavy water (100 Areas). Supporting facilities extract heavy water from natural water (400 Areas), fabricate nuclear fuel and targets (300 Areas), dissolve the irradiated materials, and chemically separate nuclear products from the radioactive by-products (200 Areas). A comprehensive research and development program, directed with increasing emphasis on peaceful applications of nuclear materials, guides the plant operations.

The major waste management facilities for radioactive liquids and sludges are adjacent to the separations areas and consist of two waste tank farms linked to the separations areas and to each other by pipelines with secondary containment. One centrally located site (burial ground) is used to store the radioactive solid waste and degraded solvent generated at the Savannah River Plant, as well as occasional special ERDA shipments from offsite.

Almost all aqueous high-level wastes are treated in some manner before storage in underground tanks. The purpose of these treatments, such as evaporation, adsorption, and/or filtration, is to minimize the volume of wastes that must be stored indefinitely. Combustible wastes, such as degraded solvent, are washed to transfer most of the radioactive constituents to the aqueous system before transferring the solvent to tank storage. Gaseous wastes are decontaminated by various processes such as scrubbing, adsorption, and/or filtration.

Aqueous (Noncombustible) High-Level Waste

Evaporators: Separations Facilities (F and H) — Various aqueous streams generated during the solvent extraction processes in 221-F and 221-H hot and warm canyons, as well as laboratory-generated wastes, are concentrated by evaporation in the separations areas (200-F and 200-H) before transfer to underground storage tanks 241-F and 241-H. All wastes are made alkaline before storage in underground, carbon steel tanks. This prevents chemical corrosion of the primary container but results in a sludge being precipitated. The sludge is composed primarily of iron and manganese hydroxide. Recovery of nitric acid during waste concentration reduces the amount of sodium hydroxide that must be added to the evaporator concentrate and consequently reduces the volume of waste to be stored. The aqueous waste evaporation processes are performed in each separation area before the wastes are transferred to their respective waste tank farm. The high-activity waste and low-activity waste processes reduce overall waste volumes by factors of about 50 and 30, respectively.

Waste Tank Farm Evaporators (242-F and 242-H) — Each waste farm has an evaporator that is used to concentrate alkaline waste that has been aged a suitable period following receipt from the canyons. The aging allows separation of

FIGURE 6.9: THE SAVANNAH RIVER PLANT SITE

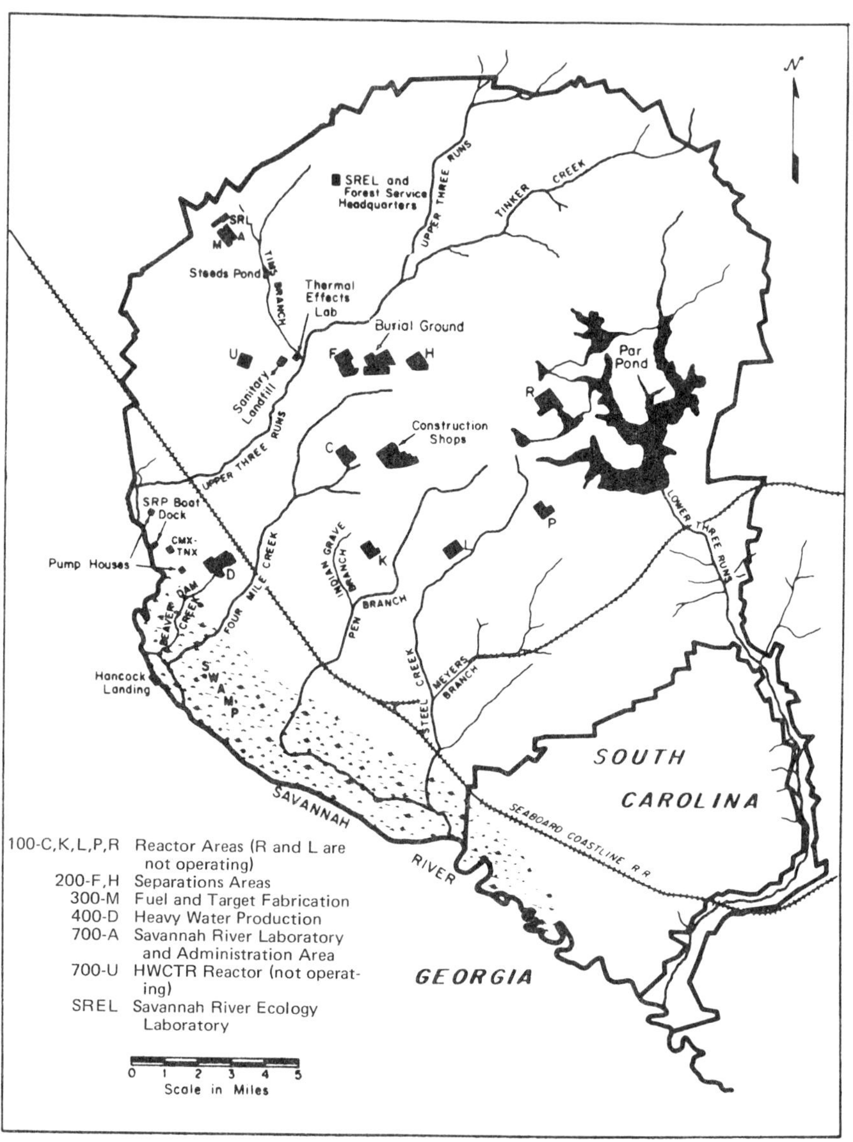

Source: SRO-TWM-76-1

the sludge and supernate and also allows the shorter-lived radionuclides to decay to acceptable levels. The tank farm waste evaporators are a very important part of the waste management facilities, and their successful operation has obviated the need for many additional storage tanks. Radioactive waste, as received and stored in the separations areas tank farms, can be reduced to about one-third of its original volume and immobilized as crystallized salt by successive evaporations of the liquid supernate. Such a dewatering operation has been carried on routinely in F Area since 1960 and in H Area since 1963. This operation is shown schematically in Figure 6.10. The evaporators used for this operation are single-stage, bent-tube units.

The evaporator was designed for simple maintenance-free operation. Shielding permits evaporation of wastes containing up to 100 Ci of ^{137}Cs per gallon. Double-contained concentrate transfer lines up to several hundred feet long are used for both gravity and pumped systems between the evaporator and the storage tanks.

Evaporator Design and Operation — The pot-type evaporators are constructed entirely of stainless steel and operate at atmospheric pressure under alkaline conditions. Each is steam-heated by a bent-tube bundle designed to facilitate remote descaling. The bowed tubes terminate at both ends in heavy tube-sheets maintained at fixed spacing by tie rods. By periodically surrounding the tubes with relatively cool and dilute waste or water and heating them intermittently with bursts of steam, alternate thermal expansion and contraction flexes the bowed tubes and aids in spalling and/or dissolving accumulated scale. An integrally mounted, three-tray, dry bubble-cap column provides de-entrainment of suspended concentrate droplets from the effluent vapor steam. The steam is condensed in a water-cooled condenser vented through a filter to the atmosphere. Future evaporators will have wire mesh separator pads in addition to the bubble caps to improve de-entrainment.

Condensate from the evaporator is checked for radioactivity carryover by a gamma monitor. If the radioactivity level exceeds a preset scale reading, the condensate is diverted manually to a waste storage tank. Currently, the condensate is pumped through a cesium removal column to reduce the ^{137}Cs content. This column is a stainless steel vessel containing an ion-exchange resin (currently a synthetic zeolite) that gives a decontamination factor for ^{137}Cs of about 200.

Effluent from the cesium removal column is alternately passed to one of two stainless steel hold tanks. When the hold tank is full, a sample is analyzed for radiocesium; after satisfactory analysis, the contents of the hold tank are pumped to a seepage basin. Piping is being installed to return the evaporator condensate to the separations plant for re-evaporation or for process use.

Concentrated bottoms are steam-lifted from the evaporator pot, continuously or intermittently, and flow by gravity directly to nearby waste storage tanks or to a below-grade concentrate transfer system pump tank in a stainless-steel-lined reinforced concrete pump pit for transfer to more remote storage tanks. The hot concentrate is continuously circulated by the main pump through double-walled piping, through the upper interior of several waste tanks in series, and back into the concentrate transfer system pump tank. Air-operated drawoff valves, specially designed to minimize holdup space, are operated intermittently under automatic control to draw a fraction of the circulating stream into a given

FIGURE 6.10:   HIGH-LEVEL LIQUID WASTE PROCESSING AT SRP

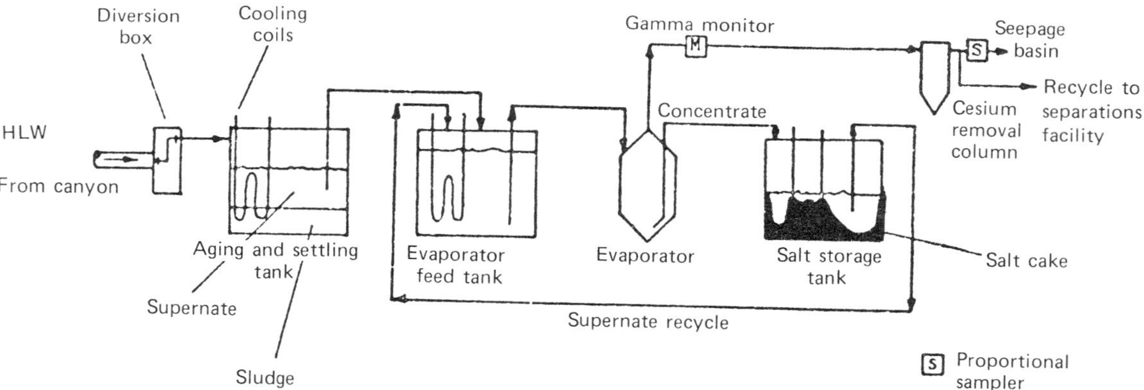

Source: SRO-TWM-76-1

receiving tank. This recirculation system is used to accommodate relatively low net drawoff rates (original design, 3 gpm) and still maintain, in practical-size piping (1½ to 2 inches), sufficient fluid velocity (greater than 2 ft/sec) to keep the fluid hot and prevent settling of suspended salt particles. A second pump is provided in the pump tank to permit prompt emptying of the pump tank if the main pump fails. This provision is necessary because salt crystallization would plug the system if the concentrate were allowed to remain stagnant and to cool off. Experience has shown that this spare pump must be run continuously to maintain its operability. During normal operation its discharge flow goes directly back into the pump tank.

Concentrate Recycling and Salt Storage — Concentrated bottoms from the existing pot-type evaporators must retain enough water to maintain fluidity during transfer through pipes by gravity and/or by the concentrate transfer system to receiving tanks hundreds of feet away. Depending on dissolved salt composition and transfer temperature, 20 to 30% water content by weight is required (hot specific gravity about 1.5 to 1.62) for reasonably trouble-free operation on a continuous basis; more dilution is required during process upsets. After cooling (thermally) in the receiving tanks, a portion of the dissolved salts crystallizes, and coherent, fairly dense salt cake is formed. This cake normally adheres to heat-absorbing surfaces, such as coils, tank walls, or concentrate transfer system piping and/or equipment. The salt mixture of one type of low-heat waste, cladding waste from the dissolution of aluminum cladding from natural or low-enriched uranium fuel elements, has a high enough temperature coefficient of solubility and absorbs sufficient water of crystallization to yield a solid salt cake devoid of free liquid.

In practice, all other waste mixtures do not solidify completely on cooling but leave a liquid phase of saturated solution around and in the interstitial voids of the salt cake. Depending on composition, temperature, and evaporator boildown, the volume of the liquid phase may range from about 50 to nearly 100%. Because the water contained in the concentrate occupies valuable storage space (even the water of crystallization) and also defeats the goal of waste immobility, the supernatant liquid phase is jetted out of the salt tank periodically and recycled to the evaporator for removal of additional water.

Absorption Beds: Reactor Areas — Suspended and dissolved radioactivity is removed from the fuel storage basin through deionization, filtration, purging, and natural decay. In normal operation, the amount of nonvolatile radioactive contamination of the basins is so small that radioactive decay and removal by the sand filters are adequate. However, larger amounts of activity can be introduced to a storage basin from fuel that has cladding penetration after discharge from the reactor. Carryover of radioactivity on the surface of the fuel tube (oxide and residual D_2O) from the reactor is confined to the vertical tube storage which is isolated from the rest of the basin during discharge operations and remains isolated until an acceptable level of activity is achieved.

No earlier than 5 days after discharge operations, removal of ionic material is begun by recirculating water in the vertical tube storage basin through a deionizer. Delay in operation of the deionizers is to allow the sand filters to remove the particulate activity.

Three types of deionizers are used; four deionizers use a mixed-bed resin, one contains a cation resin, and one contains an anion resin. Deionizers may be operated as single units or may be operated in tandem; i.e., one mixed bed and one cation bed, or one cation bed and one anion bed. Operation conditions will dictate the best type of operation. In general, the removal of nuclides is as follows: ^{134}Cs, ^{137}Cs, removed by cation resin; ^{51}Cr removed by anion resin; and ^{131}I, removed by anion resin.

When the nuclide removal ratio (effluent activity/inlet activity) reaches 0.75 on two successive 8 hour samples and the effluent activity of the deionizer (for ^{51}Cr, ^{131}I, ^{134}Cs, and ^{137}Cs selected by area supervisor as a depletion index) is greater than 25 dis/(min-ml), the deionizer is depleted.

To maintain the effectiveness of the resin, the basin pH must be kept between 5 and 8. When basin pH drifts downward, it is controlled by adding Na_2CO_3 to the recirculation pump suction. The addition rate is limited to no more than 10 lb per shift to permit uniform control. A high pH is controlled by nitric acid addition.

Normally, water is removed from the surface of the discharge canal, is passed through the deionizer, and is returned to vertical tube storage. Normal flow through the resin is about 100 gpm; this flow decreases to about 50 gpm as the bed is partially plugged through normal use. Effective throughput for the deionizers is about 470,000 gal for removal by ion exchange and 1,000,000 gal for some additional activity removal because of the filtering effect of the beds. When activity levels are below specified limits, the deionizer effluent may be returned to the transfer pits or may be purged to the 50,000,000 gallon basin (K Area) or to the building effluent (C and P Areas) to reduce tritium content (if tritium content in vertical tube storage or machine basin is greater than 0.2 μCi/ml in P Area or 0.4 μCi/ml in K and C Areas). Current control values are 2.3 x 10^{-5} μCi/ml for ^{131}I and ^{134}Cs^{135}Cs and 4.5 x 10^{-5} μCi/ml for ^{51}Cr, ^{95}ZrNb, and ^{141}Ce^{144}Ce.

All deionizers are mobile for use in different reactor areas and for transfer to RBOF for regeneration of the beds. The radioactivity removed is stored in 200 Area. The deionizer pad, on which the trailers are packed, is located outside the reactor building. The pad drains to a process sewer which flows to the river.

Receiving Basin for Offsite Fuels (RBOF) — In the event the water in any of the basins becomes contaminated, the water can be decontaminated by circulation through deionizers. Currently, basin water is circulated through the deionizer almost 100% of the time. These deionizers contain a bed of mixed cation and anion ion exchange resin. Flow rates up to 125 gpm are normal. The conductivity of deionizer effluent is typically 0.3 to 1.0 μmho. Typical feed levels of activity are 1 to 5 x 10^{-4} μCi/ml; deionizer effluent levels are 0.01 to 0.05 times 10^{-4} μCi/ml. Thus, the normal inventory of activity is 0.2 to 1.0 Ci in about 400,000 gal of basin water. Levels of activity in isolated basins have been as high as 0.1 μCi/ml.

Under typical feed conditions, the deionizer processes about 4 x 10^4 bed volumes (2 x 10^7 gal) during about 6 months of service. It contains about 20 Ci of radioactivity (mostly radiocesium) when regeneration is required.

Waste Tank Farm (241-F and 241-H) — Each separations area waste tank storage facility has a cesium removal column located on the top of an uncooled waste storage tank. These columns are used to remove cesium from the condensate that results from the concentration of waste supernates. In addition, the column in H Area is used to remove cesium from aqueous wastes discharged by the resin regeneration facility and the receiving basin for offsite fuel.

Each cesium removal column is fabricated of stainless steel and holds 10 ft^3 of zeolite (Linde Molecular Sieve AW-500, 20 to 50 mesh synthetic zeolite). The zeolite (alumina silicate) is supported by a conical, perforated metal screen. Flow is downward through the zeolite bed with overall decontamination factors for radiocesium of 100 to 200.

In F Area, where column feed consists only of evaporator (242-F) overheads, approximately 2,000 to 3,000 bed volumes may be processed until unacceptable radiocesium levels are detected in the column effluent. [The column effluent is collected in one of two hold tanks and sampled for radiocesium. After an acceptable analysis (F Area — 1.1 x 10^{-3} μCi ^{137}Cs/ml; H Area — 6.8 x 10^{-4} μCi ^{137}Cs/ml), the contents of the hold tank are discharged to the seepage basin. If unacceptable radiocesium concentrations are detected, the contents are recycled.] Occasionally, pressure drop across the column becomes excessive before radiocesium breakthrough necessitating zeolite replacement. Spent zeolite is discharged directly into the waste tank beneath the column.

In H Area, aqueous waste from Tank 23 is blended with the evaporator condensate as feed to the cesium removal column. Tank 23 contains waste discharged by the resin regeneration facility. Experience has shown only 500 to 1,000 bed volumes of this feed can be processed before zeolite replacement is necessary. This lower capacity is attributed to the deleterious effect of competing nonradioactive cations and colloidal solids present in the Tank 23 waste.

Filters: Nuclear Production Reactors — Visibility control in the fuel component storage and disassembly basins in the reactor areas are maintained using sand filters at high flow conditions. This system provides the visibility required to conduct operations in 30 ft of water. Untreated water becomes turbid because of the growth of algae and bacteria, corrosion of mild steel equipment, and spalling of oxide film from component surfaces.

Each reactor basin filter installation contains 100 ft^2 of filtering area. The filter media consists of 18 inches of anthracite (0.72 mm) over 18 inches of sand (0.36 mm) and is operated at 5 to 7 gpm/ft^2. Each storage basin and disassembly basin is provided with 1,000 gpm pumps feeding parallel filters through a common header.

The filter backwash is collected in an uncovered 22 foot diameter cone-bottomed vessel where the sludge settles for 2 to 4 hours before the supernate is automatically pumped back to the filter. The sludge is periodically pumped into a shielded trailer and transported to the 200 Area waste tank farm where it is transferred into underground storage tanks.

Receiving Basin for Offsite Fuel — A porous stone tube filter is used to prevent solids from entering the deionizers located in the basin water recirculation system. When a change in pressure drop across the filter is observed, the filter is backflushed and recoated with filter aid. This normally is done 3 to 4 times per month. Occasionally the filter tubes are treated with oxalic acid and caustic.

Fuel and Target Fabrication Facility — When the acid solutions used in the various processes reach a uranium concentration of 100 g/l and the anodic etch solution reaches 16 g/l, they are treated with caustic to precipitate uranyl hydroxide, and the slurry is processed through a rotary filter. The filter cake is removed continuously during filtration, and the depleted uranium precipitate is sent to the burial ground. A filter press is used to recover aqueous suspensions of UO_2 resulting from autoclave failures. This material is also sent to the burial ground. These filter operations recover about 99% of the uranium from solution; the remainder is released to the settling basin.

Discharge of Low-Level Liquid Wastes

Low-level liquid wastes are those waste streams that may be discharged to the environment with assurance that persons will not be exposed to concentrations in excess of those prescribed in the ERDA Manual Chapter 0524. These effluents are monitored for activity to determine and control the quantities of activity discharged. Effluents that might be radioactive are monitored or sampled continuously if the discharge is continuous; if batch discharge is used, then each batch is sampled before discharge. If the activity content of an effluent exceeds limits, corrective actions are taken; these actions include diverting the flow to contain the activity, shutting the operation down, and isolating the source of contamination.

Fuel and Target Fabrication Facility (300-M): The only significant radioactive releases from these facilities are the aqueous effluents containing a small fraction of uranium not recovered from the recovery process in 313-M and the uranium generated in 321-M Building from cleaning and etching uranium-aluminum fuel components. These aqueous effluents are routed to the 905-51G Settling Basin.

A proportional sampler inventories releases specifically from 313-M. Another sampler (nonproportional type) located on the sewer line to the seepage basin, but beyond the last point at which any other effluents tie into the sewer, samples continuously. The sampler on the effluent to Tims Branch is a proportional type. In all locations, samples are collected weekly for analysis.

Alloy Extrusion (321-M) — The aqueous wastes generated from cleaning and etching of uranium-aluminum fuel components are analyzed and then either released to the settling basin or retained for possible treatment in the separations areas waste management facilities.

Uranium Metal Element Fabrication (313-M) — Elements failing quality control requirements are reprocessed after the aluminum cladding is dissolved in caustic and the nickel plate is dissolved in nitric acid. Caustic solutions are released to a settling basin; nitric acid is sent to the 313-M uranium recovery facility.

When the acid solutions used in the various processes reach a uranium concentration of 100 g/l and the anodic etch solution reaches 16 g/l, they are treated with caustic to precipitate uranyl hydroxide, and the slurry is processed through a rotary filter. The filter cake is removed continuously during filtration, and the depleted uranium precipitate is sent to the burial ground. A filter press is used to recover aqueous suspensions of UO_2 resulting from autoclave failures. This material is also sent to the burial ground. These filter operations recover about 99% of the uranium from solution; the remainder is released to the settling basin.

Miscellaneous flows (about 450 gpm) are currently discharged, via the 300 Area wastewater system, into Tims Branch, which empties into Steeds Pond about 1½ miles from the 300 Area. Overflow from the pond runs into Upper Three Runs and after another 2 miles into the Savannah River. Every 15 minutes, the effluent is sampled in amounts proportional to the flow; samples are combined and analyzed every week to determine the weekly release of uranium.

Target Extrusion (320-M) — Aqueous releases from this facility are mixed with the 313-M effluent to Tims Branch and included in the weekly analyses.

Metallurgical Laboratory (322-M) — Aqueous releases from this facility are sent to the settling basin together with those from 321-M.

Heavy Water Production and Recovery Facility (400 Area): Low-level waste from this facility is principally tritium-contaminated liquid originating from the reprocessing of degraded reactor D_2O moderator in a separate distillation facility.

Rework Unit (420-D) — Waste campaign distillate is collected in drums and analyzed before being discharged (at 1 gpm) to the cooling water effluent (1,300 gpm), which goes to Beaver Dam Creek as a part of the total area discharge of about 40,000 gpm. The discharge rate is adjusted to meet procedural controls on tritium concentrations in the effluent sewer (currently <0.1 μCi/ml) and on total quantities released.

Drum Cleaning Facility (421-2D) — 50 gallon drums used for heavy water storage and transport are cleaned with either acid or alkaline phosphate solutions before reuse. Drums used for degraded D_2O moderator retain some tritiated water, which is the major source of activity release from this facility. Spent cleaning solutions are stored in tanks and pumped to Beaver Dam Creek in about 18,000 gallon batches after analyses to confirm the acceptability of release. The solution may be transported by tank truck to the 200-F Area waste treatment system if sufficient radioactivity is present (radioactivity is compared to prorated fractions of operating guides).

Analytical Laboratory (772-D) — The quality control laboratory in the 400 Area serves both the heavy water area and the reactor areas. Aqueous releases to Beaver Dam Creek are sampled continuously and analyzed monthly. Tritium-contaminated heavy water samples that cannot be economically reprocessed are poured into sink drains which discharge into the effluent stream with uncontaminated cooling water as a diluent.

Nuclear Production Reactors (100 Areas): Sources of low-level aqueous waste in the reactor areas are (1) heat exchanger cooling water, (2) fuel and target basin purge via deionizers, and (3) miscellaneous (sumps, etc.). All 100 Area seepage basins are inactive.

Heat Exchanger Cooling Water — The effluent cooling water from each process heat exchanger is continuously monitored before discharge to a plant stream by in-line gamma detectors. These detectors are capable of detecting heavy water leakage from a single exchanger of about 0.3 lb/hr. This monitoring is supplemented by more sensitive tritium analyses on samples drawn once each week during operation. In the event of contaminated cooling water, the stream is manually diverted to a 50 million gallon, unlined, earthen basin.

Fuel and Target Basin Purge — When the tritium content of the basin water has built up beyond a procedural control point (currently 0.2 μCi/ml) and other activities are below their procedural limits, the deionizer effluent is directed to a plant stream in P and C Areas. This purge is continually sampled for laboratory analysis throughout the purge. In K Area, the deionizer effluent is routed to the 50 million gallon earthen basin.

Separations Facilities (200 Areas): Low-level aqueous streams routinely charged to the environment from the 200 Areas consist of segregated cooling water, surface runoff from the waste tank farms, laundry waste, waste farm evaporator overheads, cesium removal column effluent, condensate from the general purpose evaporator and acid recovery units located in 211, selected liquid regeneration wastes from the resin regeneration facility in H Area, and cooling water from 232-H and 234-H.

Cooling Water Systems — Both separations areas have water-handling systems that are basically the same in concept, although different in detail. Both areas have circulating cooling water and segregated cooling water. The circulating cooling water is monitored continuously to detect any contamination from leaks in cooling coils of process vessels that contain activity. If activity is found, the water is diverted to seepage basins at moderate levels of activity and to retention basins at higher levels.

The segregated cooling water system is once-through and is used for process vessels in which the same coil may be alternately used for heating and cooling. The segregated cooling water normally discharges to Four Mile Creek. If activity appears in the segregated system, the water may be diverted to the seepage basins or to retention basins, depending on the activity content. The two areas are different in that segregated cooling water in F Area is continuously monitored and discharged through a delaying basin that provides four hours of holdup to allow time for action if activity is detected. In H Area, segregated cooling water is collected and discharged on a batch basis according to the monitoring result.

The water-handling system in each separations area waste tank farm is basically the same in concept, although different in detail. Cooling water to the shell side of the waste tank closed-loop cooling system heat exchanger is supplied from wells. The cooling water discharge from the heat exchanger is used to cool the 242 waste evaporator condensors. In both areas, a portion of the heat exchanger effluent is used as makeup water in the 285-H cooling spray tower.

Cooling water from the 242-F waste evaporator is diverted back into the 221-F circulated cooling water system upstream of the 281-4F water monitor. Steam condensate is routed to the 221-F segregated water system upstream of the 281-6F segregated water monitor.

In H area, the evaporator cooling water is monitored and routed to Four Mile Creek after one cycle independent of the 221-H segregated water system. When this monitor is inoperable, the cooling water is routed to the 221-H segregated water system upstream of the 281-6H monitor, or it can be diverted to the retention basin or seepage basin independent of the 221-H system.

The general rules for control of activity in the circulating cooling water are:

> Less than 4.5 x 10^{-6} μCi/ml beta-gamma and $<$1.4 x 10^{-6} μCi/ml
> alpha activity indicated on the alpha and beta-gamma monitors is
> considered normal.

At 4.5×10^{-6} to 1.4×10^{-4} μCi/ml beta-gamma and 1.4×10^{-6} to 4.5×10^{-6} μCi/ml alpha activity, the overflow of surplus cooling water to Four Mile Creek is stopped, and the source of activity is sought.

At 1.4×10^{-4} to 4.5×10^{-4} μCi/ml beta-gamma or 4.5×10^{-6} to 4.5×10^{-5} μCi/ml alpha, the water is diverted to seepage basins. Above 4.5×10^{-4} μCi/ml beta-gamma or above 4.5×10^{-5} μCi/ml alpha, the water is diverted to retention basins.

For the segregated cooling water, there is a corresponding set of guidelines for action:

Less than 4.5×10^{-6} μCi/ml beta-gamma or $<1.4 \times 10^{-6}$ μCi/ml alpha activity is considered normal.

At 4.5×10^{-6} to 4.5×10^{-4} μCi/ml beta-gamma or 1.4×10^{-6} to 4.5×10^{-5} μCi/ml alpha, the water is diverted to seepage basins. Above 4.5×10^{-4} μCi/ml beta-gamma or above 4.5×10^{-5} μCi/ml alpha, the water is diverted to retention basins.

The cooling water beta-gamma monitors are swirl cells in which the liquid is pumped tangentially into an upright cylindrical chamber. The liquid flows around the walls, drains out the open bottom, and forms a hollow, rotating cylinder of water. The radiation detector (a scintillation unit) is mounted in the hollow vortex of the cylinder of water.

To ensure that these monitors are functioning properly, the response of the units is tested daily with an external radiation source, the units are calibrated each year with a circulating solution of known activity, and samples of the cooling water itself are taken twice a shift and analyzed in the laboratory.

In the cooling water alpha monitors, water flows across a plastic film, which is thin enough for alpha particles to penetrate and thick enough to protect the scintillation detector on the other side. The response of the cell is checked weekly with an alpha source plate, and the unit is calibrated annually with a solution of known activity.

Storm Water System — Surface water runoff at the waste tank farms is normally routed directly to a plant stream (Four Mile Creek). However, because this sewer drainage is subject to contamination from surface spills of waste, the system is segregated and continuously monitored with swirl cell gamma monitors. The F Area waste farm terrain is divided into two zones and the H Area waste farm into four zones. All zones are monitored individually and, in the event any monitor detects radioactivity, the contents of that storm sewer system are automatically diverted to a lined retention basin for further handling. Once in the retention basin, the water may be:

Pumped to natural effluent streams if within guidelines.

Pumped to seepage basins if this would not exceed the current guide limits for such discharges. In general, as a maximum, guide limits are set such that the maximum amount of a given radioisotope of one to 30 year half-life, other than tritium, added to a series of basins in a calendar year will not exceed the quantity of that isotope destroyed in one year by radiolytic decay, based on the radionuclide inventory in that basin as of January 1.

Pumped through a filter-deionizer system for removal of radioactivity with effluent from this system recycled to the retention basin, sent to seepage basins, or released to a plant stream.
The filter-deionizer would be regenerated, and the radioactivity
collected would be sent to waste tank storage.

Laundry Wastes — Approximately 10,000 gallons of laundry washwater is generated daily. This waste stream is normally extremely low in radioactive contamination, but is discharged to the F-Area seepage basin as a precautionary
measure. During transfer, a drip-type sampler collects a composite sample for
subsequent analysis.

Effluent from the 242-F and 242-H Cesium Removal Column — The effluent
from the cesium removal column is alternately routed to one of two stainless
steel hold tanks. During filling, a sampler in the hold tank operates continuously.
When the tank is full the sample is analyzed on a pulse height analyzer and
compared to a sample containing a known amount of radiocesium. If acceptable,
the contents of the hold tank are pumped to the seepage basin. The sample is
forwarded to 772-F for additional analyses. If the pulse height analysis results
are outside specified limits, F Area — 1.1×10^{-3} μCi ^{137}Cs/ml, H Area — 6.8 x
10^{-4} μCi ^{137}Cs/ml, the waste is either recycled through the cesium removal column
or pumped back to the waste evaporator feed tank.

Resin Regeneration Wastes — Liquid wastes that result from regeneration of
deionizer resin beds at RRF are qualitatively monitored to determine the site
to which the waste must be transferred. This survey is a gamma count against
the side of a 250 ml sample bottle.

Later this year, resin regeneration wastes from 245-H that would previously
have been transferred to the seepage basin and Tank 23 may be recycled back
to 211-H as feed for the General Purpose Evaporator.

211-F and 211-H Facility Condensate — The radionuclide content of overheads
from the General Purpose Evaporators and acid recovery units in each separations area are normally within guidelines that allow discharge to a seepage basin.
A 4 ounce sample of the overheads is qualitatively surveyed for radionuclide
content using a gamma detector. Based upon experience, a count rate of <500
counts/min assures that the radionuclide concentration is within operating guidelines and accordingly may be discharged to the seepage basin. If the count rate is
>500 counts/min, the overheads are either recycled, or if time permits, held
until the analytical results are available from the laboratory.

Tritium Facilities (Cooling Water) — The large amount of water used in the
tritium processing buildings (about 250 gpm) is normally routed to Four Mile
Creek. However, if the tritium concentration exceeds the operating guide, the
cooling water from either 232-H or 234-H can be manually diverted to the
H-Area seepage basin.

Currently, 24 hour composite samples of the effluent cooling water are analyzed
daily. To forestall the potential release of tritium to the creek that this scheme
might allow before detection, a continuous monitor has been designed and is
to be constructed and installed. The instrument has a design sensitivity of
1 μCi/l (equivalent to about one curie per day in the combined effluents of
Buildings 232-H and 234-H), and will be capable of diverting the cooling water

to the seepage basins automatically in the event of tritium contamination in excess of operating guides.

Savannah River Laboratory: Cooling water and trade waste effluents are presently discharged to Tims Branch. These streams are continuously sampled. Samples are collected weekly and forwarded to 735-A laboratory for radioanalysis.

If either system became contaminated, there is no mechanism to divert the streams from Tims Branch until such time the source of contamination can be identified and corrected. A study is planned of the individual sources of cooling water and trade waste as to their potential for becoming contaminated. If certain systems are judged to be an unacceptable risk for direct discharge to Tims Branch, these systems will be routed to the low-activity waste facility at 776-A.

Contents of the waste hold tanks at 776-A are held for analysis to determine the proper disposition. Means are provided by which the contents of these tanks may be discharged to either Tims Branch, the seepage basin or to the appropriate waste trailer for shipment to 211-F for unloading and processing.

Combustible Liquid Waste (Degraded Solvent)

At present there are no facilities for treatment or disposal of degraded solvent stored in underground tanks at the burial ground. Before 1971, waste solvent was burned in open pans. Burning of solvent was suspended in 1971 because the existing incinerator did not meet South Carolina air pollution regulations and standards. Efforts to improve the burning characteristics of the existing solvent incinerator so that it meets current regulations and standards have been unsuccessful. Alternate methods for disposing of spent solvent are currently being investigated.

Sludge (Reactor Areas)

Turbidity formation (suspended solids) in the spent fuel assembly storage and disassembly basin can cause water clarity to deteriorate in 24 to 48 hours and hinder operation of the facility. Water purity also affects the extent of final decontamination required before shipping casks offsite.

Originally the basin was cleared by purging with filtered water which released these suspended solids and absorbed radioactivity to effluent streams. This release is now avoided by operating the disassembly basin as a closed system, by stopping the filtered water purges, and by removing and concentrating the suspended solids (into a sludge) by the use of Graver Mono-scour sand filters. Analysis of suspended solids from the basin by optical and electron microscope methods and radioautography showed that the particles ranged from about 0.01 to 4 μm diameter; chemical and spectroscopic analysis showed that the particles were about 80% iron oxides, about 15% aluminum oxides, and 5% miscellaneous material.

Each Graver Mono-scour sand filter contains 100 ft^2 of filtering bed consisting of an 18 inch thick layer of anthracite of 0.72 mm effective size over an 18 inch thick layer of sand of 0.36 mm effective size. Basin water at rates of from 750 to 1,500 gpm (depending on the reactor area) flows through the filters and is returned to the disassembly basin.

In P and K Reactors two sources of supply of the filters are used, one from the discharge canal and temporary storage and one from the machine basin and its adjacent final storage and transfer pits. To provide this supply, one pump (750 gpm in K, 1,000 gpm in P) is installed in the temporary storage section, and a second pump of the same size is installed in the machine basin. Both pumps are used when the gates between sections are open. When the temporary storage is isolated, only one of the pumps is used. The pumps supply a pressure of 38 feet at the filters, and permit a pressure loss through them of up to 10 feet.

A pump suction point is located in the fuel canal close to the bottom, another is located near the surface where fuel is first immersed, and a safety suction break is located about 18 inches below the surface. In the temporary storage basin, a suction entry is located in the adjustable skimmer on the surface. In the machine basin, suction entries are placed in the vicinity of underwater machinery, in addition to a skimmer and suction break. A suction break provides backup protection for level-sensing and pump shutoff devices so that if basin water level drops, suction is lost and the pump ceases to reduce the level further.

Filtrate is discharged to both main sections of the basin as well as to an isolated corner of final storage and to the transfer pits, where clarity is needed for loading the shipping casks.

In C Reactor two smaller basins are served by one 1,500 gpm pump and one 100 ft^2 filter specially designed internally for 15 gpm/ft^2 flow rate. In this case the supply pressure at the filter is 44 ft, providing for a differential pressure loss of 16 feet across the filter bed.

When the differential pressure across a filter reaches a predetermined value, the filtering process is automatically interrupted and the filter media is air-scoured and backwashed to remove the residue to a setting tank. Backwash can be initiated manually when other conditions warrant. In current practice, manual backwash is initiated each day (alternate days in reactor areas with two filters). Water for the backwash (about 7,500 gallons) is stored in the top section of the filter.

A backwash cycle consists of:

> Closing off the filter backwash storage compartment;
> Draining excess water from the top of the filter bed (5 minutes);
> Air scouring the filter media (5 minutes);
> Settling the bed (5 minutes);
> Backwashing the filter at 15 gpm/ft^2 (5 minutes); and
> Returning the filter to operation.

The backwash is collected in an uncovered 22 foot diameter cone-bottomed vessel where the sludge settles for 2 to 4 hours before the supernate is automatically pumped back through the sand filter. Precautions are taken to prevent the backwash from entraining air into the settler. Air bubbles would break through the water surface in the settler and could spread smearable radioactivity on the surrounding ground.

Auxiliary equipment includes the mixing tank and the feed pumps used when adding chemicals such as inhibitors and coagulant aids to the basin water before filtration and the level controllers which permit water to be added to the vertical tube storage basin or machine basin automatically if the level decreases.

Solid Waste

Radioactive solid waste receives no treatment before storage. Facilities are planned to incinerate solid waste and to disassemble and decontaminate non-combustible waste.

The solid radioactive waste storage site is principally for the managed storage of solid radioactive wastes in underground trenches or on covered pads on the surface. Examples of the materials handled to date are:

> Contaminated equipment—obsolete or failed tanks, pipes, and
> other process equipment.
> Reactor and fuel hardware—fuel components and housings not
> containing fuel or products.
> Spent lithium-aluminum targets—the waste target alloy after
> tritium was extracted by melting the alloy.
> Oil from gas displacement pumps in the tritium facilities—prior
> to burial, the oil is placed in drums containing an absorbent
> material.
> Laboratory and operating waste—small equipment, clothes,
> analytical waste, decontamination residues, plastic sheeting,
> gloves, etc.
> Special shipments from offsite—tritiated waste from Mound
> Laboratory; ^{238}Pu process waste from Los Alamos Scientific
> Laboratory and Mound Laboratory; debris from 2 U.S. airplane
> accidents in foreign countries.
> Spent deionizer resins—from reactor use.

The solid radioactive waste storage site is divided into sections for transuranium alpha waste, low-level waste, and high-level waste.

Operating Procedures: Procedures and job plans are written prior to initiating storage of waste in order to achieve maximum protection from radiation and contamination to personnel and equipment. Coveralls, rubber shoe covers, gloves, eye protection, and hard hats are required for personnel assisting with waste handling operations. Only essential vehicles are permitted to enter the solid radioactive waste storage site. The vehicles are surveyed for contamination before leaving. A Health Physics inspector observes burial of high-level waste and makes routine surveys to determine ground surface or vegetation contamination.

The supervisor of the solid radioactive waste storage site keeps accurate records of the contents, radiation level, and burial location of each load received. Shipments are described and recorded on a Radioactive Solid Waste Record, and permanent computerized records are maintained on magnetic tape. The exact location of the trenches is defined by use of a 100 foot grid system laid out in 1960. The 100 foot grids are further divided into twenty-five 20 foot squares. Previous to 1960 the trenches were defined with concrete markers.

Burials are made in trenches that are 20 feet deep and 20 feet wide. Low-level waste is unloaded manually or emptied directly into trenches. Where the radiation dose rate is high, the waste is handled remotely. For the highest dose rates, a shielded crane is used. Waste is covered by soil soon after burial to reduce radiation, contamination, and the possibility of fire. The minimum soil cover is 4 feet, but must be sufficient to reduce surface radiation to 6 mR/hr or less.

Past Operating Practices: Routine Burials — Radioactive waste has always been segregated into transuranium alpha, low-level, and high-level waste categories. These are described below:

(1)   Transuranium Alpha Waste—From 1964 to 1974 this waste was segregated into two divisions:

Retrievable—Waste containing greater than 0.1 curie per package was placed in prefabricated concrete containers and then buried. These containers were 6 feet in diameter by 6.5 feet high. Waste that did not fit into the prefabricated concrete containers was encapsulated in concrete. Transuranium waste from the Savannah River Laboratory (SRL) was buried in square concrete containers. Prior to 1964, this waste was not placed into retrievable containers.

Nonretrievable—Waste containing less than 0.1 curie per package was buried in a low-level transuranium alpha trench.

(2)   Low-Level Waste—Low-level waste was defined as that measuring less than 50 mR/hr at 3 inches from an unshielded package, less than 50 mR/hr at 10 feet from the truck load, and less than 0.1 Ci of transuranic alpha activity per package. Full shipments of waste, e.g., skip pans or closed container dumpsters with radiation intensities to 50 mR/hr at 10 feet, were disposed of in low-level waste trenches. Scrap uranium from the fuel fabrication operation was also placed in these trenches.

(3)   High-Level Waste—High-level waste was defined as that exceeding 50 mR/hr at 3 inches from an unshielded package.

The volume and radioactivity content of waste buried in 1974 are listed in Table 6.2. The volume and radioactivity of waste buried since startup through 1974 are summarized in Table 6.3.

Special Burials—Occasionally shipments of classified wastes are received per ERDA request. Two such shipments occurred following crashes of airplanes.

(a)   Spanish Soil—A collision during mid-air refueling on January 17, 1966, between a bomber carrying nuclear weapons and a refueling plane contaminated the ground at Palomares, Spain, with plutonium. Decontamination procedures produced 4,827 55-gallon drums of soil and vegetation. These were placed in two separate trenches in 1966. The drums were buried 10 feet below the ground surface as a precaution against local infestation with plant and soil diseases from Spain.

(b)   Greenland Ice—On January 21, 1968, a bomber that carried nuclear weapons crashed in Greenland, producing large quantities of contaminated ice and aircraft parts (3). Recovery activities required 535 containers with a storage volume of 120,000 cubic feet for aircraft parts and 680,000 gallons of water potentially contaminated with plutonium. The water was filtered, monitored, and sent to a seepage basin, except for a small fraction that was evaporated and its concentrates stored in the high level waste tanks. Aircraft parts and storage tanks have been buried in three separate trenches.

In addition to these several categories and examples of solid waste, degraded solvent is temporarily stored in the solid radioactive waste storage site in 20 underground tanks (150,000 gallons in storage in 1975). The solvent contains residual transuranics and fission products that were not removable by washing procedures. The transuranics in 1975 totaled 45 Ci of alpha radioactivity, primarily ^{238}Pu and ^{239}Pu. Most of the fission product activity in the solvent is short-lived. A facility for incineration of this solvent inventory is being designed.

TABLE 6.2: RADIOACTIVE WASTE BURIALS IN 1974

Waste Classification	Radioactivity Content, Ci	Volume, ft³
1.　Transuranium Alpha Waste		
Retrievable	5,000	7,000
Nonretrievable	200	74,000
2.　Low Level	5,000	280,000
3.　High Level	280,000	42,000

TABLE 6.3: RADIOACTIVE WASTE BURIALS FROM STARTUP THROUGH 1974

Waste Classification	Radioactivity Content, Ci	Volume, ft³
1.　Transuranium Alpha Waste		
Retrievable	500,000	70,000
Nonretrievable	20,000	1,100,000
2.　Low Level	3,200,000	6,700,000
3.　High Level	4,100,000	700,000

Source: DP-1366

Present Operating Practices: In 1974, the bases for operating procedures were modified to reflect new ERDA criteria governing retrievable storage of solid transuranic waste, to protect high-level stored waste from contact with water-saturated soil, and to define storage limits from beta-gamma radionuclides. These new bases are:

  (1)　All solid radioactive waste originating at SRP shall be stored at the solid radioactive waste storage site. Every practicable effort shall be directed toward reducing the amount of radioactivity in wastes prior to storage. Permanent records and maps shall be maintained to locate and identify stored waste.

(2) Solid radioactive waste from operations at SRP shall be stored
     in the following manner:

    (a) Transuranium wastes contaminated to greater than
          10 nCi/g, except as specifically excluded in Section
          (b) (below), shall be stored protected from contact
          with water-saturated soil in containers that can be
          retrieved free of external contamination and intact
          for at least twenty years from the time of storage.
          Combustible and noncombustible wastes shall be
          stored in separate containers.

    (b) Canyon equipment and other bulky wastes contaminated
          with transuranium nuclides to greater than 10 nCi/g
          and also intensely contaminated with gamma emitters
          shall be excluded from the packaging requirements in
          section (a) (above). These wastes shall be stored in
          earthen trenches protected from contact with water-
          saturated soil.

    (c) Concentrations and quantities of radionuclides in waste
          containers shall conform to criticality safety and heat
          load criteria.

    (d) All other solid radioactive waste shall be stored in earthen
          trenches.

    (e) All solid radioactive waste shall be stored with a cover of
          at least 4 feet of soil and with at least 10 feet of undis-
          turbed soil between the permanent groundwater table
          and the stored waste.

(3) The SRP limits on the quantity of beta-gamma radioactivity
     emplaced each year at the solid radioactive waste storage site
     are the following:

^{137}Cs	500 Ci
^{90}Sr	500 Ci
^{60}Co	3×10^5 Ci
^{3}H	4×10^5 Ci
Other nuclides ($T_{1/2} > 10$ y)	1×10^3 Ci
Other nuclides ($T_{1/2} < 10$ y)	5×10^5 Ci

(4) A comprehensive surveillance program shall be provided to monitor
     migration of radionuclides from their storage locations.

With the attention currently given to monitoring and control of migration, the
solid wastes can remain safely in their present location for as long as is necessary
for a national policy to be established for their eventual disposal. Leaching of
fission products, activation products, and transuranium nuclides has been negligible.
However, tritium is leaching from buried wastes. Because of the low movement
rate of groundwater, the dose-to-man projection from tritium leaching from
the inventory in the burial trenches is estimated to be less than 0.02 man-rem
per year. Uptake of radionuclides by vegetation growing over buried waste has
shown that deep-rooted vegetation should not be permitted to grow over the
waste. Thus, long-term (100 to 200 years) management will primarily require
vegetation and erosion control.

SRP waste management procedures for transuranium wastes are compatible with recovery and removal of buried solid wastes if national policy should so dictate. Segregation of waste according to source and radiation levels permits minimum management for much of the area and permits recovery of any one type of waste. Transuranium alpha emitters buried in concrete can be recovered without including soil. Detailed records of waste burial locations will facilitate recovery of wastes.

Gaseous Wastes

Fuel and Target Fabrication (300 Area): 321-M — Off-gas exhausts from various 321-M operations that are potential sources of radioactive release are vented through HEPA filters to the atmosphere outside the building. The exhaust downstream of the filters for the casting furnaces used to prepare uranium-aluminum billets is monitored by daily filter paper samples, and process conditions are adjusted if releases increase.

Off-gases from the perchloroethylene-caustic-nitric-acid cleaning process are released from a 100 foot stack at a rate of 32,000 cfm. Air sampling equipment is periodically used to monitor activity concentrations in the stack and at other locations in the building to provide data toward minimizing releases.

HEPA filters from off-gas exhausts are analyzed and buried in the burial ground if they contain less than 20 grams of ^{235}U. Filters with more than 20 grams are sent offsite for recovery of ^{235}U. Scrap graphite from the casting process is buried in the burial ground ($<0.3\%$ ^{235}U) or processed for ^{235}U recovery if necessary ($>0.3\%$ ^{235}U).

313-M — The 313-M building contains facilities for bonding fuel and target elements in aluminum, cleaning and testing. Off-gases from nitric acid processes are exhausted through three 100 foot stacks at 8,000 to 14,000 cfm. Most other exhaust air streams are vented directly to the atmosphere outside the building. No particulate filters are used in the atmospheric exhausts from 313 M because of low potential for radioactive release. Ventilation air to the stacks and in other building locations is sampled for particulate activity continuously, and the filter paper samplers are monitored daily.

320-M — The 320-M building contains facilities for extruding lithium-aluminum tubes for tritium production. Various off-gas exhausts in the extrusion process contain HEPA filters for particulate removal. Because the amounts of radioactive materials handled in this facility are low, potential for radioactive release is minimal.

322-M — The 322-M building handles many materials, including test specimens of all materials fabricated in the 300 Area. Some irradiated materials are also examined here. Ventilation air from various operations exhausts through HEPA filters and is vented from 25 foot and 50 foot stacks.

Heavy Water Production and Recovery Facilities (400 Area): The principal radioactive containment in the ventilation systems is tritium, and no facilities are installed to contain this radioisotope. The most significant release of tritium to a ventilation system occurs both in the Rework Unit (420-D) during reprocessing degraded reactor D_2O and in the Analytical Laboratory (772-D). Kanne chambers monitor these air streams for tritium to detect leaks and estimate releases.

Nuclear Production Reactors (100 Areas): Clean outside air is drawn through the reactor building at a design rate of 128,000 cfm. The distribution of flow through the three process areas (purification area, pump and heat exchanger area, and reactor room) is also controlled to assure that air flows from uncontaminated to potentially contaminated or contaminated areas before discharge through the stack.

During operation, all exhaust air to the stack passes through an activity-collection (confinement) system designed to collect and retain radioactive particulates and iodine vapors. The system is designed primarily to limit the consequences of unlikely reactor accidents and is on line at all times. It consists of: (1) moisture separators (crimped wire mesh and Teflon fibers intended to prevent entrained water particles from collecting on the particulate filters), (2) HEPA filters, and (3) activated, impregnated carbon absorbers designed to remove elemental iodine vapor and gaseous iodine compounds.

However, the radioactive particulates and iodine in normal operation are only minor constituents of the radioactivity released to the ventilation system. The major radioactive constituents, tritiated water vapor and the noble gases ^{41}Ar and isotopes of Kr-Xe, are unaffected by the present confinement system. Stack air is continuously monitored by drawing a small sample stream through a Kanne chamber. The Kanne signal is proportional to the sum of the tritium, ^{41}Ar, and Kr-Xe present and accordingly serves only as a qualitative monitor. Continuous monitors for noble gas activity are being installed in all three reactor areas.

Separations Facilities (200 Areas): Ventilation air flows from clean areas to potentially contaminated or contaminated areas before discharge to a stack. The air is filtered before discharge with one, two, or three stages of filtration to remove airborne particulate radioactivity. Emergency power and spare ventilation capacity are provided on air streams that normally contain particulate activity so that this air will always move through filters before discharge. Many stacks of varying heights serve the buildings of the separations areas; they provide for dispersion and dilution of any released activity before return to the ground, and they provide some emergency natural draft in case of power failure.

Canyon Building Ventilation Systems — In Building 221-F, exhaust air from the several sections and processes in the building passes through filtration and activity-removal systems before being combined into a common effluent of 225,000 cfm from a 200 foot tall stack. The main contributors to the activity in the stack effluent are the canyon ventilation air, the vessel vent system that collects off-gases from the process equipment, and the air from the plutonium powder-handling cabinets. The vessel vent air and the plutonium cabinet air are each filtered independently and then combined with the canyon air for passage through deep-bed sand filters. The sand filter effluent of about 120,000 cubic feet per minute then combines with the air from other building areas, from the dissolver off-gas, and from the acid recovery unit for exhaust from the stack.

Two small pipes are attached to the main stack and discharge at the same height; these pipes allow the off-gases from the two fuel dissolvers (after passing through an iodine absorber and filter) to be diverted from the main stack during decladding operations to prevent the formation of ammonium nitrate in the stack. Additional air is exhausted from the plutonium processing area (exclusive of the powder-handling cabinets) via separate filters and a stack 160 feet above grade mounted on top of the canyon building.

The air from the vessel vent in the H-Area hot canyon passes successively through a scrubber and fiber glass filter before combining with the canyon air. The scrubber contains water. In the H-Area warm canyon and both F-Area canyons, the scrubber is replaced with a dehumidifier.

In Building 221-H, the general pattern of ventilation is similar to that in 221-F. As in 221-F, the exhaust air from the main canyons and the vessel vent system exhausts through a sand filter (a second sand filter is under construction), and air from the central personnel areas is filtered separately. The finishing lines for ^{237}Np and ^{238}Pu also are filtered separately and do not exhaust through the sand filter, unlike the 221-F system. There are no separate small lines from the dissolvers because there is no aluminum decladding with sodium hydroxide to generate ammonia. The combined air streams of about 220,000 cfm are discharged through a 200 foot stack.

Filtration Systems: The air systems use HEPA filters, packed glass fiber filters, and two deep-bed sand filters that serve the two main canyon buildings (221-F and 221-H). Continuing attention is given to maintaining adequate filtration efficiency. It generally is not feasible to measure all filter efficiencies on a continuous basis, so normally the activity releases from stacks are monitored, and when releases increase, the efficiencies are studied. Periodic tests are made by injecting a mist of dioctyl phthalate upstream of a filter to verify that air is not bypassing the filter. In large systems where more than one filter is tested in parallel, the efficiency must be at least 99.95%.

Each separations area was originally equipped with a sand filter. A second sand filter was added for 221-F in 1975, and a second sand filter for 221-H is presently under construction. Performance of the original sand filters is of particular importance in limiting the amounts of particulate activity releases from the separations areas. The sand filters are 100 feet wide, 240 feet long, and 8 feet deep. Only the roof section is above ground level, and is supported by columns that extend through the filter bed. The bed is made from layers of sand and gravels in graduated sizes.

The largest rocks (1¼ to 3 inches diameter) are supported on special tiles which distribute the air from 12 tunnels connecting to the main supply tunnel. Progressing from the bottom to top, the air from the canyon building passes through layers of successively smaller particle size until the primary filter medium (20 to 50 mesh sand, 3 feet deep) is reached. Coarser sand and gravel are at the top of the bed to prevent loss of sand through entrainment. Air passes through exit ports into an exhaust tunnel and then to a 200 ft stack. Sumps are provided in the supply and exhaust tunnels to collect any condensed moisture.

The new sand filters are of similar construction except that they are 360 ft long, the supports for air distribution tubes over the laterals were replaced with stainless steel grates, and an additional 8 inch thick layer of sand was added as the final filter medium. The additional sand layer improves the sand filter efficiency.

Total routine flow through a bed has ranged from 100,000 to 130,000 cfm depending on pressure drops allowed and operating conditions. No single number can be given for filtration efficiency, but under routine operating conditions the release levels from the filters correspond to efficiencies of 99.8 to 99.99% for beta-gamma and 99.5 to 99.9% for alpha activity.

Canyon Support Facilities — Off-gases from the acid absorber unit and from the SRL trailer unloading system contribute a measurable fraction of the radioactivity in the air normally exhausted from the main 221-F stack; this area has no apparent large potential for abnormal releases to the air.

Tritium Facilities (232-H, 234-H, and 238-H) — Releases from the tritium process buildings can be categorized according to the controls that are imposed on the specific streams. Tritium escapes in very low concentrations when small batches of inert gas or air are discharged, when light-hydrogen isotope wastes from the isotopic separations are disposed of, and when unavoidable releases into the general ventilation system occur (from leaks, opening equipment, etc.)

The first category of release is individual waste batches of inert gas or air. Absorption beds are used where feasible to reduce the amount of tritium that otherwise would be lost in this manner. One system in use has an oxidizing bed to convert any elemental tritium into water, followed by a zeolite bed to absorb the water; another system requires only the zeolite beds.

A second category of release is the waste from purification operations in which tritium (^{3}H) is separated from protium (^{1}H) and deuterium (^{2}H). The final waste fraction is analyzed for tritium before discard and is recycled if the tritium is recoverable. However, the waste hydrogen isotopes must be discarded at some point in the cycle and will contain some amount of tritium; this point is ultimately determined by separations capacity and cost.

The third category of release includes releases to the general ventilation system, so that the tritium is mixed with very large volumes of air. The large volume of dilution air precludes simple recovery because of the resulting low concentration of ^{3}H. Examples of the sources of these releases are leaks, opening lines for maintenance work, routine opening of equipment (such as the extraction furnace), and routine disconnections as materials are loaded into and out of the process system. Successive evacuations and flushings with inert gas are the steps taken to minimize the releases for maintenance and routine disconnections or equipment openings.

Waste Tanks — The ventilation systems for the underground storage tanks are designed for purging the interior space to dilute any radiolytically produced hydrogen that may be formed and preventing any radioactivity being released to the atmosphere.

Type IV Tanks (17-20, F Area; 22-24, H Area): These waste tanks are spaced so that a single exhaust system provides interior purge through two tanks (17-19, 18-20, 21-22, 23-24). Inlet purge ducts are equipped with HEPA filters on all tanks except 20. The exhaust system consists of a tea pot filter and exhaust blower. The exhaust air is not sampled. These systems are currently being up-graded such that the purge exhaust passes successively through a de-entrainer, condenser, heater, parallel HEPA filters before finally being discharged to the atmosphere by an exhaust blower.

Type I, II Tanks (1-8, F Area; 9-16, H Area): The existing tank interior ventilation designs provide a positive air sweep across the vapor space, exhausting successively through a condenser and a heated, jacketed tea pot filter located in the exhaust duct. Purge air is supplied either by a tank annulus blower or separate purge blower. The plant is currently converting all these systems to negative. Presently, 10 of 16 systems have been converted. The new systems include HEPA filters on the inlet purge duct with the tank vapor sweep being exhausted successively through a de-entrainer, condenser, heater and parallel HEPA filters by an exhaust blower.

Type III Tanks (33-34, F Area; 29-32, H Area): The ventilation systems for Type III primary tanks are negative pressure systems designed for purging the interior space to dilute any radiolytically produced hydrogen that may be formed. In a typical installation, the air enters the tank through a HEPA filter. The exhaust air is drawn successively through a wire mesh separator to remove entrainment, a water-cooled condenser to extract potentially contaminated moisture, a heater to prevent condensation, and HEPA filters before being finally discharged to the atmosphere by an exhaust blower.

Receiving Basin for Offsite Fuel (RBOF) and Resin Regeneration Facility (RRF): All air exhausted from the basin areas, control room, and regulated portions of the office and change room section passes through HEPA filters before discharge to the atmosphere through a 5 ft diameter short stack 53 ft above grade. These filters are preceded by roughing filters to extend the serviceability of the HEPA filters. When operations such as fuel cutting are scheduled, activated charcoal absorbers are installed in series following the HEPA filters to retain radioiodine isotopes.

Process vessels in the decontamination and waste cells are vented by means of an acid-resistant vent system that is equipped with an exhaust fan and a HEPA filter. The filtered exhaust is discharged through a 10 inch diameter pipe 53 ft above grade. The process vessel vent system does not have the versatility that is desirable for maximum containment of airborne activity. The single filter housing can contain either a charcoal filter for ^{131}I retention or a HEPA filter for particulate activities. The type of filter used is determined by operating needs. There are plans to install a revised system that can mount both types for use either singly, or in series or parallel as required. Containment effectiveness will be improved.

Savannah River Laboratory (700 Area): Outside air flows into clean areas such as offices and personnel corridors and then to potentially contaminated areas such as laboratories and utility corridors. Most of this ventilation air flow is exhausted from the laboratories through a single stage of HEPA filtration to a 100 ft stack. This air is continuously monitored for radioactive contaminants, and when contamination is detected, this air is diverted to a sand filter until the source is identified and neutralized. The remainder flows through an inlet HEPA filter into contaminated process enclosures such as glove boxes or manipulator cells and then through at least two stages of HEPA filters to the sand filter and a 160 ft stack. The effluents from the three principal stacks in the Technical Area are continuously monitored for gross alpha and beta-gamma particulates and for radioiodine. Other stacks are continuously sampled. C stack in 773-A is continuously monitored for tritium releases.

Auxiliary diesel generating systems maintain continuous ventilation air flow in the event of the loss of the normal electrical power supply. All filters except those attached directly to process enclosures are tested on a regular schedule by measuring their retention of 0.3 μm diameter particles of dioctyl phthalate. A filtration efficiency of 99.97% is required for acceptance.

REFERENCES

(1)   H.T. Fullam, *Compatibility of Cesium Chloride and Strontium Fluoride with Contain-ment Materials,* BNWL-1673, Battelle, Pacific Northwest Laboratories, November 4, 1972.

(2)   Hanford Engineering Development Laboratory, TMT-3, *Waste Management Manual,* November 1973.

(3)   *Project Crested Ice,* Danish Atomic Energy Commission Report RISO 213, Research Establishment, Riso, Denmark (1970). Also published in: *USAF Nuclear Safety,* 65, No. 1 (Part 2), special edition (Jan/Feb/Mar 1970).

WASTE MANAGEMENT AT COMMERCIAL FUEL REPROCESSING FACILITIES

The following reports are the sources of the material in this chapter: PB-235 806, PB-258 316, CONF-700440-1, Docket 50-564-1, PB-260 559 and BNWL-SA-5725. See p 359 for a complete bibliography.

Three commercial nuclear fuel reprocessing plants have been constructed in the United States, each on a relatively large remote site: the Nuclear Fuel Services (NFS) plant, the General Electric Midwest Fuel Recovery Plant (MFRP) and the Barnwell Nuclear Fuel Plant (BNFP).

Up to the present, only one commercial facility has been operational, but is now shut down for modification. This is the Nuclear Fuel Services plant in West Valley, New York, with a capacity of 1 tonne of uranium per day. Processing of spent nuclear fuel was initiated in 1966 and continued with sporadic interruptions until December 1971 when the plant was shut down to permit expansion of processing capability which will increase plant capacity to 3 tonnes per day. The plant is located on a 3,300 acre tract owned by the State of New York in Ashford Township, Cattaraugus County, New York (1). Buffalo, New York, is 26 miles from the plant and several of its southern suburbs are within a 25 mile radius.

Considerable dairy farming and other agricultural activities are conducted close to the site. New York State conducts a comprehensive monitoring program around the facility including daily raw milk surveillance (2)(3). Extensive studies have been conducted at this facility by EPA and its predecessor organizations (4)(5)(6) and (7).

The Midwest Fuel Recovery Plant (now called the Morris Operation) is a tonne per day plant located at Morris, Illinois, adjacent to the Dresden Nuclear Power Station on privately owned property (8). The General Electric Company, which has received a permit to store spent fuel and process unirradiated fuel, owns and operates the plant. The MFRP was not able to start up because of problems encountered in cold checkout operation.

The closest population center is Joliet, Illinois, 14 miles from the plant. Part of the city of Chicago is included within a 50-mile radius of the plant, thus significantly increasing the local population of concern. Offsite concentrations must be evaluated in terms of a multiple source since the three Dresden reactors are in close proximity.

The Barnwell Nuclear Fuel Plant is designed to reprocess 5 tonnes of uranium fuel per day (1,500 tonnes per year), and is located adjacent to the Savannah River Laboratories (SRL) in Barnwell County, South Carolina, on privately owned property (9). Construction of this facility was begun during the spring of 1971. The main sections of the BNFP are complete but the plutonium conversion and HLW solidification facilities have not yet been finished.

Augusta, Georgia, which is 31 miles from Barnwell, is the nearest population center. The population within 50 miles of the facility is about 500,000 or a factor of 10 below the corresponding population at Midwest.

The Exxon Corporation has applied for a construction permit to build a complete fuel reprocessing center at Oak Ridge, Tennessee. Fuel reprocessing plants are essentially complex chemical plants, the complexity being compounded by the fact that the materials being processed are highly radioactive. The specific process used to separate the spent fuel element into the product streams and waste stream is dependent upon the particular type of reactor fuel being serviced.

All three present facilities used a shear (chop) and nitric acid leach method to separate the light water reactor (LWR) spent fuel from the metal cladding. Following this step, the Purex process with tributyl phosphate (TBP) as the solvent is used to extract the uranium and plutonium from the fission product waste in column contactors (10)(11)(12)(13)(14). The uranium and plutonium are separated and further purified by various means, including ion exchange, scrubbing, evaporation, etc. The design recovery rate is 99.5% for uranium and plutonium.

Plants designed for processing of spent fuel elements from LWR systems could be used to process elements from LMFBR as long as the facilities are derated (handle smaller quantities of fuel) to avoid criticality problems and not exceed constraints on effluents (15).

The processing of HTGR spent elements requires a different "head end" processing system since such fuel elements may require the burning of the graphite which contains the coated fuel particles. The decontamination of the off-gas stream resulting from the burning operation requires development of "head end" processes unique to the HTGR processing facility (16). The Thorex process, (12) will be used for the separation of uranium and thorium from the fission product wastes in spent HTGR fuels (16).

The following, three sections in this chapter give a basic description of the processes used at these fuel reprocessing plants and the control systems for reducing the radioactive discharges.

NUCLEAR FUEL SERVICES AND BARNWELL NUCLEAR FUEL PLANTS

Process Flow

The main process steps will be summarized to indicate sources and handling of radioactive waste liquids and gases.

Cask Unloading and Decontamination: Cask water is discharged to a low-level waste system.

Fuel Pool Storage: Special provisions are made for leaking fuel elements storage and resulting contaminated water and off gases. Prevention of criticality is a major design factor. Inventory control and accountability are important operating parameters.

Fuel Transfer and Mechanical Processing: NFS disassembles fuel elements before shearing the individual fuel rods. Barnwell will shear the entire element and thus is essentially limited to reprocessing fuel from light water power reactors because of the geometries involved. Lengths of sheared fuel rods range from about 1.25 to 7.5 cm (½ to 3 inches). Very little of the krypton-85 and tritium is released during the shearing process. Goode (14) reported that less than 1% of the krypton-85 is released during this step and Cochran et al (6) confirmed this conclusion based on field studies at an operating plant.

Fuel Dissolution: In this step the spent fuel is leached from the sheared cladding in nitric acid as a preparatory step to the chemical separation processes. The leached hulls are analyzed for plutonium content and returned for further leaching if necessary. Over 99% of the krypton-85 (14)(6) and about 6% (17) of the tritium is released into the off-gas system by this operation.

In addition to the noble gases, a large fraction of the halogens are also released to the off gas in this step. Most of the radionuclide particulates present in the off gas result from this step (19), although this may not be the case at Midwest where large quantities of particulates will be added from the high-level waste solidification process. Since NFS dissolves fuel on a batch basis, essentially all the krypton-85 in the batch is discharged within a 3- to 4-hour period. Present NFS Technical Specifications limit fuel dissolution to 2 metric tons per day.

Midwest and Barnwell will dissolve on a semicontinuous basis. Waste gases are processed separately through the dissolver off gas system (DOG) at NFS and Barnwell, since most of the radionuclides in the waste gases are generated in this step at these facilities. Midwest has designed their gaseous waste treatment system somewhat differently and waste gases are not segregated by source.

Chemical Separation and Purification: The nitric acid feed from the dissolver which contains the uranium, plutonium and fission products is counterflowed in a contactor column with TBP and nitric acid. The uranium and plutonium are preferentially dissolved in the TBP, and the fission products are retained in the nitric acid. Separation of the plutonium and uranium in the organic solvent is achieved by reducing the plutonium to its trivalent state where it can be stripped with a nitric acid scrubbing process in a contactor column. The uranium is subsequently also stripped from the TBP which is then recycled. The plutonium

and uranium are then purified through a series of processes. Midwest has designed
a calcining system for conversion of the uranium to UF for shipment directly to
an enrichment facility. NFS and Barnwell will ship uranyl nitrate in tank trucks
to conversion facilities. Plutonium will be stored and shipped in the nitrate
form in critically safe containers.

Recovery of Solvent and Acid: The TBP solvent and the acid are recovered and
recycled. The fission products are concentrated in the evaporator bottoms of
the high-level waste concentrator system and the acid overheads are recovered.
Additional fractions of the semivolatiles and the halogens are discharged to the
waste gas system in this step, along with some particulates.

High Level Waste: At NFS and Barnwell high-level liquid waste will be stored
in stainless steel tanks which are located underground and externally cooled.
These tanks are placed on concrete saucers to permit monitoring for leaks and
inside concrete vaults which provide a secondary containment. These acidic
wastes are not neutralized prior to storage to simplify future solidification. At
the Midwest facility, which uses a fluidized bed calcination process for solidification, high-level wastes are solidified immediately after separation.

This process has been used on a production scale with intermediate level waste
since 1963 at the Idaho Chemical Processing Plant (ICPP) (18). The solidified
waste will be sealed and stored in a cooling pool. The off gases from this process will contain some radioactive particulates (19), especially the semivolatile
fission products such as ruthenium.

NFS Tanks — The NFS West Valley plant was designed to use both the standard
Purex and the Thorex processes (20). Wastes generated by both processes are
now in storage at West Valley. There are two storage tanks for neutralized Purex
waste at West Valley, one serving as a spare.

These are similar to tanks at the Savannah River plant in that the primary tank,
including internal support columns, sits in a secondary pan 4 feet high relative
to the bottom of the primary tank. Each primary tank has a working volume
of 600,000 gallons; one tank is full. The tanks are fabricated of carbon steel
and were stress-relieved in the field to minimize stress corrosion. They are
equipped with air sparger-circulators for mixing and for purging of radiolytically
generated hydrogen. The pans are equipped with level indicators and connected
to pumps for transfer of any leakage back to the storage tank or to the low-level
liquid waste treatment facility.

Because the neutralized waste in storage at NFS is not sufficiently concentrated
to maintain self-boiling, a temporary heat exchanger has been installed in the
active-waste storage tank to concentrate the dilute waste. Off-gases and steam
from the tank are passed through a condenser and the condensate is either transferred back to the tank or treated by ion exchange for removal of cesium and
strontium before discharge to the low-level waste treatment facility. The off-
gas is then heated to 180°F and filtered before discharge.

Thorium waste was generated in processing fuel from Core A of the Indian Point
1 Reactor. The 12,000 gallons of waste (20) are stored in a 304L stainless steel
tank 12 feet in diameter and 15 feet 9 inches high. The active tank and an

identical spare are installed in the same concrete vault, which has a stainless steel bottom liner and a stainless steel sump. The tanks are equipped with cooling coils to maintain waste temperature below 140°F. The vault sump is monitored and alarmed for leakage from the tanks. Experience with the two waste tanks in operation at West Valley has been good; there has been no evidence of leakage.

It is not planned to use the existing tanks for wastes produced by future operations at West Valley. An application by NFS (21) for modifications and increased capacity at the West Valley plant is now being reviewed. On February 24, 1976, NFS advised NRC that they plan to eliminate the necessity for HLW storage by solidifying the waste immediately after generation (22). This would be made possible by storage of irradiated fuel for an average of 24 months before reprocessing. NFS was to provide details of the waste solidification facility. However, on September 22, 1976, NFS announced that it is terminating reprocessing services.

BNFP Tanks — The BNFP provides for temporary storage of liquid HLW as an acid solution in a stainless steel tank 54 feet in diameter by 20 feet high. Net capacity of the tank is 300,000 gallons, with a 10% freeboard allowance above this level. The tank is equipped with cooling coils, ballast tanks, and air lift circulators for waste circulation and for purging of radiolytic hydrogen. The tank is also equipped with transfer jets and instrumentation for temperature and liquid level measurements.

The tank is contained in a concrete vault lined with stainless steel. A sump with instrumentation to determine leakage is part of the liner system and provision is also made in the concrete structure for collection of possible leakage through the vault liner.

The cooling coils are designed to remove 3.7×10^7 Btu/hr and are connected in a primary loop from which heat is transferred to a secondary system where it is dissipated in a cooling tower. Two backup water supplies are provided for the waste tanks. If the cooling-tower system fails, well water will be circulated through the secondary system to remove heat from the primary system. If both these systems fail, diesel-powered pumps will supply water from a nearby pond directly to the primary system cooling coils.

Purge air supplied to the air-life circulators and ballast tanks is also required for dilution of the radiolytic hydrogen produced. Hydrogen concentration in the tank atmosphere is kept below 3% by volume.

The waste tank is vented through a dome on top of the tank equipped with internal baffles to minimize entrainment. The vented gases then pass through a series of cleansing processes before being released through the main stack.

An underground tank in a lined vault identical to the liquid HLW vault is provided for intermediate-level liquid waste, which contains about 0.0001 as much radioactivity per unit volume as HLW and is therefore not self-heating. The intermediate level tank is of the same design as the HLW tank except that it does not include air-lift circulators, ballast tanks, or cooling coils. Agitation is provided by an air sparging unit. Provision is made for jetting or for installing a pump to remove liquid from the tank.

The BNFP tanks incorporate improvements over previous tank designs. The secondary containment is a fully lined vault that will hold more than an entire tank of waste. There are no obstructions on the tank bottom, so that solids cannot build up. Vapor and radioactive entrained particulates evaporated from the tanks' contents are contained and treated in the plant off-gas system. A primary waste cooling system is backed up by two other water-supply systems. In addition, temporary connections can be made to add pond water to a tank if all three cooling water supplies became inoperative.

Low Level Waste: The low-level liquid waste stream consists of wastes collected from sources throughout the plant. Midwest and Barnwell have been designed to completely eliminate the discharge of low-level liquid waste by adding an evaporator to the system to process the final low-level liquid waste stream (it is discharged to the environment at NFS). The evaporator overheads are discharged through the stack and theoretically contain all remaining tritium from the fuel. The bottoms are solidified and are currently shipped to privately operated waste burial areas on site.

In the gaseous stream, almost all of the krypton-85, and varying fractions of tritium and other volatile nuclides, such as iodine, are released to the dissolver off gas (DOG) during the dissolution process. Treatment of this waste stream is complicated since it contains varying concentrations of the oxides of nitrogen. NFS and Barnwell are designed to have a separate waste system for the DOG. The various process tanks have vents for off gases which are contaminated with volatile fission products. These are routed to the vessel off-gas system (VOG). Midwest will combine the DOG and VOG systems since they have additional sources of airborne wastes from the calcining of the high level liquid waste.

The gaseous waste treatment system at all three facilities basically consists of a caustic scrubber, followed by a silver zeolite absorber and then final filtration through a high efficiency particulate air (HEPA) filter. NFS has incorporated both an acid scrubber and a caustic scrubber in their system. The scrubbers and silver zeolite adsorption systems are installed to collect the iodine in the off-gas stream. Midwest and Barnwell have installed or plan to install in series two independent particulate filtration systems which are isolated to avoid dual failure. Two HEPA filter systems are to be installed at Barnwell. Midwest has installed a HEPA filter system in the off-gas stream and a sand bed filter system for final filtration of both the off-gas stream and the plant ventilation air.

Control Points for Effluents

The cladding on the fuel normally provides the primary barrier (or containment) for preventing the release of fission product wastes. This barrier must necessarily be destroyed in the nuclear fuel reprocessing plant in order to recover the fissile and fertile material for reuse. This fact, in addition to the very large quantities of fission products present, requires that effluent control procedures must be incorporated at all processing steps. In practice, control of the effluents from each process can be achieved through the use of a common collecting system such as the vessel off-gas header system which collects the off gases from several processing vessels.

The use of such common systems has definite economic advantage. Fewer control systems are required and a higher degree of reliability for control of discharges

is possible. Since there are fewer components subject to failure, better quality
equipment can be installed. An example of a common collection system is the
low-level liquid radioactive waste system. Sources feeding this system include,
but are not limited to: cask decontamination water, leakage from the fuel storage pool and waste storage tanks, laboratory wastes, laundry wastes, high-level
waste equipment drains, and floor drains. This waste is collected and processed
through an evaporator where the overheads can be condensed and recycled, discharged, stored, or handled in a combination of these options.

NFS, Inc. has chosen the method of discharge of the condensed overheads
through a system of settling lagoons to a public waterway. Midwest has chosen
a combination of treatments where the evaporator overheads are condensed and
recycled to the maximum extent possible. Low-level liquid wastes which cannot be recycled, such as air scrubbing wastes, are collected in a low-level waste
vault which is maintained at a constant volume by use of a second evaporator.
The overheads of this evaporator are not condensed but discharged through a
second air cleaning system and the sand filter to the stack.

In general, air cleaning systems follow the same principle. Contaminated air is
collected from various processes in a common header and then treated to remove
contaminants. Reprocessing plants have sufficient chemical contaminants in some
off-gas streams to cause problems, such as overloading of various air cleaning
systems. Therefore, the off-gas streams are frequently segregated by source,
especially for initial treatments.

Specific effluent control points and the principal contaminants are: (1) Dissolver
Off Gas — noble gases (krypton-85 and xenon-133), halogens (iodine-129 and
iodine-131), tritium and particulates. (2) Solvent Extraction and Purification
Off Gas — particulates. (3) Solvent Recovery Off Gas — mixed fission products.
(4) Acid Recovery Off Gas (High-Level Waste Concentration) — halogens and
other volatile species. (5) High-Level Waste Solidification — ruthenium and
other potential volatile radionuclides including technetium, cesium, selenium,
and tellurium (19). (6) Low-Level Waste Treatment — tritium, strontium,
ruthenium, cesium and other longer-lived radionuclides (23).

Investigations are being conducted into modifying spent fuel reprocessing systems
to provide more positive control of the effluents. ORNL is presently directing
efforts to the design of equipment and process flow to obtain "near zero release"
processing of short-cooled LMFBR fuel (24). The design includes a new "head-
end" processing step (voloxidation) that is designed to release tritium from
LMFBR fuel and deactivate sodium prior to aqueous processing. The tritium,
krypton, and iodine are evolved during the head-end operations (which include
voloxidation and dissolution) and vented to their respective primary removal
systems.

The off gas from the dissolution cell and from the process equipment beyond
dissolution and feed adjustment is subjected to a secondary off-gas treatment
system which includes filters for particulates and scrubbers for the oxides of
nitrogen, halogens, and ruthenium. Argonne National Laboratory is also developing an alternate "head-end" pyrochemical process for decladding of LMFBR fuel
which potentially may produce an improved method for control of the effluents
(25)(26).

Solid Wastes

The solid wastes resulting from the recovery of uranium and plutonium can be
categorized as high-level solidified wastes, spent fuel cladding wastes, and low
and intermediate level wastes. Federal regulations (27) require that high-level
wastes, generally interpreted as self-heating, be solidified within 5 years of processing. In addition to the radiation exposure protection which must be provided
for these wastes, cooling must also be provided. The spent fuel cladding wastes
contain residual amounts of the fuel in addition to the activated metallic radionuclides making up the cladding itself. There are many sources of low and intermediate level wastes: air filters, spent resins, silver zeolite, evaporator bottoms,
sand filters, etc. It appears the greatest problem presented by these wastes is the
presence of long-lived radionuclides of health significance such as the alpha-emitting transuranics and iodine-129. These long-lived components will be present
in all solid wastes from reprocessing plants.

Discharge Control Options

In general, the fuel reprocessing industry has incorporated the most advanced
technology into their waste treatment systems. For example, the control system
for iodine, one of the limiting radionuclides in the local environs of a reprocessing
plant (28) will use silver zeolite technology which has only recently been developed. It should be noted that a control method is not available for tritium and
only the one control system is planned for iodine.

Most of the radioactive discharges to the environment from reprocessing plants
will be in the gaseous waste effluent. The two newer plants have designed their
processing systems to eliminate the discharge of liquid radioactive material. However, neither of these facilities has been operated with irradiated fuel. Thus, a
decision on whether it is preferable to discharge radioactive waste to the atmosphere only or to discharge via both the liquid and gaseous pathways should be
postponed until operating experience with both methods is obtained. Since
most of the current efforts to reduce discharges have been directed toward gaseous effluents, the methods discussed in this section are limited to the control of
airborne discharges. However, liquid waste discharges may require additional investigation in the future.

The radioactive pollutants that are most likely to be released from normally operating reprocessing plants are krypton, tritium, iodine, and the actinides. Other
fission products such as strontium, ruthenium, and cesium and induced activities
in the fuel element cladding can also be released. In addition to the anticipated
normal discharges (gaseous waste stream), miscellaneous airborne releases can
occur because of the complexity of the various processing operations and the
unproven reliability of some of the control systems. Such releases may not be
detected by monitoring of the gaseous waste stream (stack effluent). Inplant
air monitoring for contamination control can, however, indicate these possible
pollutants. The long-term operational reliability of control systems is unproven
with the exception of the HEPA (high efficiency particulate air) and sand filter
systems.

Krypton-85 Control

Up to the present, krypton-85 and other noble gases have been released directly to the atmosphere from nuclear reactors and fuel reprocessing plants. With no off-gas treatment for noble gases about 10,000 Ci of krypton-85 is estimated to be discharged per metric ton of spent fuel processed, assuming a burnup of 33,000 megawatt days per metric ton. Several methods have been suggested to limit such releases. The processes are classified as ambient temperature adsorption, cryogenic adsorption, cryogenic distillation, selective absorption, permaselective membranes, and clathrate precipitation.

The processes have been previously reviewed by several authors (29)–(33) with regard to development status, advantages and disadvantages, cost, and efficiency. Kirk (29) has prepared a comprehensive review of the radiation hazard from krypton-85. The conclusions reached in these reviews indicate that at present only cryogenic distillation, selective absorption processes, and cryogenic adsorption are worthy of consideration for control of krypton-85 discharges from reprocessing plants (30).

Systems based upon both the cryogenic adsorption and cryogenic distillation processes have been designed for and are being installed at light-water-reactors for extension of holdup times for gaseous effluents containing noble gas radionuclides. The selective absorption process has been developed for application to reactor systems, but requires further development to be applicable to fuel reprocessing plants (31).

The cost of krypton collection systems is highly dependent on the design of the dissolution process. Essentially all of the krypton present in the spent fuel is released during this dissolving or leaching process. To minimize the costs, it is necessary to minimize the total volume of off gases from this process since all of the off gas must be treated to remove the krypton. Therefore, in a new plant the cost of a krypton collection system would probably be significantly less than installation of such a system in an operating plant where no effort was made to minimize the total off gas from the leaching process. The costs are typical of costs for a krypton control system in a new facility. At the Midwest facility an effort was made to prevent large scale dilution of the off gas.

Cryogenic adsorption systems remove krypton from process gas streams by adsorption in refrigerated activated charcoal beds until the bed capacity is reached, followed by desorption into a purging gas while heating the beds. This was demonstrated on a large scale at the Idaho Chemical Processing Plant (ICPP) more than 15 years ago. The disadvantages of this process are high refrigeration costs, fire hazard potential, explosion potential due to hydrocarbons, nitrogen oxides and ozone, and impurity plugging of the adsorbers. An overall recovery fraction of 31% was obtained for krypton although with design changes and modifications in operating procedures, a recovery fraction of the order of 99% could be achieved. The system can be considered but may not be the best for application to fuel reprocessing plants. It may have more potential for application to interim holdup for effluent gases in reactors.

Experience has been gained for this method in the development of the HTGR and a decontamination factor of 10 appears to be achievable under most operating conditions. Operation of the adsorber beds at cryogenic temperatures helps overcome the problem of an occasional abrupt release of adsorbed contaminants which has been experienced with ambient temperature absorbed beds. Assigning a decontamination factor to ambient temperature systems is questionable since experience has shown that under adverse conditions it is possible to experience a negative decontamination factor for interim periods.

Total capital costs, including installation, for a cryogenic adsorption system are estimated at 3 million dollars (1973) based on general estimates for use of systems at reactors. Slansky (30) estimated 1 million dollars for capital costs which appears low. An annual operating cost of $150,000 is reasonable.

Cryogenic distillation, which is based upon separation of gases due to differences in their relative volatilities at low temperatures, has been demonstrated and operated on a significant scale at ICPP for removal of krypton and xenon from an off-gas stream (34)(35). Recovery of krypton and xenon in a form suitable for bottling in gas cylinders is possible with this process. An additional advantage is the lower capital and operating cost of this cryogenic system as compared to cryogenic adsorption. The cryogenic distillation process entails some potential for explosion. In spite of the explosion potential, cryogenic distillation is considered to be one of the two most promising processes for noble gas control at reprocessing plants.

Considerable experience has been gained in operating these systems in liquified gas (or air) plants. While decontamination factors across the cryogenic stage of 10^3 have been estimated, small leaks can occur in a system. Estimates of the recovery factor for an overall system range from 98% to 99.99%. Decontamination factors of 10^3 (99.9% recovery) should be attainable with this system and guarantees of such performance have been submitted to the General Electric Company with bids for installation of a cryogenic distillation system at their Midwest plant (36).

The General Electric Company received bids ranging from 0.75 to 1.5 million dollars (1971) for the equipment needed to install a cryogenic distillation system in their Midwest plant (36). It is estimated that installation costs would match equipment costs, thus producing a total capital cost of about 3 million dollars for this system. Annual operating costs of $100,000, postulated by Slansky (30), are probably a good estimate. Xenon is also recovered by this system and if kept separate from the krypton, has a potential market value.

Selective absorption (liquid extraction) depends upon the relative solubilities of gases in the solvent used, Freon-12 being the typical solvent under consideration. Krypton and xenon are selectively absorbed in this solvent while other materials pass through. The solvent is then processed to recover the krypton and xenon which can be stored while the solvent is recycled. Bench-scale studies have been completed (31)(37) and, as indicated previously, commercial systems for application to reactors have been developed. Because of the anticipated explosion hazard associated with the cryogenic systems, management personnel of commercial fuel reprocessing facilities (NFS, Barnwell) have indicated a preference for fluorocarbon selective absorption systems of this type.

Major disadvantages of such a system include radiation degradation of the solvent, and a requirement for pretreatment of gas streams to increase system tolerance to impurities. Babcock and Wilcox indicates that these problems can be circumvented with good engineering design. Decontamination factors of 100 to 1,000 are expected for such systems (31).

The cost of a selective absorption system for installation at a reactor was estimated at 1 million dollars (1973). Assuming a slightly greater capacity system for a reprocessing plant results in a total installed cost of $1.5 million. Annual operating costs of $100,000 predicted by Slansky (30) also appear reasonable.

Tritium Control

Processes available for tritium control have recently been reviewed (38). The techniques considered include: chemical exchange, distillation, electrolysis, diffusion and centrifugation, radiolysis, adsorption and chromatography, solvent extraction, and molecular excitation. At present the information on these techniques is inadequate to project the technical or economic feasibility of retaining tritium.

Voloxidation is another method of tritium control that looks promising (24). In this process the fuel pins, just after shearing, are heated to approximately 650°C in a stream of air or oxygen. Tritiated water is generated which should be relatively free of ordinary water and consequently occupy a much smaller volume than tritium wastes from presently planned or operating reprocessing plants. This process may collect approximately 99% of the tritium which is present in the unprocessed fuel. Since the voloxidation process requires a major change in the head-end design of fuel reprocessing plants, it would probably be impractical to back-fit existing or planned facilities.

The projected release rate of tritium from a facility having no tritium control systems is 88 Ci/MTU. Releases may be either by the air or water pathways depending on plant design. With the voloxidation head-end process, tritium decontamination factors of 100 are anticipated. Cost information on the voloxidation process is not presently available; however, General Electric considered three other methods of tritium recovery and/or disposal in connection with their Midwest plant (36). Removal from off gas was stated to cost approximately $10 million (1971) with no process or technical feasibility defined. Deep-well disposal of tritiated water was reported to run between $400,000 and $500,000. Finally, off site shipment cost was estimated to be between $250,000 and $350,000 with no estimate of feasibility or safety associated with the packaging and transport.

Radioiodine Control

A variety of processes have been developed and used for collection of radioactive iodine. These processes include (1) wet collection—aqueous scrubbing (reactive sprays, towers, wet filters); (2) adsorption (charcoal, activated charcoal, silver zeolite, metallic filters); and (3) filtration (high efficiency particulate—HEPA, sand, deep-bed fiber glass).

The particular treatment process to be selected depends upon the collection efficiency desired, quantity and chemical species of iodine involved, stream characteristics, i.e., flow rates, temperature, humidity, etc. The actual overall removal efficiency of a system is dependent upon the off-gas flow paths as well as the characteristics of the treatment techniques.

Wet Collection Techniques: These have been used to quantitatively retain iodine in the liquid phase by adding mercury salts during fuel dissolution at the Savannah

River Plant. The resulting Hg-I complex is solvent extracted and the solvent is washed in a solvent scrubber to remove any remaining iodine. Disposal of the solid Hg-I complex must be effected. However, there are no AEC guidelines for disposal of such waste at present.

A mercuric iodate precipitate can result from scrubbing the off gas with a mixture of 8 to 14 molar nitric acid and 0.2 to 0.4 molar $Hg(NO_3)_2$ (mercuric nitrate). Solid iodine can be obtained by use of concentrated 17 to 19 molar nitric acid at room temperature. The latter systems can be considered capable of removing massive concentrations of elemental iodine and trace quantities of other forms. Decontamination factors of 10^4 for all forms of iodine are reputed to be theoretically feasible in either packed or bubble-packed scrubbers at operating throughput rates considered appropriate for fuel reprocessing (39). If the off gas contains high concentrations of NO some loss in the decontamination factor occurs. Alkaline solutions (NaOH, $NaHCO_3$, Na_2CO_3) are also used in scrubber columns and have reported decontamination factors of 10 to 20. In packed columns, the separation efficiency is a function of bed height, fiber drag coefficient, fiber diameter and fiber volume fraction.

An operational facility (Eurochemi Fuel Reprocessing Plant) which uses two scrubbers [NaOH for low iodine concentrations and $NaHCO_3$ or $Hg(NO_3)_2$ for high iodine concentrations] reports decontamination factors of 500 for gaseous iodine and 2,500 for iodine in aerosols. The chemical form of iodine which emanates from these scrubbers is predominantly organic (40) although traces of hypoiodous acid have also been reported.

Reactive sprays, hydrazine and thiosulfate, have been studied for application to iodine and methyl iodide washout from reactor containment atmospheres with decontamination factors of 2,000 and 100 reported respectively. Such systems have not been applied to fuel reprocessing plants. Their performance is found to be a function of relative humidity, temperature, drop size, solution pH and concentration.

Charcoal Adsorption: This has been used for more than 10 years in reactor and fuel reprocessing off-gas systems. The iodine removal efficiencies of activated charcoal that have been reported cover a broad range (50 to 99.99%). The actual decontamination factor is dependent upon the forms of iodine, concentration, stream humidity and flow velocity, and charcoal impregnant.

Silver zeolite (AgNO_3 impregnated in an alumina-silica molecular sieve) is reputed to be superior to impregnated charcoal for the adsorption of methyl iodide (approximately 20 times that for impregnated charcoal under dry air conditions) (41). The principal design consideration is the effective residence time in the bed which is related to the face velocity. If the silver zeolite (AgZ) beds are designed to permit a mean residence time of about 0.5 second, efficiencies are greater than 99.9% under the most adverse conditions of humidity and chemical form of the iodine (41)(42). Removal efficiencies can be optimized through selection of the zeolite material with its characteristic alumina-silica ratio. However, this ratio largely determines the resistance of the AgZ to acid vapors and thus the effective life of the bed. In general, high acid resistance results in lower removal efficiencies.

Although the efficiency of the AgZ is acceptable, consideration must be given to the loading characteristics of AgZ in terms of the total iodine cleanup system. The AgZ beds will adsorb all halogens and probably cannot be used when HCl is used for fuel dissolution. In addition, there is a considerable amount of iodine (iodine-129 and -127) present in the off gas, and this may load the AgZ at unacceptably high rates. Current technology employs the alkaline scrubber to remove a large fraction of this iodine which is theoretically elemental in composition. Work is continuing on the development of other metallic zeolites which may eventually replace the scrubbers. Lead zeolite currently appears to be the most attractive of these more economical and 'ess efficient systems.

The lifetime of the AgZ system is improved by introduction of an oxidizing catalyst upstream of the sorbent (39). The use of AgZ systems has been shown to be feasible and practicable, but full scale operation has yet to be accomplished. Adsorption of iodine on metallic filters (copper, steel or aluminum, silver-coated copper) has resulted in recovery in the range of 97.4 to 99.9%. Silver-coated silica gel has a reported decontamination factor of 10 for iodine (elemental).

Filtration: This is primarily designed for control of particulates. It cannot be seriously considered as a primary technique for iodine collection, since it depends upon sorption of iodine to particulates which are then trapped by the filter. In any event, all off-gas streams from fuel reprocessing plants will be filtered through sand, deep-bed fiber glass or high efficiency particulate air filters in addition to the iodine control systems. The iodine collected by particulate filters will experience desorption at a rate which is dependent on stream conditions. While this desorption rate may be significant for the short-lived iodine-131 discharge, it will have a negligible effect on the iodine-129 discharge. Therefore, the removal efficiency for filtration of iodine is negligible.

Control Systems Evaluation: It is estimated (43) that much of the discharged iodine will be in an organic form and that most of this iodine will interact with atmospheric particles and settle out within 10 to 20 miles of the discharge point. Thus, the discharged iodine appears to be principally a local problem, although long-term environmental transport must be considered because of iodine-129 with its long half-life.

LWR spent fuel with a burnup of 33,000 megawatt days per metric ton will contain about 2 Ci of iodine-131 per metric ton after 150 days cooling. In addition, to being dependent on the irradiation history of the spent fuel, the iodine-131 content decreases rapidly with the cooling period because of its 8 day half-life. The iodine-129 content will be about 0.04 Ci per metric ton of spent fuel.

An overall system for iodine collection in fuel reprocessing plants will probably consist of a combination of wet collectors (liquids and off-gas scrubbers) followed by a catalytic decomposition system, an adsorption system, and a filtration system. The overall efficiency will be dependent upon the detailed design of the off-gas flow system. For example, if the system is assembled in the manner of the Midwest facility, the off-gas stream bearing the iodine is first processed through the caustic scrubber (eff = 90%) and then through the AgZ bed. The gas stream is then routed to the sand filter and stack.

The overall removal efficiency for this treatment should be greater than 99.9%. However, the scrubber solution, theoretically containing 90% of the iodine, is

routed to the low-level-waste storage vault. The low-level waste in the vault is
routinely evaporated for volume reduction, opening a way for a large fraction
of the iodine to be revolatilized. This second off-gas stream is processed through
a second scrubber but no AgZ bed. Therefore, about 10% of the iodine routed
to the low-level vault can be discharged to the sand filter and the stack. The
buildup of the long-lived iodine-129 in the waste storage vault must also be considered in this system since it is reasonable to assume that the iodine-129 discharge will increase with the inventory buildup.

The development work previously discussed may produce more effective iodine
cleaning systems for reprocessing plant off gases (24)(25)(26). However, even
with present technology it is difficult to speculate what the cleaning efficiencies
of actual installed systems will be. Once these systems are in operation their
performance can be monitored and documented. In the interim the use of an
overall decontamination factor of 100 to 1,000 appears acceptable.

Based on experimental evidence it is reasonable to assume that the wet scrubber-
AgZ system can be designed to achieve a minimum decontamination factor
of 1,000 under most conditions. However, as discussed previously, the performance of the off-gas iodine cleaning system may not be the controlling factor in
determining total iodine discharges from a reprocessing plant. In particular, the
system for handling or processing the scrubber solutions must minimally provide
the same cleaning efficiency as the off-gas stream since a large fraction of the
iodine is expected to be in the scrubber solutions. Therefore, it can be concluded that while the efficiency of currently planned iodine removal systems
will be a minimum of 99.9%, the total iodine waste handling system introduces
a large uncertainty.

Meager data are available regarding the costs for iodine removal systems. The
costs are dependent upon the volume of off gas which requires processing as
well as the total quantity of iodine to be removed.

Particulate Control

Radioactive particulates associated with fuel reprocessing will be found in both
the liquid and the gaseous effluent streams. The particulates arise due to the
various operations (shearing through waste solidification) and as a consequence
can include a variety of radioisotopes. They will also be in a variety of chemical
and physical forms. Easily condensable vapors should be considered as a source
of particulates bearing radioactivity. Most of the radioactive wastes available
for discharges are attached to particulates. Included in this category are the isotopes in the actinide series, many of which are highly radiotoxic since they decay
by alpha emission. Other radionuclides which potentially can contribute significantly to the waste gas stream particulate makeup include the volatile isotopes,
such as strontium-90 and cesium-137.

The particulate effluent from reprocessing plants will for the most part be soluble
and will quite probably be in the nitrate form. The particulate effluents from
the solidification process will probably be oxides and thus insoluble. Ruthenium
may be an exception, especially from the solidification process, and may be com-
plexed as an organic.

Probably the most acceptable theoretical estimates of discharged radionuclides were made by Oak Ridge scientists (19). These estimates appear conservative, however, as indicated by the gaseous discharges measured at the NFS facility (28). An annual discharge of 1.0 to 10 Ci of beta-gamma fission products, excluding tritium, noble gases and the iodines, currently is the most reasonable estimate. The routine annual gaseous discharge of actinides at the USAEC Rocky Flats Plutonium Recovery Facility (48) averaged 2.4 millicuries from 1953 to 1970. Fuel reprocessing plants may be expected to have higher discharges.

Control Techniques: Particulates are generally collected from gaseous waste streams through the use of inertial separators (cyclone, or gravity settling), filtration (fabric, glass fill, sandbeds, HEPA), precipitation (electric, thermal), sonic agglomeration, and liquid scrubbing. The specific system used is dependent upon the type of source and efficiency of control desired.

Filtration: Because of the concern regarding the inhalation of radioactive particulate material, high efficiency filtration systems for removal of particulates from air have received the primary emphasis in the nuclear industry and are widely used (43). Particulate filters and materials of a variety of types are available. The materials include cellulose-asbestos, glass, glass-asbestos, plastic fibers and ceramics. Filters are generally classified as panel (viscous impingement) filters which are coated with a tacky substance to increase particle adherence, and extended medium dry-type filters which are called "bag" or "sock" filters.

Panel filters are designated Group I and have a low efficiency for small particulates. Bag filters are designated as Group II and Group III (medium and high efficiency, respectively). HEPA (High Efficiency Particulate Air) filters are classified as a special group. Sand filters (44) in general are designed for a specific application, usually for use in cleaning corrosive materials from air streams. Their use in the nuclear field has been limited, but is expected to increase since they can be considered relatively fail-safe. Deep-bed sand filters have been employed at Hanford and Savannah River for reprocessing plant gaseous effluent filtration. The Midwest facility has also installed a deep-bed sand filter for final filtration.

The parameters affecting filter performance, in addition to face velocity characteristics are: resistance, geometrical size, pressure drop, penetration, collection efficiency, dust and holding capacity, particle sizes, loading, humidity, temperature, and chemical resistance.

Liquid Scrubbing: This method has been used in cleaning the off-gas stream from solidification processes (45)(46). Scrubbers are used as initial cleaning devices and the scrubber solutions can be recycled to the process streams. Experience is limited to the Idaho Chemical Processing Plant and the prototype work conducted under the Waste Solidification Engineering Prototype program conducted by Battelle at Hanford.

Control Systems Evaluation: The various methods of particulate control may be grouped in the order of increasing efficiency as follows: fair—Group I panel filters; good—Group II bag filters, deep-bed sand or fiber glass filters and scrubbers; better—Group III bag filters; best—HEPA filters.

The anticipated performance of fibrous filters can vary from 2 to 99.97% retention for 0.3 micrometer particulates and up to 100% for 10 micrometer size.

HEPA filters have consistently demonstrated penetrations less than 0.3×10^{-3} (99.97% efficiency). However, the overall system efficiency is highly dependent on filter integrity, proper installation, operating conditions (particularly air flow rates and loading), and aging characteristics. Adequate in-place testing is necessary since improper installation or filter damage in shipment or installation could result in leakages of up to 30%.

A typical filter installation consists of a prefilter (roughing filter) followed by two HEPA filters in series or by a HEPA filter and a sand filter. The prefilter generally has a rating of greater than 75% (Group III for 1.0 micrometer particles) and offers a considerable cost savings by reducing mass loadings on the more costly HEPA filters. The HEPA filters are rated at a minimum efficiency of 99.97% for 0.3 micrometer particulates. The reported efficiency of deep-bed sand filters for submicrometer particles is greater than 99%. Thus, the overall decontamination factor for the filters themselves should be on the order of 10^5.

However, the efficiency of the system is also dependent upon factors such as leak tightness and aging characteristics. The minimally acceptable inplace test results would probably be in the order of 99.97% for a typical HEPA system (it is emphasized this is a system test). The system would probably be assigned a credit of 99 to 99.9%, the actual credit depending on the testing frequency.

Capital and operating costs are highly dependent on volume flow rates. Based on Silverman's estimates (47) which are 1960 prices, a HEPA filter would have a capital cost between $200 to $800/1,000 cfm and annual operating costs between $100 to $450/1,000 cfm depending upon specifications, corrosion resistance, and other operating conditions. A prefilter for reducing the HEPA loading would cost between $100 to $300/1,000 cfm for capital cost and $15 to $25/1,000 cfm for annual operation.

The costs of a glass-fiber or sand deep-bed filters were reported by Silverman (47) to be between $2,860 to $5,500 capital costs per 100 cfm with associated operating cost between $400 and $800/100 cfm (based on only depreciation and air-moving costs). The reported capital cost for the sand-bed filter, discharge stack, and associated equipment is $400,000 (1970) for the Midwest facility (44).

MIDWEST FUEL RECOVERY PLANT

The General Electric Company's Midwest Fuel Recovery Plant, now called Morris Operation, is located near Morris, Illinois and is an example of a type of reprocessing plant that could be operating in the 1980s. It exemplifies the nuclear industry's continuing emphasis upon the control and confinement of radioactivity. Beginning at the time the initial design concepts for this facility were being developed in 1963, the confinement and ultimate disposal of all radioactive materials entering the plant have been major considerations. Key objectives established at that time include:

(1)   The immediate solidification of highly radioactive wastes and their confinement in a manner permitting continuous surveillance, inventorying and retrieval.

(2)   The avoidance of the discharge of any liquid waste to the environs.

(3)   An off-gas treatment system designed to be virtually incapable of

      failure and to assure that off-site dose rates were less than 1%
      of the federal regulatory limit as stated in 10CFR20.
(4)   The minimization of radiation exposure to plant personnel.

These and related design objectives had strong influence on both the design and
the cost of the Midwest Fuel Recovery Plant (MFRP).

A very simplified presentation of the major steps taking place in the plant, with
emphasis being given to the wastes produced is shown in Figure 7.1. Starting
on the left, there is the mechanical feed preparation step in which the fuel assemblies are dismantled, the relatively uncontaminated end fittings and spacers
being discharged to the cladding vault. The fuel tubes are sheared into short
lengths to facilitate chemical leaching, and the fuel sections pass from the enclosed shear directly to the vibratory conveyor which functions as the leacher.
Here the uranium oxide fuel and attendant plutonium and fission products are
dissolved from within the cladding, which is subsequently washed, monitored to
assure completeness of leaching and then transferred to the cladding vault. The
gases released during the shearing and leaching processes pass from the leacher
into the process vent system for subsequent treatment prior to release.

The solution containing the dissolved fuel is then subject to a cycle of solvent
extraction, typical of fuel reprocessing plants, which achieves a very high degree
of separation of the uranium, plutonium and neptunium products from the fission products. The waste stream from the solvent extraction process, containing
essentially all of the entering fission products, is concentrated and the resulting
concentrate solidified in a fluidized-bed calciner, the immediate solidification being a unique feature of the MFRP. The solidified wastes are subsequently sealed
in stainless steel containers and stored under water in the waste basin.

The semipurified product solution passes through two successive cycles of anion
exchange in which plutonium and neptunium are separated and purified. Three
or four times a year, the ion exchange resin used in these processes must be replaced. At these times, the resin is washed exhaustively after which it is dissolved
chemically and the resulting solution combined with the high level waste from
solvent extraction.

At this point the product stream, containing primarily uranium, is concentrated
and converted to solid uranium trioxide in a fluidized-bed calciner. The resulting uranium oxide is converted to uranium hexafluoride in a fluidized-bed reactor
employing an unreactive granular aluminum oxide as the bed material. Several
times a year this bed material is discarded to the dry chemical vault due to a
buildup of miscellaneous contaminants in the bed.

The uranium hexafluoride is frozen out of the gas stream circulating through the
fluorinator and is subsequently purified by means of sorption beds and by distillation, producing the final UF_6 product. The sorption beds are discarded in-
frequently to the dry chemical vault due to a buildup of contaminants or a de-
cline in sorption efficiency.

A portion of the UF_6-free, circulating fluorination gas stream, equivalent in volume
to the incoming elemental fluorine, is exhausted to maintain a constant system
pressure. This exhaust stream passes through a fluidized bed of activated alumina
which removes unreacted fluorine, hydrogen fluoride and other trace level fluorides
which may be present.

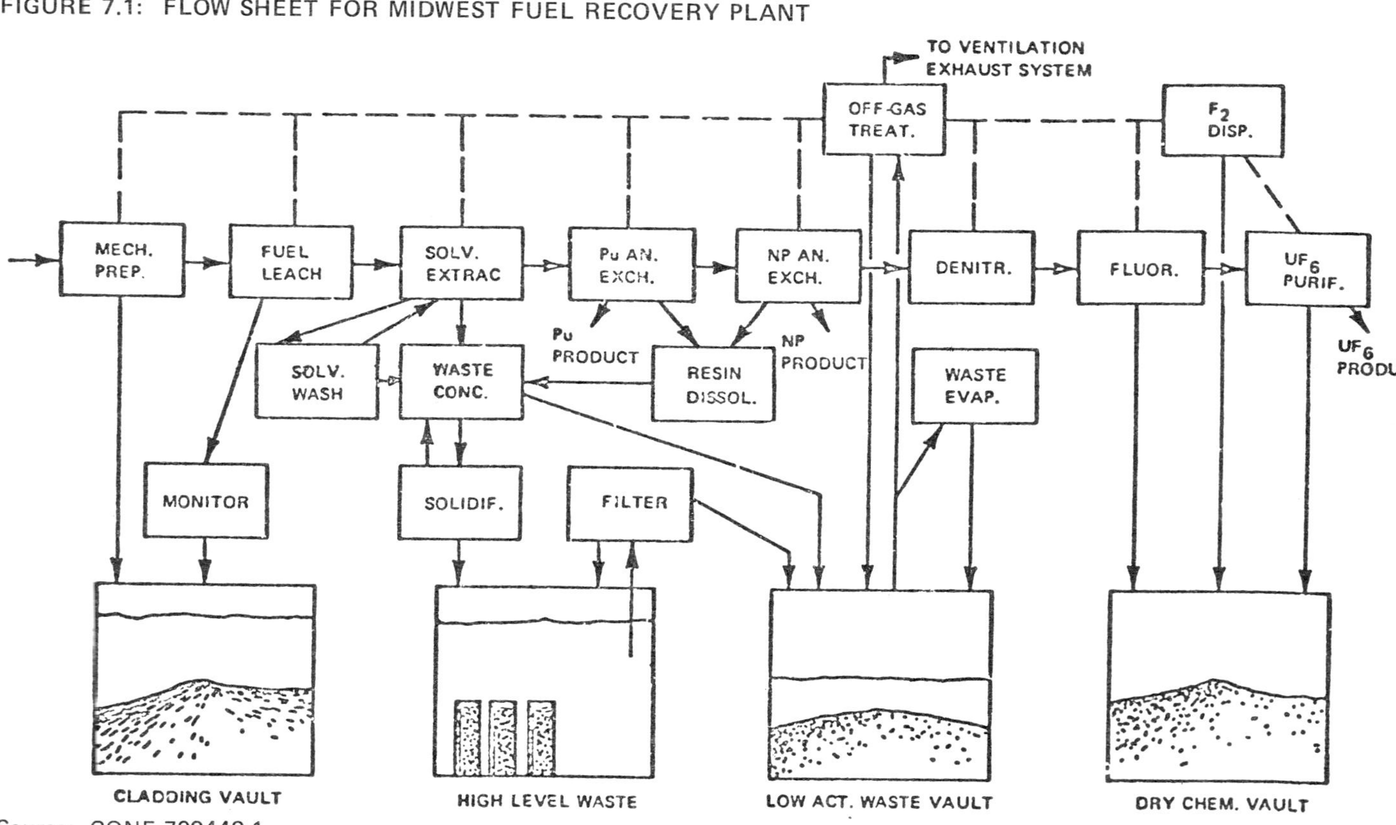

FIGURE 7.1:   FLOW SHEET FOR MIDWEST FUEL RECOVERY PLANT
TO VENTILATION EXHAUST SYSTEM
OFF-GAS TREAT.
F2 DISP.
MECH. PREP.
FUEL LEACH
SOLV. EXTRAC
Pu AN. EXCH.
NP AN. EXCH.
DENITR.
FLUOR.
UF6 PURIF.
UF6 PRODUCT
Pu PRODUCT
NP PRODUCT
SOLV. WASH
WASTE CONC.
RESIN DISSOL.
WASTE EVAP.
MONITOR
SOLIDIF.
FILTER
CLADDING VAULT
HIGH LEVEL WASTE
LOW ACT. WASTE VAULT
DRY CHEM. VAULT
Source: CONF-700440-1

The cleaned gas stream is subsequently washed with an alkaline solution and routed to the ventilation exhaust filter. The aluminum fluoride produced in the fluorine disposal reactor is periodically discharged to the dry chemical vault.

The wastes resulting from each of these processing steps are disposed of in the following manner. The cladding material, consisting of the end fittings, spacers and sheared, leached tubing from the fuel, amounts to approximately 15 cubic feet per metric ton of uranium in the fuel processed. The cladding contains substantial quantities of fairly short-lived radioactivity induced by neutron irradiation during its time in the reactor. In addition, although it has been leached essentially free of fuel material and subsequently washed, it still is contaminated with relatively small amounts of fission product radioactivity.

One approach to the disposal of such cladding scrap is to load it into containers and bury it at an AEC-licensed waste burial site. Because of the induced radioactivity in the cladding, such an operation requires a heavily shielded container for transport, virtually dictating that a reprocessing plant be located at the burial site. An added complication of this approach is the potential for radiation exposure of plant personnel during the transfer from the vehicle to the waste burial trench.

At the MFRP, General Electric decided to store the cladding waste under water in an underground vault adjoining the process building. This approach permits completely remote operation thereby avoiding potential personnel exposure and also avoiding dependence upon the acceptability of the reprocessing site as a burial facility. In addition, vault disposal affords more effective containment of the radioactive contamination than would ground burial. The vault in which the contaminated cladding scrap is stored is a large cylindrical stainless steel-lined concrete tank. The zone between the stainless steel liner and the concrete is equipped to permit detection, sampling and removal of any water that leaks into this region. The vault, as designed, will permit retrieval of the scrap for off-site disposal at some future time, if required. An alternative final disposition of these wastes, after filling of the vault, would be to fill it with concrete grout, leaving the residual contamination locked in concrete inside the stainless steel lined vault.

The conventional methods for handling concentrated, highly radioactive wastes from nuclear fuel reprocessing consist of storing the waste solutions in large metal-lined concrete tanks, set in a metal-lined concrete containment basin, all located underground. The tanks are equipped to permit reliable dissipation of heat arising from the radioactive decay of fission products. For acidic solutions, corrosion-resistant stainless steel linings are used, while for those solutions which have been neutralized and are alkaline, mild steel linings are suitable.

For either acidic or alkaline storage, spare tanks must be available to receive the contents of any tank that may begin to leak into the secondary containment basin. Eventually it would be necessary to solidify the wastes for disposal at a Federal waste repository. For waste solutions that had been stored for a period of 5 years, the rate of energy release due to radioactive decay at the time of solidification would be less than one-tenth that at the time of fuel processing.

The choice of immediate solidification of the highly radioactive wastes from the MFRP was based on the anticipated eventual requirement for solidification and the desire to avoid the complications involved in solution storage. The solidified highly radioactive wastes, occupying about 2.5 cubic feet per tonne uranium and containing almost all the radioactivity from the reactor, are stored under water in specially-designed stainless steel containers of very high integrity. These containers are designed to dissipate the heat generated as a consequence of radioactive decay and to accommodate safely any pressure which might arise from radiolytic decomposition of abnormal materials involuntarily introduced into the container. Since these containers are stored in an accessible water-filled basin, they are readily subject to surveillance and retrieval. After storage under water for periods up to 10 years, the wastes may be shipped to a Federal waste repository in accordance with government policy.

The low activity waste vault receives contaminated chemical wastes arising from the treatment of fuel storage basin water, solvent and offgases and smaller amounts of miscellaneous low activity wastes. Extensive recycling of process reagents is employed to minimize the volume of low activity wastes generated. The wastes, amounting to about 15 cubic feet per tonne of uranium processed, are segregated from the highly radioactive wastes to avoid unnecessary dilution of the latter.

These low activity wastes are introduced to the vault as warm, concentrated solutions permitting the chemical salts to crystallize upon cooling and settle out as a dense salt cake. The residual solution above the salt cake is circulated back to the process for further concentration and for reuse, to deplete the unreacted caustic present. The vault, in which these wastes are contained, consists of a large high quality steel tank enclosed within a steel-lined concrete vault. The annulus between the tank and the vault is equipped for detection, sampling and removal of any liquid that leaks into this area. The design of this vault would permit retrieval of the salt cake for off-site disposal at some future time, if required. An alternative final disposition of these wastes, upon deactivation of the vault, would be to complete the removal of free liquid and to seal the upper portion of the tank and the annulus with concrete, asphalt or other sealant such as montmorillonite clay, thus assuring that water could not penetrate to the residual radioactive contamination until extensive time for decay had transpired.

The dry chemical vault receives various chemical wastes from the fluorination and the uranium hexafluoride purification systems. These slightly contaminated wastes amount to about 10 cubic feet per tonne of uranium processed, mostly aluminum fluoride from the fluorine disposal reaction. The volume of these wastes would have been reduced significantly by releasing unreacted fluorine and fluorides up to the limit legally attainable taking maximum advantage of the 300 foot tall stack. This, however, was deemed undesirable from an environmental standpoint, so provisions were made for efficient fluorine removal.

The vault in which these wastes are confined consists of a steel-lined concrete vault with the zone between the concrete and the liner equipped to detect and sample any ground water which might intrude into this region. As is the case with the other vaults, the dry chemical vault is designed to permit retrieval of the wastes, if required. In view of the very low levels of radioactivity involved, an alternative would be to seal the wastes from early contact with ground water and leave them in the vault.

Small quantities of other solid wastes are generated as an unavoidable consequence of radiochemical processing. Some of these, such as sample bottles or failed equipment, can be decontaminated to the extent practical and confined in the cladding vault. Other wastes, such as absorbent materials used in some decontamination operations, are not suitable for inclusion in any of the waste vaults. These wastes, being limited in volume and very low in contamination level, are compacted and sealed in containers and monitored for transfer to a licensed disposal site.

A common problem facing nuclear fuel reprocessing plants is the disposal of water used in the process routinely having very low levels of contamination, but with a potential for higher levels of radioactivity due to very infrequent instances of equipment failure. The conventional approach to the disposal of process water is to release it into a fairly large stream of surface water under carefully controlled conditions to assure that the radioactivity level remains well below the allowable limits.

The continuing need for rigid controls and the concern over possible reconcentration mechanisms reduced the attractiveness of this relatively simple disposal method. An alternative means of disposing of process water is to discharge it underground into a permeable soil stratum. In this approach, the ion exchange capability of the soil provides a mechanism for trapping certain radionuclides. However, unless the discharged water is of very low activity or the discharge enters a stagnant aquifer or one whose flow characteristics are very well done, a very extensive system of wells may be required to monitor the path of the process water in the water table. The problem of liquid disposal is avoided in the MFRP by a combination of techniques including:

(1) The use of a closed-cycle process steam system which recycles rather than discards the steam condensate. Because of the possibility that radioactivity may enter this system, it is located in a shielded area. In addition, several extra precautions were taken in the design to minimize the likelihood and potential consequences of radioactive contamination should a heat exchanger fail due to corrosion.

(2) The process cooling water is also recycled, being cooled by means of air-cooled heat exchangers. By locating the coolers high on the process building, it is possible to maintain the coolant hydrostatically pressurized with respect to the process equipment, thereby providing a higher degree of assurance against the possibility of leakage of radioactivity into the cooling water than if it were discharged at ground level.

(3) The amount of water used in the process is minimized by avoiding unnecessary additions of water to the system and by extensive water recycle within the process. The water passing through the process is decontaminated from virtually all radioactivity except tritium. Since tritium is an isotope of hydrogen and is present as tritiated water in concentrations less than one part in hundreds of millions, effective separation of the tritium from the process water is impractical. Tritium is a beta-emitter which has a decay half-life of 12.26 years, and hence is not reduced appreciably by decay prior to processing. The biological effects of tritium are relatively low since it distributes almost uniformly through the body, has an effective retention half-life of only 12 days and a comparatively soft beta (less than 1% of the average energy for all fission product betas).

Since the amount of process water to be disposed of is relatively small, it is practical to release it in the ventilation air thereby taking advantage of the large dilution that is obtained through use of a 300 foot tall stack. The small volume also makes it possible to hold up the excess process water in the cladding and low activity waste vaults for periods of several months, thereby permitting substantial flexibility in avoiding periods of unfavorable meteorological conditions. Thus, it is possible to dispose of the water at the most opportune times and with very favorable and predictable dilution assured.

Gaseous effluents are a problem which is a major consideration in the design and operation of all fuel reprocessing plants. A certain amount of air, used for instrumentation, agitation or ventilation control, comes in contact with the highly radioactive process materials. These process gases are subject to sequential scrubbing, sorption and filtration to assure the removal of noxious chemicals, radioactive iodine and entrained particulate activity. Following this treatment, the gas stream is monitored for radioactivity and released with the ventilation air stream. In the case of the MFRP, the process off-gases join the ventilation air at the inlet to the ventilation air filter and are subject to additional filtration.

The ventilation filter may be a highly efficient packaged filter employing very fine glass fibers. It may be a bed of more durable coarser glass fibers compressed to achieve filtration efficiency, or a bed of graded gravel and sand may be employed to assure highly reliable and efficient filtration. The latter type of filter was selected for the MFRP, despite its higher costs, because of its assured reliability.

Gaseous effluents from reprocessing plants normally contain insignificant radioactivity other than Kr-85. This isotope has a 10.76 year half-life, and hence does not decay significantly prior to reprocessing. Since krypton is a noble gas, which means it is chemically unreactive, it is not readily removed in conventional off-gas treatment. This lack of chemical reactivity also makes the Kr-85 relatively nontoxic, since it is not retained by the body. In addition, Kr-85 is primarily a beta emitter, with less than 1% gamma radiation, thereby significantly reducing the distance over which it can cause radiation exposure to a person in contact with air containing Kr-85. The radiation from Kr-85 results principally in irradiation of the clothing and exposed skin. Normal practice in a reprocessing plant is to discharge the gaseous effluents from a stack tall enough to assure that the average Kr-85 concentration is well below the allowable limit at any point at ground level.

The variation of meteorological conditions at the MFRP site, coupled with the use of a 300 foot tall stack, assures that the maximum radiation exposure from Kr-85 that a person could receive by standing at the site boundary all year would be less than 2% of 10CFR20 limit. These same average conditions would result in tritium concentrations that would limit the maximum radiation exposure from this source to 0.1% of the 10CFR20 limit.

NUCLEAR FUEL RECOVERY AND RECYCLING CENTER

Location and Scope of Proposed Activities

The Exxon Corporation has applied for a construction permit to build a complete nuclear full recovery and recycling center (NFRRC). The Exxon Nuclear site is located in Roane County, Tennessee, about 8 miles southwest of the center of the city of Oak Ridge. The site is within the perimeter of the Oak Ridge Reservation owned by the U.S. Government and administered by the Energy Research and Development Administration.

The plant site is in an area of low population density. There is no residence within 2 miles of the site, and no major concentrations of population within 5 miles. The 1970 population within a 50-mile radius was less than 700,000. Only two cities in this area had 1970 populations exceeding 25,000; the center of the city of Oak Ridge (28,319) lies about 8 miles to the northeast, and the city of Knoxville (174,587) lies about 25 miles to the east of the site.

The NFRRC is designed to receive, store, and process, irradiated fuels from light water-moderated nuclear power reactors. The facility will separate and purify uranium and plutonium from irradiated nuclear fuel. These products are used as nuclear fuel to meet the nation's identified need for energy. The plant will process fuel at rates up to 2,100 tonnes (metric) of heavy metal (uranium plus plutonium) per year. Prior to the start of fuel reprocessing, the storage facility may accumulate as much as 7,000 tonnes of irradiated fuel. This storage may be required to avoid reactor shutdowns while adequate fuel reprocessing capacity is being established.

The nuclear fuels to be processed consist of uranium oxide or mixed uranium and plutonium oxides, enclosed in capped zirconium or stainless steel tubes which are incorporated into fuel assemblies. The plant is designed to routinely accept fuel containing fissile isotopes up to 3.5% of the uranium and plutonium present; assemblies containing up to 5% fissile material may be accepted and processed at reduced rates or under special processing conditions. The plant is designed to accept fuel for storage 90 days or more after reactor discharge and for processing after 150 days "cooling." It is planned to accept fuel after irradiation to an average exposure of up to 40,000 megawatt days per tonne (MWd/T) at specific power levels up to 40 megawatts per tonne (MW/T); single fuel assemblies irradiated to higher exposures (up to 48,000 MWd/T) and/or at higher rates (up to 48 MW/T) may be accepted under special processing restrictions.

Uranium hexafluoride and plutonium dioxide, packaged in licensed shipping containers, are the principal plant products; other product forms, such as uranyl nitrate and uranium trioxide, will be available on customer request. Solid wastes that are contaminated with radioactive materials are packaged and stored within the plant pending shipment to approved repositories. Radioactive and noxious chemical components of gaseous wastes are removed to an extent sufficient to cause negligible impact on the environment. Liquid effluents are limited to those that have not contacted radioactive materials of process origin.

General Process Description

Fuel Receipt and Storage: The technologies for handling and storing irradiated fuels in water basins are well established based on some hundred of plant years of reactor operation. The water transfers the heat of radioactive decay from the fuel to the cooling tower and protects operating personnel from the gamma and neutron radiation emitted by the fuel. The water is kept clean by use of commercially available ion exchange and filtration systems.

The design of the facility for removing the irradiated fuel from transport casks permits cask unloading without immersion in the storage pool. In the NFRRC, the cask is opened in air in a heavily shielded cell; contamination of the cask exterior is limited to the lid, and that receives only minimal quantities of airborne contaminants. Surface contamination is removed from the fuel by flushing and cleaning before the fuel is placed in the fuel storage pool. Contaminated water from the fuel rinse, fuel cask, and building flushes are routed to the radwaste evaporator.

End Fitting Removal: Before the start of chemical processing, a saw is used to separate the fuel rods from the end pieces and miscellaneous hardware that hold the assembly together. The separated hardware is packaged for storage.

Fuel Shearing: The fuel bundles are sheared into short pieces to expose the fuel materials for processing. This technique has been used by Nuclear Fuel Services (NFS) at their West Valley Plant and by Europeans at the Eurex and Windscale Plants. Shearing is conducted in an enclosure to confine dust and blanketed with nitrogen to avoid the possibility of zirconium combustion. The sheared segments fall into a cooled surge drum and are fed at a controlled rate to the fuel dissolver. A fraction of the iodine, krypton and xenon is released.

Dissolution: The fuel rod pieces fall into a basket which is immersed in hot nitric acid in an annular dissolver. Most of the uranium, plutonium, and the fission products dissolve by reaction with nitric acid. The dissolver is blanketed with oxygen and the off-gases are scrubbed with water to convert the nitrogen oxides formed back into nitric acid. This technique is known as "fumeless dissolving," because the volume of off-gases is reduced to a practical minimum.

The remaining krypton and xenon are evolved as gases during dissolution. Krypton is isolated and stored. Xenon is released to the atmosphere. While some tritium is released to the atmosphere as the elemental gas, most of the tritium is oxidized and becomes part of the process water. Efforts are in progress to develop a practicable means of isolating tritium for packaging. Radioiodine volatilizes partially during the dissolution operation and more completely in a subsequent sparging operation; the iodine is sorbed, packaged, and stored.

After dissolution of the fuel materials is complete, the solution is transferred to the accountability station. The cladding fragments are then rinsed with nitric acid to remove remaining fissile materials; the rinse solution is combined with dissolver solution. The basket of cladding fragments is then removed from the dissolver, measured to determine its content of fissile materials, and dumped into a cladding storage container.

Accounting and Feed Preparation: Solid materials (small cladding fragments, undissolved fission products and a small fraction of the PuO_2) are centrifuged from the dissolver solution and leached with a mixture of ceric nitrate and nitric acid to dissolve any plutonium present. Ruthenium, volatilized by this treatment, is sorbed on ferric oxide granules. The leach solution is centrifuged to remove solid materials and added to the dissolver solution. Separated solids are routed to high-level waste. The quantity of fuel in the solution is determined. The solution is also analyzed to obtain a high-precision measurement of the isotopic content of uranium and plutonium. The acid and uranium concentrations and the plutonium valence state are adjusted, as necessary, in preparation for solvent-extraction processing. Techniques developed in ERDA facilities and applied in commercial plants are used to sample and analyze the fuel materials.

Solvent Extraction: An adaptation of the well-established Purex process is used to isolate uranium and plutonium from each other and from other radionuclides. This process is based on the selective extraction of uranium and/or plutonium from a 2 to 3 molar acid solution into an organic solvent which consists of 30 volume percent tributyl phosphate (TBP) in dodecane. Other radionuclides remain in the aqueous phase and are routed to the waste concentration system. The uranium-plutonium laden solvent is then scrubbed with a nitric acid solution to remove contaminants and then contacted by a dilute (less than 0.1 molar) solution of nitric acid to strip the products back into the aqueous phase. The transfer of product from an aqueous to an organic to an aqueous phase constitutes a solvent extraction cycle.

Pulse columns are used to effect countercurrent flow of the aqueous and organic phases while keeping the phases well mixed to promote the desired transfer of solute from one phase to the other. This mode of operation has been used successfully in ERDA fuel reprocessing plants for more than 20 years and, more recently, in commercial fuel reprocessing plants.

In the flowsheet outlined in Figure 7.2, uranium and plutonium are separated from fission products in the Codecontamination Cycle consisting of the HA (extraction), HS (scrub), and HC (strip) Columns. After concentration, the uranium and plutonium are separated and the uranium purified in the Partition Cycle which consists of the 1A (uranium extraction) and 1C (strip) columns. Plutonium is prevented from extracting by reducing it to the trivalent state. Uranium is further purified in a Second Uranium Cycle (not shown in Figure 7.2). Plutonium is further purified in the Second and Third Plutonium Cycles (2A, 2B, 3A, 3B Columns) in which plutonium is twice extracted into solvent (A Columns) and stripped (B columns) into an aqueous phase. The 2B Column also serves to separate residual uranium from the plutonium by again reducing the plutonium so that it is stripped under conditions that keep uranium in the solvent.

The solvent extraction flowsheet contains features designed to minimize the loss of plutonium and the volume of waste streams to be processed.

    (1)   The distillate from Uranium Concentrators is used to strip uranium in the strip columns. This technique was demonstrated in the Hanford Purex Plant.

    (2)   The aqueous effluents from the 2A and 3A Columns (plutonium extraction) are combined and routed through the 2S Column to minimize the plutonium content of the aqueous phase routed to the Acid Concentrator.

    (3)   The organic effluent from the 1C Column (uranium stripping) is reused as the extractant in the HA Column. The organic effluents from the 2B, 3B and 2S Columns are combined and routed to the HC Column where uranium and plutonium are stripped from the solvent.

    (4)   Hydroxylamine is used in the HC Column to reduce plutonium to the inextractable trivalent state. This step counters the effect of dibutyl phosphate, a solvent degradation product that

Radioactive Waste Disposal

FIGURE 7.2: URANIUM-PLUTONIUM ISOLATION AND PURIFICATION BY SOLVENT EXTRACTION

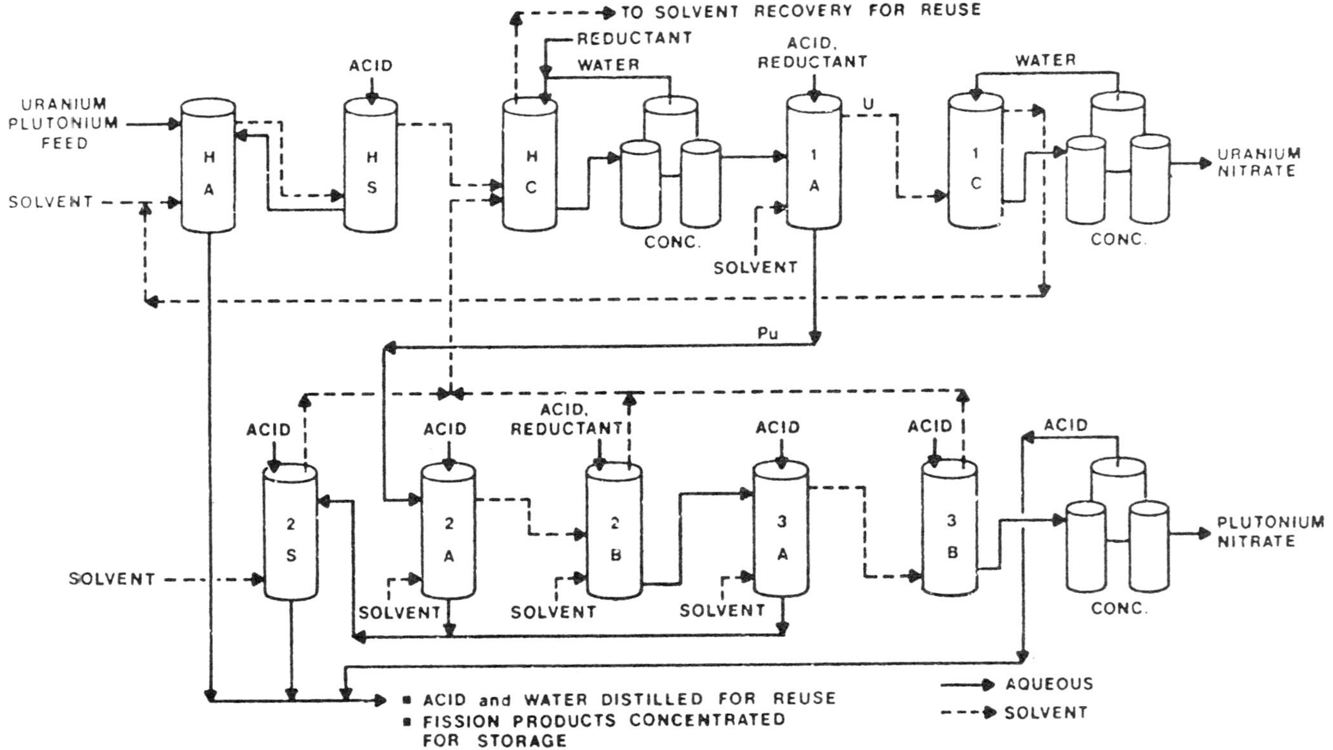

Source: Docket 50-564-1

can form a solvent-soluble complex with oxidized plutonium
thereby causing a high plutonium loss via the solvent effluent.

Plutonium Nitrate Conversion: Plutonium nitrate is converted to plutonium
dioxide by thermal denitration in a screw calciner. This technique has been
demonstrated in a pilot plant at Hanford.

Uranyl Nitrate Conversion: Uranyl nitrate is thermally decomposed in a fluidized
bed to form uranium trioxide. This process was developed at Argonne National
Laboratory and demonstrated on a large scale by the Mallinckrodt Chemical
Company at Weldon Spring, Missouri and by the Allied Chemical Corporation at
Idaho Falls, Idaho.

Uranium Trioxide Conversion: Uranium trioxide is converted to uranium hexafluoride (UF_6) by direct fluorination in a fluidized bed. The direct fluorination
of uranium oxide has also been demonstrated in a flame tower in ERDA's gaseous diffusion plant at Portsmouth, Ohio.

Uranium Hexafluoride Purification: Uranium hexafluoride is purified by a combination of distillation and sorption steps designed to remove specific impurities.
Pertinent features of this flowsheet are identified below.

(1) Plutonium, niobium, and ruthenium are sorbed on a bed of
     sodium fluoride as demonstrated at Oak Ridge and Paducah.
(2) Cold traps are used to isolate the UF_6 from fluorine and hydro-
     gen fluoride, as practiced at ERDA facilities.
(3) Neptunium is sorbed on a bed of cobaltous fluoride, as developed
     at Paducah.
(4) Technetium is sorbed on a bed of magnesium fluoride, as developed
     at Paducah and Oak Ridge.
(5) A pressurized Vaporizer and Distillation Tower are used to separate
     UF_6 from technetium and any remaining "high boilers." The
     purified UF_6 vapor is then condensed by cooling and transferred
     into licensed transport cylinders. The UF_6 solidifies upon cool-
     ing to ambient temperature.

Waste Treatment: The NFRRC is designed to release no waste liquids of process
origin and to hold the release of radioactive contaminants to the atmosphere as
low as reasonably achievable. All liquid process wastes are either purified and
reused in the process, solidified and stored as a solid waste, or vaporized and
released to the atmosphere. Solid wastes are routed to either a commercial disposal site or stored on site pending delivery to a Federal Repository, depending
on the concentrations of transuranics in the wastes.

High-Level Wastes: About 99.9% of the fission products and actinides (except
neptunium and product) are present in the aqueous effluent from the HA Column.
This stream, combined with distillates from the high-level waste calcination operation and concentrate from the acid recovery system, is concentrated about ten-
fold in the High-Level Waste Concentrator; the distillate, which contains nitric
acid and water, is routed to the acid recovery system. The concentrate is then
reacted with sugar to reduce the acid concentration, producing water and oxides
of carbon and nitrogen as reaction products. These operations have been demonstrated in ERDA and commercial fuel reprocessing plants.

The concentrate is then calcined in a fluidized bed; if desired, the resulting solids can be melted and combined with glass. Upon cooling, a tough low-solubility borate glass block is contained in a stainless steel cylinder that is welded closed and stored in a water pool. These operations have been studied by the Battelle Northwest Laboratory.

Acid Recovery: Aqueous nitric acid solutions from several process steps (HLW distillation, plutonium solvent extraction effluents, uranium and plutonium calcination) are evaporated in the acid concentrator. The concentrate, containing nonvolatile contaminants, is routed to high-level waste. The distillate is fractionated to produce moderately concentrated (about 10 molar) and very dilute (about 0.01 molar) solutions of nitric acid. These solutions are reused in the dissolution and solvent extraction operations. These operations are in common use in ERDA and commercial facilities. Excess water, containing nearly all of the tritium, is vaporized and discharged to the atmosphere.

Solvent Treatment: Purex solvent is purified and reused repeatedly in the process. Degradation products, such as dibutyl phosphate and monobutyl phosphate, are removed by washing with sodium carbonate solution, which also removes most of the residual uranium, plutonium and fission product. The aqueous wash solution is routed to the Intermediate-Level Waste Concentrator. This operation is in common use in ERDA and commercial facilities. The washed solvent is passed through a bed of anion exchange resin, which sorbs the remaining traces of contaminants and is reused. This resin "polishing" technique has been piloted at Hanford. It is anticipated that solvent will be reused without need for replacement. If solvent should become unusable, it would be burned in the High-Level Waste Calciner.

Miscellaneous Aqueous Wastes: In the Fuel Storage Pool, water may become contaminated with trace amounts of radioactive and chemical impurities. This water is cleansed by passing through filters and ion exchange beds to remove the impurities. Aqueous wastes accumulated from the fuel-receipt and cask-handling operations are concentrated and the distillate reused in those operations. The concentrate is incorporated in a solid and routed to a commercial disposal site.

Miscellaneous aqueous wastes from the Process Building arise from such operations as equipment decontamination, laboratory analyses, building washdowns, etc. These wastes are combined, made slightly acidic, and concentrated. The distillate is combined with process wastes for purification and reuse. The concentrate is combined with high-level waste concentrates and solidified.

Nitrogen Oxides (NO_x): Nitrogen oxides are formed during fuel dissolution, plutonium and uranium calcination, and the high-level waste denitration and calcination steps. Trace quantities are also formed in various other segments of the process. Water scrubbers are provided to convert the nitrogen oxides to nitric acid for reuse.

Iodine: Most of the fission-product iodine is volatilized during the fuel shearing, dissolution and feed preparation steps. After NO_x removal, the iodine is sorbed on a silver compound. The small fraction of iodine that enters the solvent extraction system is volatilized from the various aqueous waste treatment

systems or sorbed from the recirculating solvent by anion exchange resin. The
volatilized iodine is routed via the Vessel Off-Gas Vent System and sorbed on
a silver bed. The iodine-laden silver compound is packaged and stored pending
transfer to a Federal Repository.

Krypton: Krypton-85 is the only long-lived radioactive rare gas sufficiently radioactive to be of concern in a fuel reprocessing plant. Essentially all of the krypton
is released from the fuel during the shearing and dissolution steps. After NO_x
and iodine removal, the off-gases are routed to the Rare Gas Recovery Facility.
Xenon and krypton are condensed with liquid nitrogen, and xenon is separated
from krypton by cryogenic distillation. Krypton is packaged in commercially
available high-pressure (about 650 psig) cylinders and stored in the Fuel Storage
Pool.

Ruthenium: Gaseous ruthenium tetraoxide formed during the leaching of fuel
residues and the calcination of high-level wastes is immobilized by passing the off-
gases at about 250°C through annular beds of adsorbent material such as ferric
oxide granules. When a bed has reached saturation, the vessel is removed, capped,
and stored in the Fuel Storage Pool to allow the ruthenium to decay to innocuous
levels.

Fluorination Wastes: The solid sorbents used to confine radioactive contaminants
in the fluorination system are packaged for interim storage on site and later transferred to a Federal Repository. These solids include sodium fluoride (plutonium),
cobaltous fluoride (neptunium), and magnesium fluoride (technetium).

Fluorine and hydrogen fluoride are sorbed from the process off-gases and held
in beds of soda lime and activated alumina. A backup liquid scrubber (potassium
carbonate) is provided to trap fluorine in the unlikely event of malfunction in
the soda lime and alumina beds. Alumina and soda lime that are no longer effective in trapping fluorine are packaged and stored for delivery to a Federal
Repository.

Fuel Residues: Fuel hardware is packaged in metal containers for on-site storage and transfer to a Federal Repository. Leached cladding fragments from the
fuel dissolution step are monitored to determine the quantity of fuel remaining
with the cladding fragments. If significant quantities of fissile materials are
present, the fragments are recycled to the dissolver for dissolution. Otherwise,
the fragments are placed in metal containers for storage in the Fuel Storage Pool.

Failed Equipment: Failed equipment is decontaminated in place. Techniques
to be used were developed at ERDA sites and include flushing with solutions of
such chemicals as nitric acid, sodium hydroxide, oxalic acid, sodium tartrate,
and potassium permanganate (alkaline). After transfer to the Decontamination
and Hot Shop within the Process Building, the equipment is further decontaminated and repaired if practical. Unrepairable equipment is packaged, stored on
site and later transferred to a Federal Repository.

Miscellaneous Solid Wastes: Ion exchange resins are used to purify water in the
Fuel Storage Pools and to purify the process solvent for reuse. These resins are
discarded when they become unusable. The resins used to clarify pool water are
packaged and sent to commercial disposal sites as are similar wastes from commercial power reactors. Since resin wastes from solvent polishing steps contain

measurable concentrations of transuranics, they are packaged and stored pending delivery to a Federal Repository. Combustible solid waste materials are
baled or placed in drums for interim storage on site.

ERDA COMMERCIAL WASTE MANAGEMENT PROGRAM

Waste management R&D activities for both ERDA (weapons) and commercial
wastes have been centralized in the Division of Nuclear Fuel Cycle and Production
of ERDA. The breakdown of the Commercial Waste Management Program is
summarized as follows (49):

| | ------ $ Millions ------ | | |
	FY 1976	FY 1977	Growth, %
Waste processing development	5.4	19.9	370
Terminal storage	4.6	33.7	735
Supporting studies	2.0	6.4	320
	12.0	60.0	

The Commercial Waste Processing Development Program is further broken down
into work in each of the major technologies as follows:

| | ----- $1000s ----- | |
	1976	1977
High-level waste	$4,190	$10,840
TRU and low-level	540	8,030
Gaseous and airborne	0	1,000
	$4,730	$19,870

The program milestones for achievement of ERDA technology demonstration,
and an estimate of the earliest operation date for a commercial supplier are
summarized by Perge and others (50) and Table 7.1. Detailed recent technical
descriptions of program alternatives, accomplishments, and highlights of process
details are summarized in the references listed at the bottom of the table.

There is an important distinction to be made between the programs in Waste
Processing Technology and the programs in Long-Term Management. The Waste
Processing Technology programs are designed to develop multiple viable options,
from which commercial suppliers of reprocessing or waste technology will select
a particular alternate for application. Thus, ERDA is not developing detailed
designs for these components, but is working cooperatively with potential users
such as Nuclear Fuel Services, Allied-General Nuclear Services and Exxon.

In contrast the long-term waste management facilities will be under the operational control of ERDA, and the detailed design and demonstration activity is
included in the program.

ESTIMATED COSTS FOR WASTE MANAGEMENT FOR FUEL REPROCESSING PLANT

At the AIF Fuel Cycle Conference—76, at Phoenix, Arizona, March 23, 1976,

TABLE 7.1: SUMMARY – WASTE MANAGEMENT MILESTONES

	. . . . . . High-Level Waste		. . . . Plutonium and TRU Waste			Non-TRU and Low Level .		. . . Gaseous and Airborne . . .	
		ERDA Development Complete	Earliest Commercial Application		ERDA Demonstration Technology	Earliest Completion Date		Completion Date		Completion Date
Waste Processing	**Solidification Encapsulation**			**Waste Treatment**			**Waste Treatment**		**Waste Treatment**	
	Calcination	1979	1983	Incineration	1978	1981	Incineration	1979	High temperature and res filter	1980
	Calcination/ Vitrification	1979	1983	Vol reduct on of contaminated metals	1980	1983	Liquid waste treatment	1979	Iodine collection and retention	1981
	Direct vitrification	1980	1984	Solidification of liquid waste	1979	1983	Solidification and immobilization	1979	Kr fixation	1981
	Metal matrix	1981	1985		1979	1983			Tritium separation	1980
Long-Term Management	**HLW Terminal Storage**			**TRU Terminal Storage**			**Disposal by Land Burial**		**Terminal Storage**	
	Package acceptability criteria	1978		Waste form and package acceptability criteria	1977		Generic monitoring	1976		
	Select site pilot plant #1	1978		Site selection	1977		Shallow land burial	1982		
	Initiate operation plant #1	1984		Start operation	1983					
Programmatic References	**Commercial**			**TRU and Low Level**			**ERDA Low Level**		**ERDA and Commercial**	
	Kuhlman (49) Finnigan (51) Platt			R. Wolfe (52) R. Wolfe (53)			GAO (54) Kuhlman (55)		Dempsey (56)	

Source: PB-260 559

J.W. Finnegan of Battelle Northwest Laboratories (BNWL-SA-5725) made the following statement regarding the cost of radioactive waste management for fuel reprocessing:

> . . . there have not been any detailed engineering designs and operational concepts developed to the point where firm cost estimates are available. However, many groups here in the United States and in Europe have estimated costs for waste management associated with the Purex-type fuel reprocessing plant. Similarly, estimates have been made of transportation costs and the cost at federal repositories. The highest total cost that I have seen would approximate 0.2 of a mill per kilowatt hour for the conditioning or fixation of the waste, transportation and ultimate storage. There is no doubt that this would amount to a large amount of money; however, in terms of total costs of electric power the 0.2 of a mill would be approximately 1% of 20 mills of kW hour of power. Even for the most elaborate and complex high level waste management schemes that we can conceive, it is hard to visualize this cost rising to much more than 1.5 to 2% of the cost of nuclear power.

REFERENCES

(1) Nuclear Fuel Services, Inc., *Safety Analysis Report,* AEC Docket No. 50-201, July 1962.

(2) W.J. Kelleher, "Environmental Surveillance Around a Nuclear Fuel Reprocessing Installation, 1965-1967," *Radio. Health Data Rep.,* 10, August 1969.

(3) J.E. Logsdon and J.W.N. Hickey, "Radioactive Waste Discharges to the Environment from a Nuclear Fuel Reprocessing Plant," *Radio. Health Data Rep.,* 12, 6, July 1971.

(4) B. Shleien, *An Estimate of Radiation Doses Received by Individuals Living in the Vicinity of a Nuclear Fuel Reprocessing Plant in 1968,* BRH/NERHL 70-1, May 1970.

(5) P. Magno, et al., *Liquid Waste Effluents from a Nuclear Fuel Reprocessing Plant,* BRH/NERHL 70-2, November 1970.

(6) J.A. Cochran, et al., *An Investigation of Airborne Radioactive Effluent from an Operating Nuclear Fuel Reprocessing Plant,* BRH/NERHL 70-3, July 1970.

(7) D.G. Smith, et al., *Calibration and Initial Field Testing of ^{85}Kr Detectors for Environmental Monitoring,* BRH/NERHL 70-4, November 1970.

(8) General Electric Company, *Midwest Fuel Recovery Plant Safety Analysis Report,* AEC Docket No. 50-332, September 1969.

(9) Allied Gulf Nuclear Services, *Barnwell Nuclear Fuel Plant Safety Analysis Report,* AEC Docket No. 50-332, September 1969.

(10) W.O. Haas, Jr., "Solvent Extraction: General Principles," Chapter 4, *Chemical Processing of Reactor Fuels,* Flagg, ed., Academic Press, New York, 1961.

(11) T.H. Siddall, III, "Solvent Extraction Processes Based on Tri-n-Butyl Phosphate," Chapter V, *Chemical Processing of Reactor Fuels,* Flagg, ed., Academic Press, New York, 1961.

(12) M. Benedict and T.H. Pigford, *Nuclear Chemical Engineering,* McGraw-Hill, New York 1957.

(13) S.M. Stoller and R.B. Richards, eds., *Reactor Handbook, Volume II, Fuel Reprocessing,* 2nd Edition, Interscience, New York, 1961.

(14) J.H. Goode, *Hot Cell Evaluation of the Release of Tritium and $^{85}Krypton$ During Processing of ThO_2-UO_2 Fuels,* ORNL-3956, June 1966.

(15) Oak Ridge National Laboratory, *Aqueous Processing of LMFBR Fuels, Technical Assessment and Experimental Program Definition,* ORNL-4436, June 1970.

(16) Oak Ridge National Laboratory, *National HTGR Fuel Recycle Development Program Plan,* ORNL-4702, August 1971.

(17) J.A. Cochran, W.R. Griffin, Jr., and E.J. Troianello, "Characterization of Tritium Stack Effluent from a Fuel Reprocessing Plant," *ANS Transactions,* 15, 1, Las Vegas, June 1972.

(18) R.E. Commander, et al., *Operation of the Waste Calcining Facility with Highly Radioactive Aqueous Waste,* ICPP, IDO-14662, June 1966.

(19) Oak Ridge National Laboratory, *Siting of Fuel Reprocessing Plants and Waste Management Facilities,* ORNL-4451, July 1970.

(20) *Alternative Processes for Managing Existing Commercial High-Level Radioactive Wastes,* NUREG-0043, USNRC, April 1976.

(21) *Final Safety Analysis Report, West Valley Fuel Reprocessing Plant,* Section V-4, Docket 50-201.

(22) Letter to NRC from Nuclear Fuel Services, Docket 50-201, February 24, 1976.

(23) P.J. Magno, et al., *Liquid Waste Effluents from a Nuclear Fuel Reprocessing Plant,* BRH/NERHL 70-2, November 1970.

(24) Oak Ridge National Laboratory, *Chemical Technology Annual Progress Report,* ORNL-4794.

(25) Argonne National Laboratory, *Reactor Development Program Progress Report,* ANL-7872, October 1971.

(26) M. Krumpelt, et al., "The Containment of Fission Product Iodine in the Reprocessing of LMFBR Fuels by Pyrochemical Reactions," *Nuclear Technology,* 15, September 1972.

(27) *Code of Federal Regulations,* Title 10, Part 50, Appendix F, "Policy Relating to the Siting of Fuel Reprocessing Plants and Related Waste Management Facilities," November 1970.

(28) J.L. Russell and F.L. Galpin, *A Review of Measured and Estimated Offsite Doses at Fuel Reprocessing Plants,* NEA/IAEA Symposium on the Management of Radioactive Wastes from Fuel Reprocessing, Paris, November 1972.

(29) W.P. Kirk, ^{85}*Krypton - A Review of the Literature and Analysis of Radiation Hazards,* EPA, ORP, Washington, January 1972.

(30) C.M. Slansky, "Separation Processes for Noble Gas Fission Products from the Off-Gas of Fuel Reprocessing Plants," *Atomic Energy Review,* 9, 2, IAEA, Vienna.

(31) H.R. Merriman, M.J. Stephenson and D.I. Dunthorn, "Recent Developments in Controlling the Release of Noble Gases by Absorption in Fluorocarbons," *ANS Transactions,* 15, 1, Las Vegas, 1972.

(32) G.W. Keilholtz, "Krypton-Xenon Removal Systems," *Nuclear Safety* 12, 6, November-December 1971.

(33) J.P. Nichols and F.T. Binford, *Status of Noble Gas Removal and Disposal,* ORNL-TM-3515, August 1971.

(34) C.K. Bendixsen and G.F. Offutt, *Rare Gas Recovery Facility at the Idaho Chemical Processing Plant,* Idaho Nuclear Corp., IN-1221, April 1969.

(35) C.L. Benixsen and F.O. German, *Operation of the ICPP Rare Gas Recovery Facility During Fiscal Year 1970,* Allied Chemical Corp., ICP-1001, October 1971.

(36) General Electric Company, *Midwest Fuel Recovery Plant, Applicant's Environmental Report,* Supplement 1, NED 14504-2, November 1971.

(37) M.J. Stephenson, J.R. Merriman and D.I. Dunthorn, *Experimental Investigation of the Removal of Krypton and Xenon from Contaminated Gas Streams by Selective Absorption in Fluorocarbon Solvents: Phase I Completion Report,* Union Carbide, K-1780, August 1970.

(38) Environmental Protection Agency, *The Separation and Control of Tritium State-of-the-Art Study,* Pacific Northwest Laboratories, BMI, April 1972.

(39) ORNL-4794 Chem. Tech. Div., *Annual Progress Report Period Ending March 31, 1972.*

(40) Personal Communication, O.O. Yarboro, Oak Ridge National Laboratory Chemical Technology Division.

(41) D.T. Pence, F.A. Duce and W.J. Maeck, "Application of Metal Zeolites to Nuclear Fuel Reprocessing Plant Off-Gas Treatment," *ANS Transactions,* 15, 1, Las Vegas, 1972.

(42) J.G. Wilhelm and H. Schuttelkopf, "An Inorganic Adsorber Material for Off-Gas Cleaning in Fuel Reprocessing Plants," *12th AEC Air Cleaning Conference, Proceedings,* August 1972.

(43) C.A. Burchsted, and A.B. Fuller, *Design Construction and Testing of High Efficiency Air Filtration Systems for Nuclear Application,* ORNL-NSIC-65, January 1970.

(44) R.A. Juvinall, R.W. Kessil and M.J. Steindler, *Sand-Bed Filtration of Aerosols: A Review of Published Information on Their Use in Industrial and Atomic Energy Facilities,* Argonne National Laboratory, ANL-7683, June 1970.

(45) L.T. Lakey and B.R. Wheeler, "Solidification of High-Level Radioactive Wastes at the Idaho Chemical Processing Plant," presented at NEA/IAEA Symposium on the Management of Radioactive Wastes from Fuel Reprocessing, November 1972.

(46) J.L. McElroy, A.G. Blaservits and K.J. Schneider, "Status of the Waste Solidification Demonstration Program," *Nuclear Technology,* 12, September 1971.

(47) L. Silverman, "Economic Aspects of Air and Gas Cleaning for Nuclear Energy Processes," *Proceeding of a Conference on the Disposal of Radioactive Wastes,* Vol. 1, IAEA, Vienna, 1960.

(48) U.S. Atomic Energy Commission, *Environmental Statement Plutonium Recovery Facility, Rocky Flats Plant, Colorado,* WASH-1507, January 1972.

(49) C.W. Kuhlman, "The ERDA Program for Commercial Radioactive Waste Management," AIF Fuel Cycle Conference—76, Phoenix, Arizona, March 21-24, 1976.

(50) A.F. Perge, V.G. Trice, Jr., and R.D. Walton, Jr., "Long-Term Waste Management Goals and Programs," IAEA/NEA International Symposium, March 22-26, 1976, IAEA-SM-207/96.

(51) J.W. Finnigan, "The U.S. Domestic High-Level Waste Management System," AIF Fuel Cycle Conference—76, Phoenix, Arizona, March 23, 1976, Battelle Northwest Laboratories, BNWL-SA-5725.

(52) R.A. Wolfe, "The Research and Development Program for Transuranic Contaminated Waste Within USERDA," AICHE, Boston, Massachusetts, September 8-10, 1975.

(53) R.A. Wolfe, "Problems and Prospects in the Management of Solid Radioactive Waste," *Journal of Environmental Sciences,* July/August 1975, pp. 9-15.

(54) "Improvement Needed in the Land Disposal of Radioactive Wastes—A Problem of Centuries," GAO Report to Congress, January 12, 1976, Report No. Red 76-54, B-164105.

(55) C.W. Kuhlman, re ERDA's use of land burial for disposal of solid low-level waste and GAO Report B-164015, testimony before the Subcommittee on Conservation, Energy and National Resources.

(56) J. Dempsey, "FY 1977—The ERDA Airborne Waste Development Program Plan," personal communication March 16, 1976.

RADIOACTIVE WASTE MANAGEMENT IN FOREIGN COUNTRIES

The sources of the material in this chapter are the following
reports: PB-260 559, BNWL-TR 198, AAEC/TM628 and
NP-19966. A complete bibliography appears on p 359.

INTERNATIONAL ACTIVITIES IN HIGH-LEVEL WASTE DISPOSAL

All nations in the world that have made a significant commitment to nuclear
electric power, with the exception of Canada, have assumed that reprocessing
of the spent fuel will take place. The motivation is to reclaim the unused uranium and the plutonium for future use as a reactor fuel. In the nations where
processes have been selected for treatment of spent fuel, the basic Purex process has been chosen.

Furthermore, those nations that have chosen a reprocessing scheme have also
elected to adopt a HLW management process that involves immobilizing the
radioactivity in a glass. Thus, with one exception, all nations actively working
on HLW management use a system that first converts the liquid to a solid by
forming a calcine, then immobilizing the radioactive particles in a vitrification
process. The one exception is the United Kingdom which is aggressively pur-
suing one concept that goes directly from a liquid to a glass product.

The processes and activities by country are summarized in Table 8.1. Although
the basic system is the same for all nations including the U.S., differences do
exist in the manner that each process is implemented. What follows is a brief
summary of each nation's activity. For greater detail, the listed references
should be consulted.

United Kingdom

After 20 years of reprocessing experience, the HLW is still being stored in high-
integrity tanks. For the longer term, it is recognized that liquid storage will
have to be replaced by a solidification process. The present efforts are devoted

TABLE 8.1: INTERNATIONAL HIGH-LEVEL WASTE PROGRAMS

	United States	U.K.+Scotland	France	Germany	Belgium	Other Nations
Major Facilities Planned	Hanford,SNL ORNL, INEL AGNS-1500 MT Exxon-1500 MT	Windscale I,II Downreay I,II United Repro	La Hague Marcoule La Hague 800MT	WAK VERA-1500 MT	Eurochemic Mol. Belg.	Italy-Eurex I, Itrek, 1970-76 Norway
Plant Years Repro Exp.	256	58	30	6	8	
High Level Liquid Waste Storage	Yes	Yes	Yes	Yes	Yes	Canada-Spent fuel storage only
Commercial	600,000 gal					
Military	70,000,000 gal					
Calcination	Yes	Combined Calcination & Vitrification	Yes	Yes	Yes	Yes USSR, Sweden, Japan
Vitrification	Yes		Yes	Yes	Yes	
Final Waste Form	Borosilicate Glass		Glass in metal matrix	Glass	Phosphate glass beads in metal matrix	
Surface Storage	No-Backup		Yes, Marcoule	No	Yes-50 year on site	
Terminal Storage	Bedded Salt + others		No-under study	Salt dome at Asse II	Under study	
HLW Commercial Demo Date	1984-5	Mid-'80s Solidification Demo.	1978 in Surface Storage Vault	1978 in Asse II		

Source: PB-260 559

to development of a solidification process referred to as HARVEST. It consists of feeding the HLLW slurry and a slurry of glass-forming materials (both metered) into a heated container. In the container the mixture is continuously dried, calcined, and incorporated into a glassy matrix. During the filling of the container there is simultaneously a pool of liquid from which nitrate solids are crystallized, a layer of nitrates being calcined, and a layer of calcine being converted into the glassy product. When filled to a predetermined level, the feed to the container is stopped and heating is continued for some time to ensure conversion of all the contents to the glassy state. The container is then cooled and removed from the furnace.

This concept is a continuation of solidification work started in the 1950s at Harwell and called FINGAL. The whole process was successfully demonstrated on a pilot plant scale which produced 50 kg batches of glass from radioactive waste in the late 1960s.

Present plans call for a demonstration solidification plant to be in operation in the mid-1980s. At the present time about 600 m^3 of HLLW are in storage. By the year 2000, it is projected that the waste inventory would be 1,800 m^3 of solidified waste and 4,500 m^3 of liquid waste. All of this would be contained in about 7,000 m^2 of land area. No plans for ultimate disposal have been formulated as of now. (1)(2)(3)

France

The commercial nuclear fuel processing activity in France is concentrated at La Hague where an 800 MT/year reprocessing plant is under construction. An HLW solidification demonstration plant was under construction at Marcoule and scheduled for radioactive operation in mid-1977. This system consists of feeding the HLLW to a rotary kiln calciner.

The calciner will handle 30 l/hr and consists of a drum 27 cm in diameter and 3.6 meters long. The drum is rotated at 30 rpm and inclined at 3°. An idling rod inside is used to break up the calcine as it forms into large fragments. As the calcine leaves the drum, glassmaking additives are added and the mixture enters a separate furnace. In the furnace the mixture is heated to about 1150°C with an induction heater to melt the material. Periodically, the molten glass is poured into metallic containers. After cooling and solidification occurs, a lid is welded on the container and it is transferred to vaults for interim storage. Typical container dimensions are 50 cm diameter and 1 meter height. The furnace capacity is 15 to 18 kg/hr. (4)(5)

Federal Republic of Germany

The coupled spray-calcination/vitrification process has been selected as the reference HLW solidification process to be used in the 1,500 MT/year commercial spent fuel processing plant targeted for 1985 operation. In this process being pursued at the Karlsruhe Center, which is designated as VERA: (1) the nitric acid and the nitrates associated with heavy metals are destroyed by reaction with formic acid; (2) the resulting solution is dewatered and calcined in a steam heated spray calciner from which the steam is condensed and recycled; (3) the calcine is converted to a borosilicate glass formulation in a suitable metallic melter; and (4) the molten radioactive glass material is cast into metallic containers which

are welded shut and placed in storage. The containers are mild steel bottle-
shaped vessels 21 cm in diameter and 80 cm long. The filling neck is about
8 cm long and 2.5 cm in diameter. The radioactive demonstration unit (VERA
II) will be operable about 1977-1978. It will be located at the WAK plant which
has an average reprocessing rate of 20 to 25 MT/year.

While final decisions have not yet been made, the major German effort has been
on permanent disposal in salt, and this is the reference disposal process for the
waste from the commercial plant of the mid-eighties. The German government
plan is to locate spent fuel processing plants and recycle fuel fabrication plants
on land which overlies a salt deposit with the intent that waste will be disposed
in this salt deposit. The basis for the German enthusiasm with regard to salt is
the actual non-HLW radioactive waste disposal accomplished on both an experimental and production basis at the old ASSE II salt mine near Wolfenbuttel,
Germany. HLW will be stored on an interim basis once the VERA II process
is in operation in about 1978.

The mine is considered to be experimental only, but it is operated as a production facility. Fifty people now operate the mine. Up to 80 will be required in
a few years. Operation is expected until at least the mid-to-late eighties. At
present the waste is trucked 7 km from the nearest rail head to the mine.

At the Julich Research Laboratories in Germany two other nitrification processes are under investigation. One (called KFA) uses a drum dryer to remove
the moisture followed by melting in a graphite crucible to form a silicate type
glass. The second process (PAMELA) takes liquid waste and converts it directly
in a melter to form a phosphate glass. Both processes have been operated on
a laboratory scale since late 1974. Although successful, these processes do not
seem to enjoy the same degree of development as the VERA process. (6)(7)

Belgium

The major Belgian processing and waste management effort takes place at Eurochemic, a joint undertaking of 13 nuclear energy agencies of the OECD member
countries, and The Center for Studies of Nuclear Energy, both located at Mol,
Belgium.

Eurochemic's reference system for immobilization of high-level liquid wastes is
one where the aqueous HLLW is converted to oxide granules by a fluidized bed
calciner (like that used at INEL, U.S.) or to phosphate granules by the LOTES
process (Low Temperature Solidification Process) developed by Eurochemic.
Eventually the oxide or phosphate granules will be made into phosphate glass
beads. One feature different in their approach is that the final solidification
step will uniformly disperse the glass beads throughout a metal matrix such as
lead. The metal matrix is being used to provide better heat transfer characteristics in order that the glass not be overheated.

In addition, the three alloy types being considered, lead, zinc, and aluminum,
all have very low leach rates. The final storage/disposal scheme has not been
defined, but it is generally assumed that the solidified waste can be safely stored
on site for 50 to 100 years. (8)

Other Nations

The Japanese have not made firm decisions about reprocessing and waste management as of now. It is expected that they will make an appropriate commitment as they now have a pilot scale reprocessing plant in operation and are doing work on solidification and storage of waste.

It is reported that the USSR stores HLW in tanks and vitrification processes are under development. Other nations are also doing laboratory scale work on various aspects of fuel reprocessing and waste management. However, their requirements for this technology are distant enough to permit them to capitalize on the major development efforts underway elsewhere.

No reprocessing is planned in Canada because their CANDU reactor system uses natural uranium. As a consequence the economic factors do not generate incentives to reprocess the fuel today. The method now being used to handle the spent fuel is to place it in large storage basins. Work is underway to evaluate other storage concepts but they feel current techniques are adequate and present an acceptably low risk value. (9)

FRENCH EXPERIENCE IN STORING NON-HIGH-LEVEL WASTES ABOVE GROUND OR IN CONCRETED DITCHES

In the first quarter of 1965 the French Atomic Energy Commission (CEA) decided to commission a commercial company, Infratome, a subsidiary of the Potasse & Engrais Chimiques Co. (PEC = Potash and Fertilizer Co.), to perform "a project to investigate processes of treatment, packing and storage of radioactive wastes from the inventory of existing stocks and from the annual outputs of such material by the CEA and all other public or private organizations [EdF, (French National Electric Company), laboratories, hospitals, etc.] ."

In May 1966 the CEA decided to follow up the conclusions of the completed study, which it deemed satisfactory, and requested PEC Engineering to plan and create an industrial company responsible for the specific application of the conclusions of this study.

As a result, the Infratome Co. was created at the start of 1967, and was invited by the CEA to install its first storage depot on land belonging to the CEA and situated in the immediate proximity of the Hague Center. This site was chosen because it seemed to the CEA that it would simplify the psychological and health problems arising from such an establishment; at the same time it took account of the vigorous opposition of local politicians to the creation of a second nuclear facility in the province of la Manche, which would be completely independent of that of the Hague Center.

This construction on land belonging to the CEA, as well as the insistence of local government officials on seeing that the responsibility of environmental safeguards would be guaranteed by the CEA itself, led the CEA to apply in its own name for authorization for the basic nuclear facility planned. This decree, which took effect in 1969, defined the general conditions under which this facility could be operated. Construction of the center was started in 1968 and operations commenced effective on January 1, 1969.

Factors in Choosing the Site

The factors determining the choice of the Hague as the storage site are as follows:

low and disperse surrounding population;

very limited and little developed agricultural use of the soil;

domestic water supply assured by distant wells and by isolated networks;

nuclear maintenance and technology assured by the proximity of the Hague Center, which in particular permits very precise monitoring of the environment;

proximity of the sea;

the easy monitoring of effluents.

General Description of the Facility and the Operations

The facility consists of:

Storage Installations —

above-ground storage areas;

concreted ditches;

in the course of construction, planned decay and storage shafts, with the corresponding unloading installation with which 50 to 60 liter containers of highly active wastes could be received.

Pre-Storage Installations —

temporary storage hall capable of containing around 20,000 200-liter barrels;

storage tank for oils and solvents.

Treatment or Packing Installations —

a building housing a 400-ton compacting press and a machine shop;

a decontamination building with recovery of the waters in a reservoir;

a concrete batching and mixing plant.

Auxiliary Installations —

a building which houses essentially laboratories, counting rooms, locker rooms and rest rooms in the active zone and stores, electrical heating system, boiler room and cafeteria in the inactive zone;

an administration building situated in the inactive zone.

In addition to these installations, PEC-Infratome uses heavy-duty equipment for on-site transfer of wastes—lift trucks and cranes equipped for remote handling. The wastes received are essentially solid wastes of low and medium activity, packed in 200-liter metal barrels. Some of the older stock is received in the form of 100-liter and occasionally 225-liter barrels. Certain wastes have been packed in concrete blocks of 95 or 130-140 cm diameter, with an average

weight of 2 or 4-5 tons, respectively. The acceptable limit for specific activities
are as follows: 100 Ci/m^3 for ^{90}Sr; 1,000 Ci/m^3 for the other β and γ emitters;
1 Ci/m^3 for noncoated ^{239}Pu; and 10 Ci/m^3 for ^{238}Pu and other coated alpha
emitters (except radium).

It should be noted, however, that these limits, particularly for Pu, are not necessarily fixed, but that it may be possible to modify them as a function of experimental results and the storage possibilities which will become available in the
future. Packing and the method of storage are determined by the properties
of the wastes concerned:

> the packed wastes (encased in cement, bitumen, etc.) so as to
> have a very low leaching rate (maximum 10^{-3}) are dumped on
> prepared areas;
>
> the low-activity wastes contained in ordinary barrels are placed
> in "surface spaces" between the packed wastes just mentioned.
> The entire area is covered with a layer of clay to guarantee an
> impermeable layer, then with earth and turf;
>
> the wastes in barrels other than of low activity, and the loose
> wastes are stored in concrete casks. These casks are filled by
> connection to a liquid-cement grouter and covered by a bitu-
> men coating. The first concreted casks had been packed with
> sand between the barrels, and the cement grout had been applied
> only above the sand. This method, which would have been more
> suited to eventual recovery, had the drawback of keeping the
> waste barrels in contact with the moisture in the sand, and has
> now been abandoned;
>
> the special wastes (large volume, very high activity, decaying
> wastes) are placed on areas or in ditches specially designed for
> these particular cases;
>
> the very irradiative wastes, i.e., those for which the dose rate
> in contact with the packing is greater than 10 rad/hr, are placed
> in special shafts to wait until the dose rates are sufficiently low
> to permit storage on the ground surface or in ditches, without
> risking irradiation of the personnel.

Thus, between the wastes and the environment outside the site, there are a certain number of more or less effective barriers: encasement, and covering with
earth and ditches. The efficiency of these barriers is continuously observed by
the permanent monitoring system of the site.

Monitoring Methods

The monitoring methods of the site are permanent. They are assured by the existence of a series of piezometers located at different points of the depot, thus
permitting periodic sampling in the phreatic table as well as simultaneous measurement of the level of the table. Such measurements are used to verify that
the waters of the table are not contaminated. All the surface precipitation at
the site is collected in a network of drains and ditches and led to a settling
tank where possible contamination of these waters is also monitored by regular
measurements. If initial contamination were observed, a system of ducts and

pumps would then permit recovery by the station for processing the liquid effluents of the Hague Center in order to prevent the discharge of these waters into
the neighboring streams.

Possible atmospheric discharges from the Hague Center as well as from the storage site can be detected by air samplers both inside and outside the site. This
monitoring duty is performed directly by Infratome specialists and is regularly
checked by the Radiation Protection Service of the Hague Center, which in case
of need has the authority to bring its technical collaboration and its resources
to this monitoring duty.

Medical supervision of the Infratome personnel is strictly governed by the same
conditions as for the CEA personnel, i.e., periodic medical check-ups, wearing
of dosimetric films and pencils, verification of noncontamination after workshifts,
etc.

Obviously, it is not possible for Infratome to verify the exact contents of each
of the barrels or packages shipped to them by a waste producer who has de-
manded their services. In accordance with the rules governing transportation of
radioactive products, it is incumbent upon each producer who orders wastes to
be removed to define the properties of this product in a shipping bill. These
properties per barrel or per lot are, on arrival, carefully inventoried and filed,
and the storage points of these barrels or lots of barrels recorded on charts which
make it possible and will make it possible in time to know accurately the origin
and nature of the products deposited underground or aboveground at the storage depot.

Thus, although the storage measures taken can be considered as definitive, they
are not irreversible because, by virtue of the permanent information (which continues to be valid in time) given by the files and charts, it is always possible to
recover any particular stored product if such appears convenient, useful or necessary. To facilitate location in the depot and possible recovery, a system of
signpost or metal-plate markers has been designed.

Volumes Treated

A specialized company was created not only for the needs of organizations spe-
cializing in nuclear energy, such as CEA and EdF, but also to respond to the
needs of different public or private producers.

In 1975, for example, although the CEA was responsible for supplying wastes
which accounted for around 89.5% of the activities of the company, EdF was
also a large and regular client, and obviously will contribute more and more to
future development, considering the program of construction of nuclear power
plants, not to mention numerous producers: hospitals, research laboratories,
various industrial plants (more than 550 different clients) which are disencum-
bered of their radioactive wastes by the services.

To put an order of magnitude on the stored products, the following figures can
be given: by December 31, 1975 the CEA will have ordered the removal to the
Infratome storage center of 293,900 200-liter barrels or equivalents; EdF, 25,230
200-liter barrels or equivalents; and the various public or private clients, 11,220
barrels.

With regard to the third activity, it is worth noting that it is constantly increasing, because the yearly totals have been as follows:

Year	Packages, (sources or barrels)
1969	325
1970	756
1971	823
1972	899
1973	2,212
1974	3,012
1975	3,194

As of December 12, 1975 these 330,350 barrels represent a volume of around 65,000 m^3, corresponding to an approximate tonnage of 87,000 tons. The total activity thus stored, estimated from the declarations of the shippers, is on the order of 2,000 Ci of α emitters and 189,000 Ci of β and γ emitters.

If these quantities are examined with regard to the methods of storage which have been used for them, it is seen that around 272,000 barrels have been deposited directly on the surface, 30,000 have been first compacted in the press and deposited in concrete blocks, and around 25,000 have been placed directly in concreted ditches.

If these quantities are related to the estimated storage possibilities of the Manche Center, the following figures are obtained: the area available to Infratome is slightly more than 12 hectares (ca 30 acres). Thus, allowing for the construction of buildings, roads and the press, around 10 hectares (ca 25 acres) are available for actual storage. Taking account of storage practices currently in vogue, it can be estimated that the capacity of this depot is on the order of 1,200,000 to 1,300,000 barrels.

STATUS OF RADIOACTIVE WASTE MANAGEMENT IN AUSTRALIA

Estimates by the Australian Atomic Energy Commission (AAEC) of the rate of growth of nuclear generating capacity, assuming installation of a series of LWRs commencing in the early 1980s, indicate that in Australia the irradiated fuel requiring reprocessing would be about 30 MT per year in 1983, and would rise by the year 2000 to about 480 MT per year (10). These rates are equivalent to 0.1 and 1.6 MT per day respectively. The capital cost of reprocessing plants is very sensitive to scale and with the above increases there would not appear to be a strong economic incentive to build a reprocessing plant before the 1990s. However, before a firm decision was reached one would have to consider:

the cost of buying reprocessing (and waste management) services in another country;

the desirability and cost of establishing a local fuel fabrication facility to allow the separated fuel to be recycled;

the cost of allowing irradiated fuel to accumulate (replacing it with new fuel) until the accumulated amount and the projected growth rate of the nuclear power system indicated that a local reprocessing plant would be economically viable;

the rate of development of fast breeder reactors and the consequent world demand for plutonium;

the likely rate of installation of LMFBRs in Australia; and

the possibility of building a plant larger than needed and selling the excess capacity to other nations.

The AAEC has judged that it is not yet necessary to, nor indeed, desirable for it to recommend a specific technical policy for the management of high level wastes. It is the general policy of the AAEC to ensure that all operations for which it is responsible are carried out with the maximum provision for the safety of persons and equipment, and to insist that all unnecessary exposure of staff and the general public to the hazards associated with any use of radiation be avoided. Similarly, its general policy on waste management will be dominated by the need to protect man from radiation damage, and the governing factor in the choice of the waste management system used or recommended will be that the system should not expose this or future generations to any significant risk of radiation damage.

In the meantime, consideration is being given to waste management planning, and one of the most important questions is who shall have the continuing responsibility for waste management. The federal system apportions various rights and responsibilities of government among the States and the Commonwealth. Among those falling to the States are the control of matters relating to public health and safety and thus it would seem that the States would have a major responsibility for waste management.

However, the latter cannot be viewed in isolation from fuel reprocessing facilities which ideally, to minimize transport difficulties and associated hazards, would be located reasonably close to the waste facilities. At the same time a fuel reprocessing facility may best be located centrally in relation to the power plants it serves. The site of a storage facility, if this were the method chosen, would depend on geological and climatic factors. All these factors have a bearing on the decision as to who shall have the final responsibility. However, the overriding consideration must be to arrange that the responsibility for ensuring that the wastes continue to offer no threat to man continues to be recognized and accepted in effect in perpetuity.

In connection with the choice of a method of waste management to be adopted, Australia is fortunate in that she can observe and, hopefully, profit from the experiences of other nations which have had to face the problems earlier and in more acute forms. Developments in other nations in the philosophy and in the technology of waste management are under continual observation and the possibilities of their application to the Australian environment are being considered. Already it is fairly obvious that in the isolation of these wastes from this and many future generations a necessary step at some stage will be their conversion into stable immobile forms.

It also seems likely that the chances are good of finding in Australia a storage site having the necessary characteristics of: suitable geological formations; high seismic stability; remoteness from population centers; and low precipitation. Certainly, the chances are better than in many other countries. Discussion of immobilization and of storage sites leads to an important consideration in the application of the "isolation from man" policy. This is the interaction between the characteristics of the storage site and the physical characteristics of the waste to be stored there. The objective of the waste manager is to provide a guaranteed protection for man at the lowest cost. It is not impossible that wastes stored as liquids at a site in a remote desert area could be demonstrated to be as effectively isolated from man as wastes stored as solid glassy materials at a less suitable site closer to centers of population. However, the remote site concept would almost certainly raise transport problems and, probably, involve high operational costs.

Thus, while the choice of a specific method of waste management should be made primarily on the grounds of safety, economic factors must also be taken into account. Here it is necessary to note that the critical-nuclide/critical-pathway/critical-population-group approach to the assessment of the potential hazard to man provides public health authorities and waste managers with a first-class tool for an objective assessment of the level of risk associated with a particular proposal for a waste management scheme.

RADIOACTIVE WASTE MANAGEMENT PRACTICES IN JAPAN

Atomic energy facilities in Japan can be divided into two categories: namely the facilities using radioisotopes on one hand and the nuclear power plants, fuel reprocessing plants and so forth on the other. The former are controlled by "the Law concerning prevention from radiation hazard due to radioisotopes etc.". The lattter are controlled by "the Law for the regulation of nuclear source materials, nuclear fuel material and reactors".

At the facilities using radioisotopes, the processed sludge and condensed liquid wastes are temporarily stored in the storage disposal facility and are transferred to the Japan Radioisotope Association which is legally authorized as the radioactive waste disposal organization. The Japan Radioisotope Association keeps these wastes at four storage sites set up in different places in the country and requests the Japan Atomic Energy Reserach Institute at Tokai-mura since 1965 to treat these wastes.

At nuclear power plants, spent resin, sludge, condensed liquid wastes are solidified with concrete after temporary storage. Current practice in radioactive waste management in each facility are given below.

Facilities Using Radioisotopes

The liquid and solid radioactive wastes are treated in accordance with the principles given in "the Law concerning prevention from radiation hazard due to radioisotopes etc." and its enforcement regulations. But the practical method of treatment differs according to the scale and environmental situation of the facilities. Generally speaking, the low-level liquid wastes are diluted or precipitated by coagulation and the intermediate-level wastes are evaporated. The high-level liquid wastes are stored in polyethylene bottles or others. The solid wastes are enclosed in carton boxes. Large equipment is covered by polyethylene sheets. These wastes are finally enclosed in various bottles or drums to be transferred to the Japan Radioisotope Association.

Japan Atomic Energy Research Institute, Tokai

At the waste treatment plant of the Japan Atomic Energy Research Institute at Tokai, the treatment, storage and disposal are carried out for liquid and solid radioactive wastes produced from research facilities and reactors of the Institute and also for the radioactive wastes of other atomic energy facilities and the wastes collected by Japan Radioisotope Association from various places in Japan. The outline of radioactive waste management at JAERI (Japan Atomic Energy Research Institute), Tokai is described in the following paragraphs.

Low-Level Liquid Wastes ($\sim 10^{-3}$ $\mu Ci/ml$): Most of the low-level liquid waste is the washing waste water from laboratories. The liquid below 10^{-6} $\mu Ci/ml$ is diluted with general sewerage water to reduce the concentration below the maximum permissible concentration and released consequently into the sea. The liquid above 10^{-5} $\mu Ci/ml$ is transported via liquid waste pipelines or by the liquid waste transporting car to the low-level liquid waste storage tanks (six steel tanks of 36 m^3), this is followed by the ion-exchange, evaporation or precipitation treatments. The ion-exchanger has a processing capacity of 1 m^3/hr. The liquid waste from regeneration of the ion-exchange resins is treated by the evaporator.

The evaporator has an evaporating capacity of 1 m^3/hr. The vapor is condensed into liquid and transported into the storage pond for treated liquid wastes. On the other hand, condensed liquid waste is temporarily stored in a tank and subsequently solidified with concrete.

Though the precipitating apparatus has a processing capacity of 3 m^3/hr, it is not fully operational, because the amount of the waste to be treated by this equipment has not increased much and there have been some technical problems.

Intermediate-Level Liquid Wastes (10^{-3} to ~ 1 $\mu Ci/ml$): The intermediate-level liquid wastes are mainly produced from plant facilities. These liquids are transported by the transporting car to the intermediate-level liquid waste storage tank (three underground tanks of stainless steel of 5 m^3; two of these are for inorganic liquids and the other for organic liquids) in the waste treatment plant and is subsequently treated by the evaporator.

The evaporator has a processing capacity of 0.6 m^3/hr and most of it is located in a cell enclosed by a shielding wall. The condensed liquid waste is solidified with vermiculite and concrete by a remote handling operation. After it is enclosed in the shipping cask, it is stored in the high-level waste disposal facility.

Low-Level Solid Radioactive Wastes (<50 mR/hr): The low-level solid radioactive wastes which are produced in the controlled area of the Institute are classified into combustible and incombustible wastes and are enclosed in carton boxes (generally a paper bucket of 20 liters) or in a polyvinyl sheet. These are temporarily stored in the radioactive wastes storage facility.

Generally, the incombustible wastes are pressed by a pressing machine and the combustible wastes are burnt to ashes in the incinerator. The incinerator ash is solidified with concrete in the drum. Pressing machines with a total pressure of 50 kg/cm^2 and of 200 kg/cm^2 are available. Both machines are installed in a closed room to eliminate dispersion of radioactive liquid wastes and are equipped with a ventilation system. Drums containing treated wastes are kept in a radioactive waste storage facility.

Incinerator No. 1 has a processing capacity of 30 kg/day. The NaOH solution which was used for washing the gas from the incinerator is finally treated by an evaporation process. The incinerator ash is solidified with concrete. Incinerator No. 2 has a processing capacity of 50 kg/hr (as combustible wastes) and has the same mechanism as that of incinerator No. 1. In addition, the plasma cutting machine is used for volume reduction by cutting the materials which cannot be pressed such as racks, structures and pipes. The cutter is used for metals (stainless steel, lead or aluminum) with a thickness reaching 40 mm.

Intermediate and High-Level Solid Wastes ($\geqslant 50$ mR/hr): The intermediate and high-level radioactive solid wastes are generally produced by hot laboratories. These wastes are contained in paper buckets or in metal cans. The wastes in paper buckets are further enclosed in metal cartridges. They are then stored in the various facilities of the "graveyard."

Japan Atomic Energy Research Institute, Oarai

For the treatment of radioactive wastes from the division of development for radioisotope utilization and JMTR (Japanese Materials Testing Reactor) of the Japan Atomic Energy Research Institute at Oarai, the facilities in Oarai technology center of the Power Reactor and Nuclear Fuel Development Organization, and university research facilities, new treatment plants were constructed at Oarai in the frame of a five-year project which began in 1967. The detals of the plants are as follows.

The low-level liquid wastes are transported by pipelines and stored in low-level liquid waste storage tanks (six concrete tanks of 200 m^3). These are treated by precipitation, the processing capacity being 10 m^3/hr. The waste sludges are condensed by centrifugation and treated by the freezing-melting process. These are finally solidified with asphalt as intermediate and low-level solid waste. The solution remaining from the above process is temporarily stored in waste tanks and released with general sewerage water via the wastewater inspection ponds (two concrete ponds of 375 m^3 and one pond of 750 m^3).

The intermediate-level liquid wastes are stored in the intermediate-level liquid storage tanks (4 concrete tanks of 70 m^3) are treated by the evaporator with a processing capacity of 1 m^3/hr. The concentrated materials are solidified with asphalt.

The low and intermediate-level solid wastes are pressed by a pressing machine with a capacity of 5 m^3/day after temporary storage. The combustible wastes are then burnt to ashes in the incinerator.

The high-level radioactive solid wastes consist of the fragments of spent irradiation capsules, irradiated testing samples and others. These wastes are enclosed in steel cans in the hot cell and then enclosed in concrete blocks at the high level solid treatment plant. These are placed in the storage center (concrete floor of 2,500 m^2).

Nuclear Power Plants

The Tokai power plant (Revised Calder Hall type, 166 MWe) of the Japan Atomic Power Company has been operating since July 1966 and the Tsuruga power plant (BWR, 322 MWe) of the same Company has been operating since March 1970.

No. 1 reactor of Fukushima nuclear power plant (BWR, 460 MWe) of the Tokyo Electric Power Company and No. 1 reactor of Mihama nuclear power plant (PWR, 340 MWe) of the Kansai Electric Power Company was put into operation in October 1970. Many other nuclear power plants of several electric power companies are under construction and programmed.

The liquid and solid radioactive wastes which are produced from these nuclear power plants are treated in accordance with the principles specified in the Law for the regulation of nuclear source material, nuclear fuel material and reactors and the related regulations.

Tokai Power Plant of the Japan Atomic Power Company: The cooling water of the spent fuel cooking pond is treated by coagulation-precipitation, sand filter and ion-exchanger. The waste sludges from the coagulation-precipitation process and the spent resins are stored in a bunker. The regeneration wastewater of the ion-exchange resin are treated by the evaporator and its solid residues are stored in a bunker. Though the low level decontamination water from equipment is released with the cooling water of the condenser, the higher level decontamination water is treated by the evaporator. Other miscellaneous solid wastes are generally stored in a bunker.

Tsuruga Plant of the Japan Atomic Power Company and Fukushima Nuclear Power Plant of the Tokyo Electric Power Company: In both plants, the primary cooling water in the clean-up system is cleaned by the filter and demineralizer, and the condensate water from the turbine is cleaned by the demineralizer. Though the ion-exchange resins of the clean-up system are not regenerated, the resins of the condensate system are.

The filter sludges and spent resins are dehydrated by the centrifuge process and solidified with concrete. The radioactive wastewater from the spent resin regeneration is treated by the evaporator and the condensed waste is solidified with concrete. The solidified wastes and other miscellaneous solid wastes are temporarily stored in the solid waste storage facilities.

Mihama Nuclear Power Plant of the Kansai Electric Power Company: The primary cooling water of the reactor is cleaned continuously by the demineralizer. The boron recovery system which extracts and cleans the cooling water intermittently is furnished for the control of reactivity of the reactor.

Among the ion-exchange resins in the purification-demineralization systems and the boron recovery systems, only the deborating anion exchange resins are regenerated. The regeneration wastewater is treated by the evaporator and its condensed wastes are solidified with concrete. Other spent resins are dehydrated by centrifuge and solidified with concrete. The solidified wastes and other miscellaneous solid wastes are temporarily stored in the solid waste storage facility.

Fuel Reprocessing Plant

The Power Reactor and Nuclear Fuel Development Organization undertook construction of a nuclear fuel reprocessing plant and intends to operate it probably around 1977. The liquid and solid radioactive wastes from this plant will be treated as follows in accordance with the principles specified in the Law for the regulation of nuclear source material, nuclear fuel material and reactors and the related regulations.

High-Level Liquid Wastes: The high-level liquid wastes which are produced mainly during the first separation process are treated by the evaporator and stored in the high-level liquid storage cell tank.

Intermediate-Level Liquid Wastes: The intermediate-level liquid wastes which are mainly produced during the second separation process, are washed by the diluent or evaporated, and the condensed wastes are transferred to the high-level liquid waste treatment plant.

Low-Level Liquid Wastes: The low-level liquid wastes such as off-gas washing water, solvent washing solution, etc., are treated by the evaporator or the coagulation-precipitation apparatus. These condensed liquid wastes and sludges are stored in the storage tank.

High-Level Solid Wastes: The high-level solid wastes such as fuel accessories, jacket and other equipment, which are produced from decladding and dissolution processes, are enclosed in the disposal containers and disposed in the high-level solid waste storage facility. The jackets for natural uranium fuel are stored intact. The hull of the condensed uranium fuel and the zirconium fragments of natural uranium fuel are stored in the wastes.

Low-Level Solid Wastes: The low-level solid wastes consist of combustible wastes such as wood, paper and cloth, the compressible wastes such as bulbs and plastic equipment, and the incompressible wastes such as tools and metal wastes. These three categories of waste are ashed, compressed or cut respectively in the waste treatment plants and solidified with concrete. Solidified waste drums are placed in the low-level solid waste storage.

Collection of Radioactive Waste by the Japan Radioisotope Association

Among the radioactive wastes whch are produced by the reactor facilities and the facilities using radioisotopes other than the Japan Atomic Energy Research Institute, the wastes which are difficult to treat in each facility such as solid (combustible and incombustible), liquid (acid, base and organic), filters and animal carcasses, are collected by the Japan Radioisotope Association from various places in the country and stored in the local storage facilities on four different sites.

These wastes are transported in accordance with the regulation on the transport of radioactive materials by vehicles. Wastes are collected as follows: (a) The containers are rented by the facilities from the Japan Radioisotope Association.

(b) The radioactive wastes are placed in the specified containers, care being taken to avoid surface contamination. 200 liter drums for combustible wastes, 50 liter drums for incombustible wastes, 20 liter ceramic bottles or 25 liter polyethylene sheets and bags for filters and 20 liter ceramic bottles for animal carcasses immersed in 10% formalin, are used for this purpose.

(c) The Japan Radioisotope Association collects the wastes when requested by each facility. The caps of ceramic bottles are sealed tightly with gypsum and the surface contamination of the containers is measured. The containers for liquid wastes and for animal carcasses are enclosed in the additional containers (50 liter drums).

(d) The Japan Radioisotope Association transports the containers in especially designed trucks to the nearest local storage facility.

High-Level and Alpha-Bearing Radioactive Wastes

In 1972, high-level and alpha-bearing wastes were generated by four institutions in Japan: the Tokai Research Establishment of the Japan Atomic Energy Research Institute (JAERI), the Oarai Research Establishment of JAERI, the Tokai Works of the Power Reactor and Nuclear Fuel Development Corporation (PNC) and the Oarai Engineering Center of PNC. In these four institutions, the generated amounts of such wastes are low enough to be stored without volume reduction or solidification processes; i.e., high-level liquid waste is stored in stainless steel tanks in liquid form and solid wastes are packed into airtight containers. But, in the future, the treatment, storage and disposal of these radioactive wastes will become one of the most serious problems.

High-Level Wastes: The only existing high-level liquid waste in Japan is the Purex-Law solution generated by the fuel reprocessing demonstration facility in the Tokai Research Establishment of JAERI, using JRR-3 spent fuels. About 10 m^3 of the liquid waste is kept in stainless steel tanks installed within underground caves. The waste was not concentrated by evaporation, therefore, the concentration is low. The fuel reprocessing demonstration facility has been closed since 1970. It is planned to use the storage tanks installed at the JAERI-Tokai site for liquid waste if required in the future.

The first fuel reprocessing plant has been constructed at the PNC-Tokai site. This Purex plant has 0.7 to 1.0 ton/day of nominal capacity and will generate about 200 m^3/year of concentrated high-level liquid waste. The liquid waste will be stored in underground stainless steel tanks installed on site. A certain amount of high-level metal wastes is also expected to be generated from the decladding process and will be kept in concrete silos lined with steel plates.

It is estimated that a second fuel reprocessing plant (capacity: 5 tons/day or 1,500 tons/year) will need to be installed by 1985, due to the progress of nuclear power generation in Japan (6 x 10^4 MWe in 1985, 1 x 10^5 MWe in 1990). In the second plant, about 50 m^3 of high-level liquid waste will be generated daily before preconcentration or about 10 m^3 after evaporation, on the assumption that the Purex process is adopted. The radioactivity content is estimated as 2.4 x 10^7 Ci in the daily discharge. If the waste is vitrified, about 200 m^3 of glass materials will be produced annually.

Concerning high-level solid wastes, irradiated materials are the principal source of this type at present together with solidified liquid wastes generated by laboratories. These wastes include metallic fragments and fuel samples irradiated in the cores of the Japan Materials Testing Reactor at the JAERI-Oarai site and other research reactors installed in JAERI-Tokai and other sites. The production rate is about 20 m^3/year. At the JAERI-Oarai site, these wastes are sorted, compressed (if applicable) and enclosed in concrete blocks or concrete-lined drums with or without lead or iron shielding for keeping their surface dose rate below 200 mR/hr. At the JAERI-Tokai site, these wastes are kept in vertical holes in the ground.

Alpha-Bearing Wastes: Wastes with a Low Surface Dose Rate — These wastes are produced mainly by three institutions: the plutonium fuel fabrication facility (fuels for the fast breeder experimental reactor, and the fast breeder prototype reactor, estimated start-up in 1978) installed at the PNC-Tokai site, the plutonium laboratory installed at the JAERI-Tokai site and the plutonium experimental installations at the PNC-Oarai site. These wastes are packed into airtight containers without volume reduction and kept in special warehouses on site.

The production rate of the wastes is considered to be about 40 m^3/year and will increase rapidly in proportion to the amount of alpha-emitting materials estimated to be produced in the near future. With the first fuel reprocessing plant in operation, the amount of alpha-bearing solid wastes will rise steeply, not only due to wastes directly from the fuel reprocessing plant but due to the wastes generated by the fuel fabrication facilities using plutonium recovered in the reprocessing plant.

For dealing with this situation, PNC is making preparations in cooperation with JAERI to install certain facilities at the Oarai site, whereby the wastes will be compressed into small volumes (between $\frac{1}{5}$ and $\frac{1}{10}$ of original volume) and solidified into 200 liter-drums with bitumen or concrete.

Relatively small amounts of alpha-bearing liquid wastes are discharged from laboratories and concentrated into small volumes by evaporation or chemical treatment. Concentrate or sludge is solidified with cement, concrete or bitumen and managed as solid waste.

Wastes with a High Surface Dose Rate — Production of alpha-bearing solid wastes with a high gamma radiation level is anticipated from the operations of the fast breeder experimental reactor at the PNC-Oarai site and the fast breeder prototype reaction of PNC as well as from the postirradiation experiments conducted at the PNC-Oarai site, mainly from the fuels irradiated in two abovementioned reactors. It is estimated that about 30 m^3 of such wastes will be produced at the PNC-Oarai site in 1977. At the Oarai Engineering Center, PNC, it is planned in collaboration with JAERI to treat the wastes as follows:

The wastes will be sealed into stainless steel containers and kept in vertical cylindrical holes arranged within large dense concrete structures installed in the ground. At a later stage, the wastes packed in the containers will be transferred elsewhere for further treatment or storage/disposal according to the national policy which is to be established at a later date. There are no plans in hand for the construction of a fuel reprocessing plant for fuels of the fast breeder reactors.

Ultimate Storage/Disposal of Low-Level and Alpha-Bearing Radioactive Wastes — Japan's final policy is not yet decided in regard to ultimate storage/disposal of high-level and alpha-bearing wastes. At the present time, the policy on ultimate storage/disposal for low-level wastes is under examination. Two possibilities are being investigated for this purpose; one is based on dumping into the Pacific Ocean and the other is based on storage in remote sites such as uninhabited islands. Marine disposal will be adopted for low-level wastes but not for high-level and alpha-bearing wastes for the present.

Expensive and precise investigations will be necessary to settle the policy for high-level and alpha-bearing wastes, taking account of Japanese characteristics;

i.e., dense population, finely divided earth strata, shallow underground water table, vivid volcanic action, frequent earthquakes and so on. Unfortunately, any bedded salt, salt domes or equivalent geological formations have not been discovered in Japan so far. The preferred character of media for storage has not yet been found specifically in Japan.

The following criteria should be taken into account: for the solidification media, high capability to incorporate waste, high chemical stability and stability against radiation, very low leachability; for geologic media, geological stability (ranging from hundreds to millions of years), natural plasticity, good chemical stability, high impermeability, good thermal conductivity, isolation from human activity.

Krypton-85 and Tritium

According to the development and utilization of atomic energy, the increasing quantities of radioactive substances in the environment will become one of the major problems for the future. Therefore, it is necessary to study seriously the effects of radiation on man and the environment. Two possible sources of radiation to man are tritium and krypton-85 produced mainly at reprocessing plants.

Krypton-85: Krypton-85 included in off-gas arising from chopping and dissolution processes is discharged to the atmosphere through high stacks after dilution with ventilation exhaust air. Off-gas prestorage systems are installed in order to hold the above off-gas in case of unfavorable weather conditions.

According to calculations made for the PNC-Tokai plant, the annual exposure dose resulting from kryton-85 released at the maximum ground concentration point is estimated as 30 millirem for the skin and 0.4 millirem for the whole body with a sufficient safety margin. The above estimates are well below the recommended figures.

Considerable studies have been carried out concerning the estimation of dose from long-term build-up of krypton-85 through world-wide nuclear power production. The range of estimates is well under one millirem for the whole body annually. Although these estimates are relatively insignificant, R & D is underway in various countries to develop methods of recovering krypton-85 released from fuel reprocessing.

Several years ago, the Japan Atomic Energy Research Institute, JAERI and PNC conducted a feasibility study on two systems, solvent-absorption and charcoal-adsorption. Research for the development of a perm-selective membrane system has begun recently at Mitsubishi Atomic Power Industry under government sponsorship.

PNC will be starting a product for the development of krypton recovery techniques. The final goal is to construct a pilot plant adjacent to the reprocessing plant and to demonstrate its capability using actual off-gas from reprocessing. If the system could become available, the recovered product would be stored on site, contained in the form of compressed gas in gas cylinders. Concerning the long-term storage of krypton, basic research to develop methods more suitable for krypton containment will be initiated in the near future.

Tritium: At the PNC-Tokai plant about one-fourth of the tritium produced will be discharged through the stack, the remaining portion being discharged with treated low-level liquid waste into the Pacific Ocean about one kilometer from Tokai. The inhalation dose of tritium released from the stack is estimated to be well under 1 millirem annually. Tritium radioactivity in seawater at the exit point of the discharge pipeline is of the order of 10^{-1} μCi/cc. As a consequence of oceanographic studies conducted by the Nuclear Safety Research Association of Japan, the dilution factor off Tokai is estimated as follows:

Dilution factor = distance from the outfall (m)/0.35, where the dilution factor is defined as the ratio of the concentration at the point of interest and the discharge point. Thus, the concentration becomes one-thousandth of that of the released effluent when it is 350 meters away from the outfall.

The above formula does not apply to the concentration distribution near the outfall. An elaborate design of the outfall will make it possible to dilute the effluent to attain a tritium concentration well below the drinking water level within a few hundred meters from the outfall. The radiological influence of the tritium discharge on the general public and on the aquatic products is estimated to be negligible.

Tritium is a weak β-emitter, being less energetic than krypton-85, therefore, the global radiological effects might be less important than those caused by krypton-85. However, for future tritium recovery, investigation of the technological status and fundamental experimental work has been initiated.

REFERENCES

(1) D.W. Clelland, "High-Level Radioactive Waste Management in the U.K.," BNFL, presented to Symposium on the Management of Radioactive Waste from Fuel Reprocessing, Paris, France, November 27-December 1, 1972.

(2) D.W. Clelland, "Present Methods of Storing Highly Radioactive Waste In the United Kingdom and Proposals for the Future," BNFL, presented to Ninth World Energy Conference, Detroit, Michigan, September 22-27, 1974.

(3) N.L. Franklin, "Irradiated Fuel Cycle," BNFL, presented to European Nuclear Conference, Nuclear Energy Maturity, Paris, France, April 21-25, 1975.

(4) P. Auchaft, R. Bonniaud, J.A. Coste, L. Rozand, C. Sombret, and A. Barbe, "French Industrial Plant 'AVM' for Continuous Vitrification of High-Level Radioactive Wastes," CEA and St-Gabain, presented to AICHE meeting, Los Angeles, California, November 16-20, 1975.

(5) A. Leseur, P. Miquel, M. Pelras, P. Auchaft, J.J. Fabre, "Concentration of Nitric Acid Fission Product Solutions Containing Fluoride Ions," CEA, presented to AICHE meeting, Los Angeles, California, November 16-20, 1975.

(6) Dr. Wolf J. Schmidt-Keuster, "The Waste Disposal System in the Nuclear Fuel Cycle," BMFT, presented to Reactor Conference of the German Atom Forum, Berlin, Germany, April 5, 1974.

(7) F. Kaufmann, H. Koschorke, W. Guber, W. Hild, H. Krause, "Recent Experience with the Steamheated Spray Calcine Units for HLW-Solidification," Karlsruhe Nuclear Research Center, presented to AICHE meeting, Los Angeles, California, Nov. 16-20, 1975.

(8) J. Van Geel, H. Eschrich, E. Detilleux, "Conditioning of Solid High-Level Waste Products by Dispersion into Metal Matrixes," Eurochemic, presented to AICHE meeting, Los Angeles, California, November 18-20, 1975.

(9)   K.J. Schneider, "High-Level Waste," Chapter 8, *Human and Ecologic Effects of Nuclear Power Plants,* edited by L.A. Sagan, Charles C. Thomas, Springfield, Illinois, 1974.

(10)   J.M. Costello, AAEC, private communication, May 1972.

BIBLIOGRAPHY

The reports listed below which were used in preparing this book are available from:

National Technical Information Service
U.S. Department of Commerce
5285 Port Royal Road
Springfield, Virginia 22151

AAEC/TM628   *Management of Wastes from Irradiated Nuclear Fuels,* E.D. Hespe and C.J. Hardy, Australian Atomic Energy Commission, presented to the Symposium on The Role of Nuclear Energy in Australia's Development, Canberra, June 1st and 2nd, 1972.

ARH-SA-269   *Management of Radioactive Low-Level Liquid, Gaseous and Solid Wastes in the 200 Areas,* A.T. White, Atlantic Richfield Hanford Company, Richland, Washington, June 24, 1976.

BNL-21571   *Some Techniques for the Solidification of Radioactive Wastes in Concrete,* P. Colombo and R. Neilson, Jr., Brookhaven National Laboratory, Upton, New York, June 1976.

BNWL-1876   *Nuclear Waste Management and Transportation Quarterly Progress Report, July–September 1974,* G.J. Dau, Battelle, Pacific Northwest Laboratories, Richland, Washington, November 1974.

BNWL-1913   *Nuclear Waste Management and Transportation Quarterly Progress Report January Through March 1975,* A.M. Platt, Battelle, Pacific Northwest Laboratories, Richland, Washington, June 1975.

BNWL-2092   *High Level Waste Vitrification by Spray Calcination/In-Can Melting,* D.E. Larson and W.F. Bonner, Editors, Battelle, Pacific Northwest Laboratories, Richland, Washington, November 1976.

BNWL-2138   *Fluidized Bed Calcination Experience with Simulated Commercial High-Level Nuclear Waste,* W.J. Bjorklund, Battelle, Pacific Northwest Laboratories, Richland, Washington, November 1976.

BNWL-SA-5725   *The U.S. Domestic High Level Waste Management System,* J.W. Finnigan, Battelle, Pacific Northwest Laboratories, March 1976.

BNWL-SA-5757   *Ceramics in Nuclear Waste Management,* J.E. Mendel, for presentation at the 12th State-of-the-Art Symposium on Ceramics in the Service of Man, Washington, D.C., June 7-9, 1976.

BNWL-SA-5895   *Fixation of Radioactive Waste in Glass,* C.C. Chapman and J.E. Mendel, oral presentation at Gordon Research Conference in Plymouth, New Hampshire, August 1976.

BNWL-TR-198   *Seven Years' Experience of Storage of Solid Radioactive Wastes of Low or Medium Activity Above Ground or in Concreted Ditches,* G. Bardet, July 1976.

BNWL-TR-203   *Embedding of Radioactive Wastes by Thermosetting Resins,* A. Baer, A. Traxler, A. Limogi and D. Thiery, August 18, 1976.

CONF-700440-1   *Nuclear Power Reactor Wastes and Our Environment,* R.G. Barnes, presented at The Third Annual National Pollution Control Conference, San Francisco, California, April 3, 1970.

DOCKET 50-564-1 *Nuclear Fuel Recovery and Recycling Center—Preliminary Safety Analysis Report,* Exxon Nuclear Company, Inc.

DP-1366   *Storing Solid Radioactive Wastes at the Savannah River Plant,* J.H. Horton and J.C. Corey, June 1976.

DP-MS-76-39   *Interim Storage of Spent Fuel Assemblies,* F.D. King and W.H. Baker, Savannah River Laboratory, E.I. du Pont de Nemours and Co., Aiken, South Carolina, paper presented at the International Symposium on the Management of Waste from the LWR Fuel Cycle, Denver, Colorado, July 11-16, 1976.

DP-MS-76-66   *Durability of Containers for Storing Solidified Radioactive Wastes,* C.L. Angerman and W.N. Rankin, Savannah River Laboratory, E.I. du Pont de Nemours and Co., Aiken, South Carolina, for presentation at Corrosion/77 NACE Meeting, San Francisco, California, March 14–18, 1977.

ERDA-1538   *Final Environmental Statement—Waste Management Operations, Hanford Reservation, Richland, Washington,* Vol. 1, United States Energy Research and Development Administration, December 1975.

HEDL-SA-851   *Treatment Technologies for Non-High-Level Wastes (U.S.A.),* C.R. Cooley and D.E. Clark, presented at the International Symposium on the Management of Wastes from the LWR Fuel Cycle, Denver, Colorado, July 11–16, 1976.

HEDL-SA-856   *Technologies for Recovery of Transuranics and Immobilization of Non-High-Level Wastes,* G.L. Richardson, paper for presentation at the International Symposium on the Management of Wastes from the LWR Fuel Cycle, Denver, Colorado, July 11–16, 1976.

ICP-1088   *Emergency Fluidization and Bed Removal During Fluidized-Bed Solidification of Commercial Wastes,* R.E. Schindler, Allied Chemical Corporation, Idaho Chemical Programs, April 1976.

LA-6252   *Incineration Facilities for Treatment of Radioactive Wastes: A Review,* B.L. Perkins, Los Alamos Scientific Laboratory, Los Alamos, New Mexico, July 1976.

LA-UR-76-1722   *Burial Grounds,* M.L. Wheeler, presented to the International Symposium on the Management of Wastes from the LWR Fuel Cycle, July 11–16, 1976.

NP-19966   *Radioactive Waste Management Practices in Japan,* Y. Nagai, prepared for Nuclear Energy Agency, Organisation for Economic Co-operation and Development, Paris 1974.

NVO-410-38   *Deepwell Disposal of Transuranic Contaminated Liquid Waste at the Nevada Test Site,* A.W. Western and D.H. Hall, Reynolds Electrical & Engineering Co., Inc. and J.S. Coogan, Environmental Monitoring and Support Laboratory, U.S. Environmental Protection Agency, Las Vegas, Nevada.

ORNL-TM-3965   *Projections of Radioactive Wastes to Be Generated by the U.S. Nuclear Power Industry,* J.A. Blomeke, C.W. Kee and J.P. Nichols, Oak Ridge National Laboratory, February 1974.

ORNL-TM-4902   *Correlation of Radioactive Waste Treatment Costs and the Environmental Impact of Waste Effluents in the Nuclear Fuel Cycle for Use in Establishing "as Low as Practicable" Guides— Fabrication of Light-Water Reactor Fuel from Enriched Uranium Dioxide,* W.H. Pechin, R.E. Blanco, R.C. Dahlman, B.C. Finney, R.B. Lindauer and J.P. Witherspoon, Oak Ridge National Laboratory, May 1975.

ORNL-TM-4903   *Correlation of Radioactive Waste Treatment Costs and the Environmental Impact of Waste Effluents in the Nuclear Fuel Cycle for Use in Establishing "as Low as Practicable" Guides— Milling of Uranium Ores,* M.B. Sears, R.E. Blanco, R.C. Dahlman, G.S. Hill, A.D. Ryon and J.P. Witherspoon, Oak Ridge National Laboratory, May 1975.

PB-221 467   *A Study of Hazardous Waste Materials, Hazardous Effects and Disposal Methods,* Vol. III, Booz-Allen Research, Inc., Bethesda, Maryland, July 1973.

PB-224 588   *Recommended Methods of Reduction, Neutralization, Recovery or Disposal of Hazardous Waste, Volume IX. National Disposal Site Candidate Waste Stream Constituent Profile Reports—Radioactive Materials,* TRW Systems Group, Redondo Beach, California, August 1973.

PB-235 804   *Environmental Analysis of the Uranium Fuel Cycle—Part I. Fuel Supply,* U.S. Environmental Protection Agency, Office of Radiation Programs, Washington, D.C. 20460, October 1973.

PB-235 805   *Environmental Analysis of the Uranium Fuel Cycle—Part II. Nuclear Power Reactors,* U.S. Environmental Protection Agency, Office of Radiation Programs, Washington, D.C. 20460, November 1973.

PB-235 806   *Environmental Analysis of the Uranium Fuel Cycle—Part III. Nuclear Fuel Reprocessing,* U.S. Environmental Protection Agency, Office of Radiation Programs, Washington, D.C. 20460, October 1973.

PB-244 928   *The Separation and Control of Tritium—State-of-the-Art Study,* L.L. Burger, Battelle, Pacific Northwest Laboratories, Richland, Washington, April 14, 1972.

PB-254 737   *Proceedings of Nuclear Regulatory Commission Workshop on the Management of Radioactive Waste: Waste Partitioning as an Alternative,* held at the Battelle-Seattle Research Center, Seattle, Washington, June 8–10, 1976.

PB-255 502   *Environmental and Safety Aspects of Alternative Nuclear Power Technologies—Fusion Power Systems,* B.J. Mann, U.S. Environmental Protection Agency, Office of Radiation Programs, Washington, D.C. 20460, May 1976.

PB-258 316   *Environmental Survey of the Reprocessing and Waste Management Portions of the LWR Fuel Cycle,* Nuclear Regulatory Commission, Washington, D.C. 20555, October 1976.

PB-260 559   *Status of Commercial Nuclear High-Level Waste Disposal,* G.J. Dau and R.F. Williams, Electric Power Research Institute, Palo Alto, California, September 1976.

RFP-2471

Fluidized Bed Incineration of Radioactive Waste, D.L. Ziegler, Rockwell International, Rocky Flats Plant, Golden, Colorado, May 14, 1976.

SAND 75-6093

Thermomechanics Problems in Nuclear Waste Disposal, L.E. Bertram, Sandia Laboratories, Albuquerque, New Mexico.

SAND 76-0224

Report to the Radioactive Waste Management Committee on the First International Workshop on Seabed Disposal of High-Level Wastes, Woods Hole, Massachusetts, February 16–20, 1976, D.R. Anderson, C.D. Hollister and D.M. Talbert, Editors, April 1976.

SAND 76-5229

Radioactive Waste Disposal Pilot Plant Concept for a New Mexico Site, W.D. Weart, Sandia Laboratories, Albuquerque, New Mexico, report to IAEA Headquarters, Vienna, Austria, March 22–26, 1976.

SRO-TWM-76-1

Integrated Radioactive Waste Management Plan, Savannah River Plant, Aiken, South Carolina, June 1976.

TID-27341

The Shallow Land Burial of Low-Level Radioactively Con-taminated Solid Waste, Panel on Land Burial, Committee on Radioactive Waste Management, Commission of Natural Resources, National Research Council, National Academy of Sciences, Washington, D.C., 1976.

UCRL-51713

Economic and Environmental Evaluation of Nuclear Waste Disposal by Underground In Situ Melting, J.J. Cohen, R.L. Braun, L.L. Schwartz and H.A. Tewes, Lawrence Livermore Laboratory, Livermore, California, November 30, 1974.

NOTICE

Nothing contained in this Review shall be construed to constitute a permission or recommendation to practice any invention covered by any patent without a license from the patent owners. Further, neither the author nor the publisher assumes any liability with respect to the use of, or for damages resulting from the use of, any information, apparatus, method or process described in this Review.

RADIATION SPECTRA
OF RADIONUCLIDES 1976

by T. B. Metcalfe

Physicists, physical chemists, spectroscopists, and other research scientists, who use radiation spectra in their work, should find this volume of immense value. It probably constitutes the most convenient reference for anyone wishing to identify gamma spectra. All those interested in activation analyses, as well as the analysts who carry out such work, will find these tables uniquely advantageous.

Utilizing the physical data of pioneering investigators such as Heath, Hollander, Kinsman, Seaborg, and others, the radiation spectra of the nearly 900 currently known gamma emitters have been arranged so as to facilitate the identification of the contributing nuclides in **mixed spectra.**

In general, preference was given to primary sources, i.e. to publications and papers in which the investigator reports actual laboratory observations either for the first time or as confirmation of previously reported discoveries made by himself or by others. Standard reference works, known to be reliable, have also been consulted, but usually only to detect discrepancies. Since newer and better instrumentation has often resulted in more accurate data, a painstaking search of the most recent journal literature was also made.

PRIMARY LISTING: The listing of radioactive isotopes is in order of the magnitude of their gamma energies. Reading forward, the earliest appearance of each isotope corresponds to the lowest gamma energy present in its spectrum.

SUBTABLES: The entire book is divided into subtables, each subtable being associated with a separate radioisotope. The subtables are numbered consecutively to permit cross-referencing, and each subtable is identified with the isotope printed in the **NUCLIDE** column.

GAMMA ENERGY: Each subtable includes the entire gamma spectrum of its radioisotope in the ascending order of magnitude of gamma energies. The first gamma energy listed in the subtable is the lowest for the spectrum of that isotope. Moving through the tables, as the higher energies appear, repetition of a previously printed subtable is avoided by cross-reference to the earliest subtable for the particular radioisotope—in those instances in which abundance has been reported in the literature, it is given here in per cent. In many cases the sum of the percentages for a given isotope is not 100. This is due to the multiplicity of sources of information concerning a single isotope. While this may seem confusing, the actual numbers have been included as the abundance observed by the individual investigators in their work. This same comment holds true for the energies of the other forms of radiation reported in this book.

BETA ENERGY: If an isotope emits beta radiation, the energies are listed in the subtables in ascending order of magnitude under the column heading **BETA ENERGY.**

ALPHA ENERGY: If an isotope emits alpha radiation, the energies are listed in the subtables in ascending order of magnitude under the column heading **ALPHA ENERGY.**

POSITRON ENERGY: If an isotope emits positrons, the energies are listed in the subtables in ascending order of magnitude under the column heading **POSITRON ENERGY.**

PARENT-NUCLIDE-DAUGHTER: The radioisotope identified with each subtable is indicated in the column headed **NUCLIDE,** giving in order the atomic number, the atomic symbol and the mass number. If the nuclide is produced from another atom by a radiation process, the parent is indicated in the column headed **PARENT.** If, through a radiation process, the nuclide produces a daughter atom, that atom is indicated in the column headed **DAUGHTER.**

HALF LIFE: The half life of the radioactive decay of each radioisotope is indicated in the column headed **HALF LIFE** in units indicated by letters: Y = years, D = days, H = hours, M = minutes, S = seconds. The magnitude is indicated by the first number to be multiplied by 10 raised to the second number as an exponent (using the sight which precedes it), e.g. **24.36 x 10³Y.**

The unique feature of this presentation of all gamma spectra provides the interpreter of mixed radiation emitters with the cross-referencing essential to the identification of each contributing isotope. This reduces the search required, saves time and increases the assurance of complete identification.

The improvement in convenience over the usual catalogs of isotopes is apparent at once, since this book makes possible the identification of an unknown without a laborious search through all spectra for observed peaks. This usually requires considerable work of cross-referring, if more than one isotope is present. This volume provides such cross-referencing automatically with ease.

Thus each energy represented in an observed spectrum (which may be from a mixture of many isotopes) can in turn be found by its magnitude and then cross-referred to all isotopes displaying that energy. This unique arrangement makes it possible to identify, in an observed spectrum of gamma energies, each "peak" as to all possible contributing isotopes. Thus, by elimination, only those isotopes compatible with the entire spectrum are recognized as actually present.

ISBN 0-8155-0620-1

394 pages

FUEL CELLS
FOR PUBLIC UTILITY
AND INDUSTRIAL POWER
1977

Edited by Robert Noyes

Energy Technology Review No. 18

Fuel cells are generators of electricity containing no moving parts except small extraneous pumps for the movement of fuel and oxidant into the cell and the products of oxidation out of the cell.

Public utilities and industrial consumers require high voltage, three-phase alternating current. In this application fuel cells must compete with turbine-driven generators which provide such current. While the output of a fuel cell is low voltage DC power, cells may be connected in various series and parallel arrangements to give whatever voltage is desired, but mechanical rotary converters or delicate electronic inverters must then provide conversion to AC (60 cycles for each phase in the USA).

The advantages of a fuel cell system over turbine-driven generators lie in greater efficiency at full load which even increases as the load diminishes, so that inefficient peaking generators are not needed. There are considerable pollution control advantages to be gained as well. Because the various suitable fuels react electrochemically rather than by burning in air, no nitrogen oxides are formed. For the same reason, emissions of unburned or partly burned gaseous and particulate products are practically nil.

Fuel cell installations using inverters have a long life with relatively little maintenance. It is now entirely feasible to have small, completely unattended fuel cell power plants using waste fuels on location (such as hydrogen from chlorine production) to produce convenient power.

This book, based on information derived from U.S. government-contracted studies, contains considerable practical down-to-earth technical information relating to fuel cells for power plants. A partial and condensed table of contents follows.

ISBN 0-8155-0676-7

325 pages

SOLAR HEATING AND COOLING 1977
Recent Advances

by J. K. Paul

Energy Technology Review No. 16

The technology for solar energy utilization is becoming increasingly available, as indicated by the large number of patents issued in the past several years. Recent developments encompass a number of areas. The emphasis in this book is on low temperature (to +90°C) solar collector construction and heating and cooling systems which use these low temperature collectors. The material discussed here is based on 175 U.S. patents, issued since 1970, which illustrate 157 processes.

In its simplest form a collector consists of a sheet of glass or other transparent material situated above a flat plate so constructed that it acts as a black body to absorb heat. The sun's rays pass through the glass and are trapped in the space between cover and plate. The heat may then be utilized by passing a fluid through a conduit system located between the cover and absorber plate; the heated fluid subsequently being used to heat a home, water supply, or swimming pool, or even run a heat pump for cooling.

Focusing collectors use curved or combinations of flat devices to reflect solar rays onto an absorber surface to achieve greater concentration of energy (higher temperatures).

Information has been included describing suitable coatings used to improve absorption properties and detailing a number of devices which employ liquids, crushed rock or other media for the storage of absorbed energy when the weather is cloudy or hazy, or at night. A partial table of contents follows here. Numbers of processes are in parentheses.

1. FLAT PLATE COLLECTORS (53)
Finned Heat Transfer Plate
Panel with Tapered Reflector
Flat Plate Panel Array
   with Complementary Reflectors
Insulation Means
Conduit Layout
Thin Film Systems
Modular Systems
Adjustable Convection Plates

2. FOCUSING COLLECTORS (23)
Parabolic Concentrators
Combination of Curved Surfaces
Tubular Types
Liquid Medium Collectors
Flat Lens Panel
Low Profile Panels

3. UPRIGHT COLLECTORS (11)
Window-Mounted Systems
Wall-Mounted Systems
Porous Barrier Across
   Fluid Flow Path
Minimum Mechanization

4. VARIOUS COLLECTORS (6)
Roof Deck Construction
Heat Recovery from Black
   Particles in a Gas Stream

5. ABSORBER COATINGS (4)
Selective Black Coatings
Anodized Aluminum Alloys
Black Nickel Coatings

6. HEAT STORAGE (11)
Eutectic Solutions
Heat Storage Cylinder
Crushed Rock System
Heat Sinks—
   Encapsulated in Resin
Granular PVC Heat Sink
Underground Installations
Liquid Aquifer Method

7. HEATING & COOLING SYSTEMS (21)
Combination Heat Pump
   & Low Temp. Collector
Plate Absorber Freezes
   or Heats Water Sheet
Solar Heat + Oil or Coal
Circulation Systems
Enclosed Patio Installation
Conduits in Rock Layer
   below Concrete Floor
Combination Water Heater
   and Water Chiller
Hydration ←→ Dehydration
   with $Ca(OH)_2$ + LiCl

8. DOMESTIC HOT WATER INSTALLATIONS (10)
Closed System for Thin Film
   Heating of Transfer Fluid
Collector Awning Device
   for Mobile Homes

9. SWIMMING POOL HEATERS (18)
Floating Pans
Heat Exchange Panels
Heating System Built
   into Pool Border
Heat Absorbers Using
   Sand or Iron Oxide
Panels With Heat Collecting
   Lamina for Pods

ISBN 0-8155-0674-0

485 pages

HYDROGEN TECHNOLOGY FOR ENERGY 1976

by David A. Mathis

Energy Technology Review No. 9

Hydrogen is attractive as a fuel because it is abundant, relatively inexpensive and ecologically clean. When hydrogen is burned in air, it forms water vapor only, there are no solid combustion residues and no soot particles or noxious gases to contaminate the atmosphere.

The use of hydrogen as a universal fuel necessitates development of methods for storing, handling, and transferring, and these are the main subjects treated in this book. Future volumes in this series will be reserved for the technology of the myriad of hydrogen production schemes now under consideration viz. electrolysis of seawater by ocean-derived thermal energy, by solar cells, certain algae, by nuclear powered direct thermochemical conversion, or from municipal waste, etc.

The first chapter describes the hydrogen economy and suggests how it can be integrated into the USA energy system. The next three chapters are concerned with handling the various forms of hydrogen: gas, liquid, and solid (in the form of metal hydrides). The fifth chapter describes some of the work which has been done or is under way in using hydrogen as a fuel or in an energy storage system. Another chapter delves into safety and the political, socioeconomic, and environmental implications of a hydrogen economy. The final chapter provides a list of hydrogen technology experts and includes a brief description of each individual's expertise in the various aspects of hydrogen technology.

A partial and condensed table of contents follows here.

1. THE HYDROGEN ECONOMY
Primary Energy Sources
Hydrogen Energy Conversion
Hydrogen Production
Transportation and Storage
Utilization of Hydrogen
Advantages and Disadvantages

2. GASEOUS HYDROGEN
Comparison of Storage Methods
Hydrogen Handling Systems
Economics of Transmission
Pipeline Systems
Embrittlement & Compatibility
High Pressure Systems
Pressure Vessels & Pipelines
Compressor Requirements
Regenerative Compressors
Cost of Compression Equipment

3. LIQUID HYDROGEN
Technology of Liquefaction
Costs (Capital & Running)
Recovery of Liquefaction Energy
Para-Hydrogen
Storage and Transfer
Storage Dewars
Losses & Fire Hazards
Pumping Problems
The "Energy Pipe"
Peak Shaving
Energy Transmission Costs for
    Near-Urban Environments

4. SOLID HYDRIDES
Hydride Storage
Heat Transfer
Deterioration
Metal Hydride Suitability
Iron-Titanium-Manganese Alloy
Fixed Bed Metal Hydride Storage
Transmission of H_2 from Metal
    Hydride "Compressors"

5. FUEL & ENERGY STORAGE & USE
Water-Modified Aphodid Burner
Hydrogen Fueled Engines
Energy Storage for Utilities
H_2-Fuel for Transportation
Vehicle Fuel Storage
Metal Hydrides for Vehicles
Hydride-Dehydride Power
    System for Refrigeration

6. NONTECHNICAL ASPECTS OF A HYDROGEN ECONOMY
Safety Implications
Energy Law
Regulatory Law
Environmental Law and Implications
Unlikely NO_x Production
    at High Temperatures
Possible Production of Reject Heat
Underground Pipeline Transmission
International Implications

7. HYDROGEN TECHNOLOGY EXPERTS
A Listing of 263 Technological Experts,
    their Affiliations and Addresses,
    arranged by their Specialties

BIBLIOGRAPHY

ISBN 0-8155-0629-5

285 pages

THERMAL ENERGY FROM THE SEA 1975

by Arthur W. Hagen

Energy Technology Review No. 8
Ocean Technology Review No. 5

Recent advances in heat transfer research and thermal power plant design suggest that sea thermal power can now be made competitive with more conventional generating methods.

Utilization of ocean thermal gradient systems appears relatively attractive from many points of view. The ocean acts as a large solar energy heat reservoir which reduces energy storage requirements and permits the system to be operated the year round, 24 hours per day.

The purpose of this book is to provide a condensed data base to aid in proof-of-concept experiments and continued R & D to prove the technical feasibility and economic viability of generating either electricity or hydrogen by harnessing the temperature gradients in the sea.

The optimum locations for such generating plants appear to be in latitudes from 23°N to 23°S. Transmission and storage problems may limit the amount of power delivered by ocean-based plants. Construction of power production facilities must consider design problems associated with the hazards of marine environment.

Reports of proposed solutions to those problems are also contained in this book which is based on government-sponsored studies by engineering firms and university research teams. A partial and condensed table of contents follows:

1. AN OVERVIEW OF SEA THERMAL POWER
Introductory Material and NASA Reports
Site Analysis
Cost
Systems Analysis

2. OPEN CYCLE THERMAL GRADIENT OCEANIC POWER PLANT
Introductory Material
   NASA and PB Reports
Preliminary Design Investigation
Overall System Design
Spray Evaporator Design
Turbine Design
Deaeration Losses
Condenser Design
Cold Water Pump Considerations
Fresh Water Production
Feasibility Study of a
   100 Megawatt Plant

3. CARNEGIE MELLON UNIVERSITY DESIGN
Introductory Material and PB Reports
Design Overview
Water Circulation
Heat Transfer Fluid Circulation
Seawater Inlet-Outlet Hydrodynamics
Boiler Technology Considerations
Cold Water Pipe vs. Ammonia Pipe
Vertical Tube Heat Exchangers
Antifouling Costs
SSPP (**S**olar **S**ea **Po**wer **Pl**ant) Topology

4. UNIVERSITY OF MASSACHUSETTS DESIGN
Introductory Material and PB Reports
Overall Design Concept (Mark I)
Cold Water Supply & Suction Pipe
Anchor & Mooring Systems
Energy Umbilical
Control of Biofouling
Heat Exchanger Design (Mark I)
Variations and Improvements (Mark II)
Evaluation of Turbomachinery
Technical & Economic Feasibility Studies

5. COMPARATIVE STUDY OF CLOSED CYCLE TECHNOLOGIES
NASA Reports
CMU vs. UMASS Designs
Cost Evaluations

6. ADDITIONAL TECHNOLOGIES
Hydrogen Utilization
Hydrogen Energy
NITINOL (Ni-Ti-Naval Ordnance Lab.)
#55 Alloy Utilization
Further SSPP Proposals
OTEC Proposal (Ocean Thermal
   Energy Conversion Proposal)
The Danish System
A Japanese System

ISBN 0-8155-0597-3

150 pages

OFFSHORE DRILLING

TECHNOLOGY 1975

by Frank R. Carmichael

Energy Technology Review No. 5
Ocean Technology Review No. 3

The pecuniary interests, impacts, and high stakes involved in offshore drilling for petroleum are clearly revealed in the patent applications of the last few years.

Indeed, many of the concepts and design parameters now being implemented by the major construction companies around the world for use in the North Sea and other offshore locations are fully described and illustrated in the U.S. patent literature.

This book presents over 190 different processes and equipment designs for all phases of modern offshore drilling techniques including descriptions and uses of drilling ships, platforms, wellhead completions and a considerable number of very complex subsea facilities. Many of the processes in this area are described as complete systems, thus consolidating multiple functions in this emerging technology. However, for continuity, these processes are presented in the sections deemed most appropriate for their overall contributions to offshore exploration and exploitation.

A partial and condensed table of contents follows here. Numbers in parentheses indicate a plurality of processes per topic. Chapter headings and some of the more important subheadings and subtitles are given here.

1. DRILLING FROM A
FLOATING VESSEL (40)
Suspended Drill Strings
Torqued Drill Strings
Drill String Steering Units
Elimination of Riser Pipe
Equipment Guide Systems

2. DRILLING PLATFORMS (33)
Bottom-Supported Platforms
Platforms from Concrete
Pile Type Anchoring
Use of Buoyancy Tanks
Floating Platforms
Constructions for Arctic Regions

3. WELLHEAD APPARATUS
AND CONNECTORS (58)
Risers & Conductors (Pipes)
Expanded Truss Risers
Christmas Trees
Wellbase Structures
Wellhead Designs
Flow Lines & Connectors

4. WELL COMPLETION (19)
Multistring Tubeless Techniques
Packoff Designs
Casing Packoffs
Safety Valve Installations
Orientation Techniques for Tube Hanging
Hangers for Casings and Tubings
Telescoping Posts
Guide Members & Lines
Wellhead Cappings (Removable)
Casing Head Designs
Production Body
   Containing All Flow Lines
Pressure Controls
Pipeline Handling
Pipeline Guidance Systems
Cable Guidance
Vertical Sliding Connections

5. SUBSEA FACILITIES (42)
Satellite Systems
Submersible Wire Line Robot Units
Foundation Unit & Satellite Body
Rigid Short Connector Units
Production & Storage Systems
Production Satellites
Submersible Vessels & Chambers
Submersible Drilling Vessel
Hull Vessels
Combination Drilling and
   Production Vessels
Inflatable Seals
Caissons & Buoyant Apparatus
Extensible Caissons
Tethered Buoyant Structures
Oil Production Capsules
Spheres
Subsea Processing Facilities
Service Modules
Reinforced, Prestressed Concrete
   Frustoconical Structures
Base Units and Production
   Fluid Handling Systems

ISBN 0-8155-0566-3

392 pages

GEOTHERMAL ENERGY 1975

by Edward R. Berman

Energy Technology Review No. 4

This book describes in detail the nature of the geothermal resources, their extent, and the currently available technology by which these natural sources of energy can be exploited and utilized to the greatest advantage. There is little pollution resulting from the use of geothermal energy, and with a little care all disturbance of natural ecological systems can be avoided.

Earth heat can be used most practically where hot volcanic rocks are comparatively near the surface, and emerging circulating ground waters act as heat collectors, either by producing steam or by serving as heat transfer media. Suitable sites have been discovered in the Continental U.S. and Hawaii, the U.S.S.R., Japan, New Zealand, Iceland, and Italy. In view of the rise in the price of oil and the intensive search for new sources of energy, geothermal power appears to be in for a period of rapid development.

This Energy Technology Review is based on international studies conducted by industrial and engineering firms or university research teams under the auspices of various governments and governmental agencies.

A partial and condensed table of contents follows here.

1. U.S. RESEARCH & EXPLORATION TECHNIQUES

Historical Background
Recent Explorations
U.S. Agencies Involved
World Survey of Major Geothermal
   Installations (except USSR)

2. RESEARCH AND PRACTICE IN THE USSR

Soviet Geothermal Electric
   Power Generation
Paratunka Geothermal Electric
   Power Station (using Freon®)
Pauzhetka Power Station
Bol'she-Bannaya Station
Makhachkala Station
Additional Sites
Non-Electric Uses of Thermal Waters
Equipment and Instruments

3. DRY GEOTHERMAL RESERVOIRS

Conventional Drilling
Ultra-Deep Drilling

Limitations of Current Technology
Areas of Further Research
Novel Techniques
A Proposed Commercial System
Man-Induced Fracturing of Hot Rocks
Artificial Geothermal Reservoirs

4. THE PLOWSHARE CONCEPT

Geothermal Heat Extraction
   Using Nuclear Explosives
Site Guidelines & Geology
Heat Source Developments
System Thermodynamics
Seismic Design Considerations

5. EXPERIMENTAL STUDIES

Scale Formation in Simulated
   Geothermal Brine
Wairakei (New Zealand) Well Analysis
Aquifer Chemistry in Geothermal Systems
Plowshare Geothermal Steam Chemistry

6. GEOTHERMAL RESOURCES OF CALIFORNIA

The Geysers
Imperial Valley
Mono Lake
Desalination of Geothermal Water
Supplemental Water for Reinjection
Land Subsidence and Cave-Ins
System Synthesis and Costs

7. ENERGY RECOVERY FROM NATURAL HOT BRINE

Chemical Composition
Fluid Pressures
Turbine Systems
Corrosion and Scale
Sperry Rand Corp. Method

8. FEASIBILITY STUDIES FOR SPECIFIC U.S. AREAS

Texas
   Port Mansfield Site
California
   Point Mugu Site
   Twenty-Nine Palms Site
   China Lake Site
Idaho
   Raft River Valley Site

9. PROPOSED RESEARCH

ISBN 0-8155-0563-9

336 pages